KU-029-379

Edinburgh
Information Technology Series

# ASPECTS OF SPEECH TECHNOLOGY :

S. Michaelson and Y. Wilks
Series Editors

Edited by
M. JACK and J. LAVER

# ASPECTS OF SPEECH TECHNOLOGY

EDINBURGH UNIVERSITY PRESS

Edinburgh University Press
22 George Square, Edinburgh

Set in Linotronic Times by
Edinburgh University Press and
printed in Great Britain by
Redwood Burn Ltd, Trowbridge

British Library Cataloguing
in Publication Data
Speech Technology: a survey—(Edinburgh
information technology series; 4)
1. Speech synthesis
I. Jack, Mervyn A. II. Laver, John
006.5′4 TK7882.S65

ISBN 0 85224 568 8

CONTENTS

## PREFACE

Speech Technology is a multi-faceted and dynamically changing technology, within the general framework of Information Technology. In such a fast-moving technology, it serves speech technology well to create a series of snapshots of recent developments and current directions in order to define the status of the technology. This text is an attempt to create an album of such snapshots, prepared by a group of expert contributors who are currently active in the field of speech technology, such that each contribution might benefit from the individual specialisation of its authors.

It is of further significance, and personal satisfaction to us, that all of the contributors work under one roof, as members of the Centre for Speech Technology Research at Edinburgh University. The Centre was established in 1984 (under the University's Information Technology Committee) as a forum to promote collaborative research, and to develop projects in human/computer interface techniques. The academic departments that collaborate within the Centre include the Department of Artificial Intelligence (intelligent knowledge-based systems), the Department of Electrical Engineering (signal processing), and the Department of Linguistics (speech and natural language processing).

Installed in new, modern University premises in 1985, the Centre is particularly interested in working with industry on an international dimension in all areas of speech technology from speech recognition to natural-language processing, text-to-speech synthesis, and speaker verification.

This text, intended primarily as a reference work and an extensive bibliographical resource, also offers a degree of tutorial discussion, as well as a critique of many aspects of speech technology as they currently exist.

Chapter 1 presents a general treatment of the techniques available to perform speech recognition using stored whole-word reference patterns. Such techniques represent the basis of presently available commercial speech-recognition products.

Chapter 2 anticipates the emergence of knowledge-based speech-recognition systems, to complement the existing pattern-matching approaches covered in chapter 1. In essence, chapter 2 provides a bridge between the published literature on acoustic cues for recognition of consonants, and a framework for using such cues in future automatic speech-recognition systems.

Chapter 3 surveys the development of speech-synthesis strategies, commenting on their suitability in specific applications, and discusses how speech synthesis might benefit from greater use of linguistic and phonetic knowledge.

Chapter 4 considers the use of speech as a biometric technique in speaker-verification systems. The operation of a speaker-verification system is described and a range of speech parameters which have been considered for use in such systems is discussed. Results are presented on the effectiveness of automatic speaker-verification systems.

Finally, chapter 5 explores the speech distortion problem known as helium speech. This phenomenon is encountered by divers when working in high-pressure, underwater atmospheres, and is due to the unusual respiratory gas mixtures necessary to sustain life in the deep-ocean environment. The chapter provides a complete overview of this field covering the phenomenon, its signal-processing options, and systems which have been developed to overcome the distortions present in such speech signals.

Readers are referred to a companion volume in the EDITS series, *Machine Translation Today,* for coverage of natural language.

Speech technology can be represented as a catalytic technology in the future of Information Technology products, a future that will affect society in general and each individual in it. This text therefore represents an attempt by speech technologists to catalogue their subject as a reference source for its fullest, and most responsible, exploitation.

Mervyn Jack and John Laver

ONE : F. MCINNES & M. JACK

# AUTOMATIC SPEECH RECOGNITION USING WORD REFERENCE PATTERNS

Most currently available automatic speech recognition systems rely on comparison of the input speech to be recognised with stored reference patterns, with each reference pattern representing one of the words of the vocabulary being used. The stored pattern for each word may be a template (Rabiner 1978), derived from one or more training utterances (tokens) of the word obtained before the recognition session begins; or a statistical model of the word's characteristics, derived from multiple training utterances (Levinson, Rabiner and Sondhi 1983). In either case, it is usually necessary to time-align a series of input vectors, each containing spectral information derived from a short section (frame) of the input speech, with the reference pattern. In this way it is possible to compute a measure of the similarity between the input word and the word which that reference pattern represents.

In many word-based speech recognition systems, the alignment of input and reference patterns is accomplished using the optimisation technique known as dynamic programming (Bellman 1957). Where the reference patterns are templates, the dynamic programming alignment procedure (Vintsyuk 1968) is known as dynamic time warping (DTW: Myers, Rabiner and Rosenberg 1980). (The phrase *dynamic time warping* is not universally approved among speech recognition researchers (Levinson 1985); however, it will be retained here as the most generally understood term for this particular application of dynamic programming.) The corresponding procedure for hidden Markov models (a commonly used class of statistical models) is called the Viterbi algorithm (Forney 1973; Levinson, Rabiner and Sondhi 1979).

The basic idea involved in template matching is that each word in the system vocabulary is represented by a template (in some

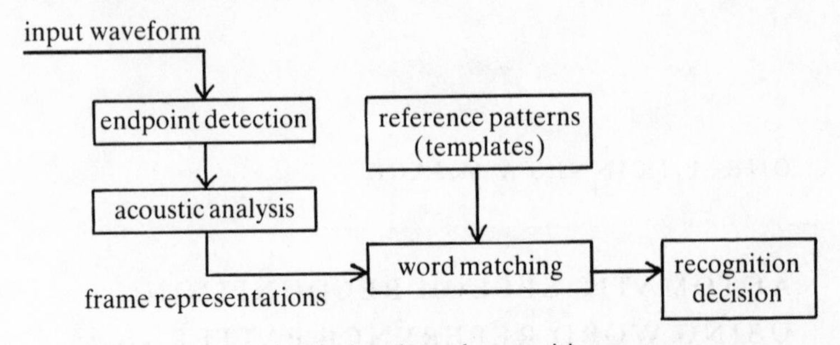


*Figure* 1.1. A typical isolated word recognition system.

cases more than one: Rabiner 1978, Rabiner *et al.* 1979), which is a reference pattern created from speech data and stored in the machine's memory. Each unknown input word to be recognised is compared with the stored templates and identified as an instance of that vocabulary word whose template best matches the unknown input.

Each word template consists of a sequence of representations of short time segments or frames of the reference speech waveform. The representation for each frame may be a vector of bandpass filter outputs, or a set of autocorrelation and/or linear prediction coefficients, or some other set of parameters such as cepstral coefficients. The unknown input speech waveform is similarly processed into input frame representations.

The simplest application of DTW for speech recognition is to the recognition of a single word spoken in isolation. The components of a typical isolated word recognition system are shown in figure 1.1. The input waveform is digitised, the beginning and end of the word are identified and an acoustic analysis is performed in each frame of the (endpoint-detected) word. The resulting sequence of frame representations is matched against each stored word template in turn, so as to identify the word which has been spoken.

In comparing the input with a template, each input frame is matched with a reference frame from the template, and a frame distance is computed, which is a measure of how different the two frame representations are. (An alternative formulation is in terms of similarities instead of distances; the details of the algorithm are the same in either case, except that, wherever a distance is to be minimised, the corresponding similarity is to be maximised.) Then an overall distance is computed from the frame distances for all the matched pairs of frames. The sequence of matched pairs of input

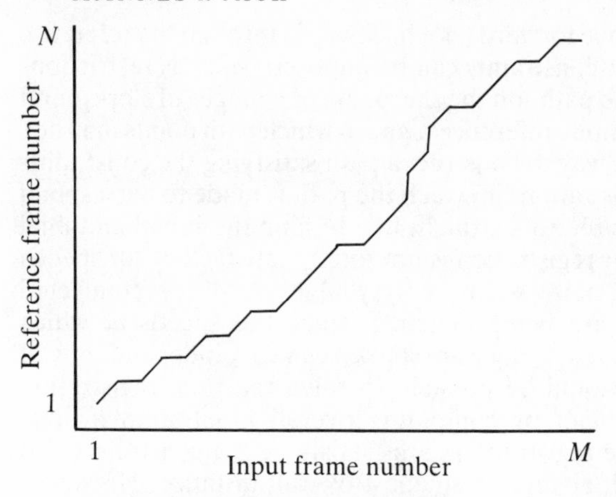


*Figure* 1.2. A time registration path.

and reference frames forms a time registration path, which can be depicted as a graph of reference frames against input frames, as shown in figure 1.2. The point $(m, n)$ on the path, where $m$ and $n$ are integers, corresponds to the matching together of the $m$th input frame and the $n$th reference frame. The slope of the path represents the degree of compression (expansion, where the slope is less than 1.0) applied to the template in aligning it with the input frames. In particular, a vertical step in the path corresponds to the matching of two successive reference frames to the same input frame, and a horizontal step corresponds to the matching of the same reference frame to two successive input frames. The overall distance is a weighted sum of the individual frame distances for the pairs of frames on the path. The weight given to each frame distance can be made to depend on the slope of the time registration path near the point defined by the pair of frames in question.

The time registration problem is that of finding the best possible time registration path for given input and reference data, i.e. the path that minimises the overall distance subject to appropriate constraints. Endpoint constraints require that the beginning and end of the input data be matched with the beginning and end, respectively, of the reference data. This requirement may be relaxed to allow for inaccuracies in the identification of endpoints when the frame sequences were created. Continuity constraints require that successive points on the time registration path be close together (in both the input and the reference dimensions). Monotonicity constraints require that as the path progresses in input time, it should move forward (not backward) through the reference template. Further constraints can be imposed, such as restrictions on the slope of the path, on the sharpness of changes of slope or on the region of the input-reference plane in which path points may lie.

The simplest way to construct a path satisfying the constraints is linear time registration, in which the path is made to correspond as closely as possible to a straight line joining the initial and final points. This linear registration is not totally satisfactory for speech recognition, especially where polysyllabic words or connected strings of words are being matched, since the speeds at which different parts of an utterance are spoken can vary independently of one another. It would be possible to solve the time registration optimisation problem by computing overall input-template distances for all time registration paths satisfying the constraints and selecting the path giving the smallest overall distance. However, this would involve an excessive amount of computation.

A much more efficient procedure relies on the application of dynamic programming (Bellman 1957). (Despite the use of the word *programming,* this is essentially an optimisation-algorithm term, not a computer-science term.) Dynamic programming is relevant to a whole class of problems where a path is to be chosen from an initial point to a final point so as to minimise or maximise a sum of quantities which correspond individually to the individual segments of the path. The essential principle involved is that if a point $P$ lies on the optimal complete path, then the partial path (i.e. the part of the optimal complete path) from the initial point to $P$ is the optimal path from the initial point to $P$. Thus, if the optimal path from the initial point to $P$ has been found, any other path from the initial point to $P$ can be discarded, and excluded from further consideration, in the attempt to construct the optimal complete path.

This dynamic programming procedure when applied to the time registration problem takes the form of a DTW algorithm. The algorithm proceeds along the input one frame at a time and, for each successive input frame, computes a frame distance $d(m, n)$ and an accumulated distance $D(m, n)$ for each value of $n$ permitted by the search area constraints (where $m$ is the input frame number and $n$ is the reference frame number). The accumulated distance is the weighted sum of the frame distances on the optimal partial path from the initial point to $(m, n)$ and is found by optimising over the possible second-last points of such partial paths. Thus, in the simple example illustrated in figure 1.3, where the constraints permit $(m-1, n-2)$, $(m-1, n-1)$ and $(m-1, n)$ as previous points, and

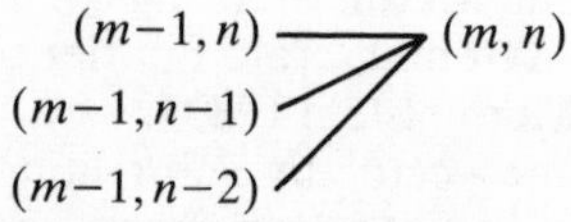

$$(m-1,n) \rightarrow (m,n)$$
$$(m-1,n-1) \nearrow$$
$$(m-1,n-2) \nearrow$$

*Figure* 1.3. Simple path constraints.

the weighting on each frame distance is 1.0 (i.e. the accumulated distance is simply the sum of the frame distances on the path so far),

$$D(m, n) = \min [D(m-1, n-2), D(m-1, n-1), D(m-1, n)] + d(m, n) \quad (1.1)$$

When the final point $(M, N)$ is reached, the overall distance between the input and the template (with optimal time alignment) is simply $D(M, N)$.

The next few sections give details of the design options occurring in various aspects of the recognition process. The ideas introduced here apply primarily to the isolated word recognition problem, but will also be applicable (with appropriate modifications) to the more complex problem of recognising words in connected speech.

### *Frame Representations and Distance Measures*

The first requirement for a DTW system is that there should be a method of representing the speech waveform in each time frame, with an associated distance (or similarity) measure for comparing frame representations. Various types of representation have been reported, notably those based on bandpass filtering (White and Neely 1976; Sakoe and Chiba 1978; Davis and Mermelstein 1980; Dautrich, Rabiner and Martin 1983a,b,c) and those based on linear predictive coding (LPC) analysis (Itakura 1975; White and Neely 1976; Tribolet, Rabiner and Sondhi 1979; Davis and Mermelstein 1980; de Souza and Thomson 1982).

A bandpass filter system (Dautrich, Rabiner and Martin 1983c) has a filter (analogue or digital) for each of a number of frequency bands covering the range of frequencies being used. The output from each filter is usually passed through a full-wave rectifier followed by a low-pass filter, to extract an averaged level which is proportional to the total signal energy within that filter band. The filter outputs (energy levels) can be transformed logarithmically and then normalised in each frame by subtracting the overall log energy in that frame. Filter outputs can be used either directly as frame representations or to compute some other representation such as cepstral coefficients (Davis and Mermelstein 1980). These representations can also be computed from discrete Fourier transform (DFT) coefficients (Davis and Mermelstein 1980).

The normalisation of each frame vector by subtracting the overall log energy in the frame entails loss of information, since all parts of a word are adjusted to the same loudness. This can be corrected by including in the frame representation a measure of the overall energy in the frame, or by some other normalisation procedure, taking into account the energy level of the utterance as a whole (Das 1980). An alternative form of normalisation involves each filter energy in a frame representation being normalised by the corresponding component of the long-time average spectrum (Kuhn and Tomaschewski 1983). These normalisation procedures have been shown (Das 1980, Kuhn and Tomaschewski 1983) to yield improvements in recognition accuracy over the normalisation of each frame independently. The filter energies may also be smoothed over adjacent frames and this too has been observed to improve the performance of the recogniser (Kuhn and Tomaschewski 1983).

The number of bandpass filters used has varied considerably from one system to another, but has usually been in the range 6–20. Increasing the number of filters used generally improves the rate of correct recognition achieved by the system (White and Neely 1976; Dautrich, Rabiner and Martin 1983b,c), although it has been observed (Dautrich, Rabiner and Martin 1983b,c) that recognition rates for female speakers decline when the filter bandwidths become too narrow, apparently because females, with high values of fundamental frequency (F0), produce for voiced sounds a spectrum where the inter-harmonic spacing can exceed the filter bandwidth, leaving a gap in the speech spectrum where the effect of noise becomes significant. The frequency spacing of the filters can be linear (with a constant frequency difference between adjacent filter bands: Dautrich, Rabiner and Martin 1983c), logarithmic (e.g. octave or third-octave filters: White and Neely 1976; Dautrich, Rabiner and Martin 1983c) or some other non-linear spacing – such as the mel frequency arrangement (Davis and Mermelstein 1980), in which the filters are linearly spaced up to about 1 kHz, and logarithmically spaced over higher frequencies, to accord with the distribution of significant information in speech. Cepstral coefficients derived from filters on a mel frequency scale have been found to yield good recognition performance (Davis and Mermelstein 1980, Chollet and Gagnoulet 1982, Gagnoulet and Couvrat 1982, Jouvet and Schwartz 1984). The filter bands can be non-overlapping or overlapping (Dautrich, Rabiner and Martin 1983b,c).

The distance between two frame representations consisting of filter energies or cepstral coefficients is often taken to be the absolute value norm of the difference of the two vectors, i.e.

$$d(m, n) = \sum_k |x(m, k) - y(n, k)| \quad (1.2)$$

where $x(m, k)$ is the $k$th component of the $m$th input frame representation and $y(n, k)$ is the $k$th component of the $n$th reference frame representation (White and Neely 1976). The Euclidean norm, or its square, can be used as a distance measure (Davis and Mermelstein 1980), but this results in a substantial increase in the amount of computation required (White and Neely 1976). Depending on the other characteristics of the system, the Euclidean metric may give recognition results slightly better than those obtained with the absolute value metric (White and Neely 1976, Das 1980), or in some cases slightly worse (Dautrich, Rabiner and Martin 1983b). These two distance measures are the most widely used, but various others have been devised (Olano 1983; Dautrich, Rabiner and Martin 1983b; Nocerino *et al.* 1985; Greer, Lowerre and Wilcox 1982). In particular, for filter energies, distance measures using dynamic programming for non-linear alignment in the frequency domain have been found to yield enhanced recognition rates, with a considerably increased amount of computation (Hunt 1985, Blomberg and Elenius 1986); and, in the case of a cepstral representation, applying different weights to the individual coefficients before the distance computation has been observed to improve the performance (Tohkura 1986; Juang, Rabiner and Wilpon 1986).

If the system is to operate in noisy conditions and the spectrum of the noise is known or can be measured, a noise masking or compensation technique can be applied (Bridle *et al.* 1984), which will result in modified filter energy vectors or a modified frame distance function. Such a technique can greatly improve performance at low signal-to-noise ratios.

The Linear Prediction (LPC) approach (Itakura 1975; Gray and Markel 1976; Gupta, Bryan and Gowdy 1978; Tribolet, Rabiner and Sondhi 1979) involves sampling the (time domain) speech waveform (typical sampling rates being from 6 to 20 kHz) and then estimating prediction coefficients $(a(1), a(2), \ldots, a(p))$ for the sequence of sample values $(\ldots, s(t-1), s(t), s(t+1), \ldots)$ so that the mean squared value of the error term $e(t)$ in the equation

$$s(t) = -\sum_{i=1}^{p} a(i)s(t-i) + e(t) \quad (1.3)$$

is minimised. The mean squared error $(\sigma^2)$ is called the prediction residual, and is given by

$$\sigma^2 = \mathbf{a}R\mathbf{a}' \tag{1.4}$$

where $\mathbf{a}$ is the vector $(1, a(1), ..., a(p))$, and $R$ is the $(p+1) \times (p+1)$ autocorrelation or covariance matrix of the signal samples. Before the prediction coefficients are calculated, pre-emphasis (Tribolet, Rabiner and Wilpon 1982) is usually applied to the sequence of sample values; this reduces the energy at low frequencies, thus compensating for the spectral tilt which is typical of voiced sounds in speech, and can result in an improvement in the performance of the recogniser (Tribolet, Rabiner and Sondhi 1979). (Pre-emphasis may be applied, also, as preprocessing for a filter bank: White and Neely 1976; Dautrich, Rabiner and Martin 1983b.)

The number of prediction coefficients calculated, $p$, has varied from system to system: some LPC recognisers have had $p$ as small as 6 (Itakura 1975); others, as large as 14 (White and Neely 1976). The windows of sample values for consecutive frames are overlapped, to produce typically 45 ms windows at 15 ms separation, with 30 ms overlap between adjacent windows (Itakura 1975) (although shorter windows have been tried: Davis and Mermelstein 1980; Tribolet, Rabiner and Wilpon 1982). Two methods for estimating prediction coefficients, the autocorrelation and covariance methods, are compared in Tribolet, Rabiner and Sondhi (1979). In estimating the vector $\mathbf{a}$ by the autocorrelation method, a windowing function is used – one frequently used example being the Hamming window (Tribolet, Rabiner and Sondhi 1979; Rabiner and Levinson 1981).

A possible distance measure between LPC representations is the Itakura metric or log likelihood ratio (Itakura 1975; Tribolet, Rabiner and Sondhi 1979), defined by

$$d(m, n) = \log \frac{(\mathbf{a}(\text{ref}, n)R(\text{in}, m)\mathbf{a}(\text{ref}, n)')}{(\mathbf{a}(\text{in}, m)R(\text{in}, m)\mathbf{a}(\text{in}, m)')} \tag{1.5}$$

where $\mathbf{a}(\text{in}, m)$ and $\mathbf{a}(\text{ref}, n)$ are the vectors of prediction coefficients for the input and reference frames and $R(\text{in}, m)$ is the autocorrelation or covariance matrix for the input frame. This is the log ratio of the prediction residuals when the input signal is predicted using the coefficients derived from the reference data and when it is predicted using the (optimal) coefficients derived from the input itself. The word *metric* is used loosely here in describing this function, since it does not satisfy the requirements of the mathematical definition of a metric. In particular, it is not symmetric, i.e. it does not necessarily take the same value if the reference and input data are swapped. Other LPC distance measures, involving the prediction coefficients and autocorrelations (or covariances) of the reference and input samples, have been devised (Gray and Markel 1976, de Souza and Thomson 1982, Olano 1983, Nocerino *et al.* 1985). Some of these, which are symmetric with respect to the two frames being compared, have been found to have better properties for distinguishing correct and incorrect frame matches than the Itakura metric (de Souza and Thomson 1982). The performance of LPC-based recognisers in noisy conditions can be improved by filtering the noisy signal before carrying out the LPC analysis (Neben, McAulay and Weinstein 1983; Wohlford, Smith and Sambur 1980). Other representations such as reflection coefficients and cepstral coefficients can be derived from an LPC analysis (Gray and Markel 1976; Davis and Mermelstein 1980; Wohlford, Smith and Sambur 1980), and for these the absolute value and Euclidean distances described above can be used. A theoretical and experimental study of several LPC-based distance measures is given in Gray and Markel (1976), where it is shown that the Euclidean distance on cepstral coefficients and a symmetric likelihood-ratio (cosh) distance are approximations to the root mean square distance between log spectra. The autocorrelation coefficients can be used directly as a speech representation, rather than to compute prediction coefficients; a distance measure allowing for autocorrelation lag offsets between the two frame representations compared has been formulated (Aktas *et al.* 1986). Comparative studies of LPC and bandpass filter representations (White and Neely 1976; Dautrich, Rabiner and Martin 1983b,c) suggest that an LPC representation gives better results on telephone-quality speech, which is band-limited so that high frequencies are lost, but not on speech without this band-limiting. In a comparison of several LPC-based and filterbank-based representations for recognition of monosyllabic words (Davis and Mermelstein 1980), the best recognition accuracy was obtained using cepstral coefficients derived from filters on a mel frequency scale. Among the LPC representations, the prediction coefficients with the Itakura distance yielded the best results (only a little poorer than those with the mel cepstral representation), and cepstral coefficients were better than reflection coefficients.

Speech representations based on models of human auditory processing have been proposed (Blomberg *et al.* 1984, Dologlou and Dolmazon 1984, Hunt and Lefebvre 1986). In some cases these have been found to yield improvements in recognition performance over the more conventional representations (Hunt and Lefebvre

1986). In experiments into the representation of the dynamic characteristics of speech (Furui 1986b), better recognition results were obtained by the use of a linear combination of each cepstral coefficient and its time derivative (estimated by regression analysis) than with either the instantaneous value or the derivative alone. A similar improvement was achieved by using the cepstral coefficients and their time derivatives as separate parameters (Furui 1986a); but the linear combination technique has the advantage that it does not increase the dimension of the frame representations, and so does not add to the computation for each frame distance. Some further improvement was obtained by using the log-energy derivative as an extra component of each frame representation. It has also been shown (Bocchieri and Doddington 1986a,b) that improved recognition performance can be obtained by using a concatenation of the feature vectors from two frames with a 40 ms time separation and applying a transformation to adjust the distance for inter-frame covariance. The recognition accuracy was increased even more when the transformations were made specific to reference frames, and when components of the vectors were selected to optimise the discrimination of different words; but this was possible only by the use of a large number of training utterances, and these techniques also increased the computational load significantly.

*Training and Template Creation*

A speaker-trained word recogniser, in which the templates are created from speech data from the specific speaker who will speak the input words to be recognised, can operate using just one template, derived from a single reference utterance, for each word in its vocabulary. The speaker has to utter a full list of the words in the vocabulary to train the system (i.e. provide it with reference data) before using it to recognise subsequent speech input. It has sometimes been found beneficial to have several templates for each word (Rabiner and Wilpon 1979b; Myers, Rabiner and Rosenberg 1980; Kuhn and Tomaschewski 1983; Bridle, Brown and Chamberlain 1983), or to derive each template from two or more utterances of the word (Rabiner and Wilpon 1979b; Zelinski and Class 1983; Kuhn and Tomaschewski 1983; Bridle, Brown and Chamberlain 1983); in either of these cases, the speaker must provide more than one utterance of each word for training.

This requirement of training to each new speaker is acceptable if the vocabulary is small or if only one speaker is to use the recogniser. However, for a system with a larger vocabulary and a high turnover of different speakers, it becomes a limitation. There are at least two possible options for avoiding the training requirements of this fully speaker-dependent mode of operation. These are: (a) the speaker-adaptive mode, in which the recogniser is trained to each new speaker using a selected subset of the full vocabulary (from which speaker-adapted templates are deduced for the remaining words using previously stored speaker-independent information); and (b) the fully speaker-independent mode (Rabiner 1978, Rabiner *et al.* 1979, Rabiner and Wilpon 1979a), in which templates are created from the speech of a limited number of training speakers, and are then used in the recognition of words spoken by whatever speakers may use the system.

The first of these possibilities is difficult because of the complexity of the pattern representing a word. Using any of the types of frame representation described above, large amounts of previously stored information and computation are likely to be required to adapt every frame of every vocabulary word to the characteristics of a new speaker (as determined from the smaller training vocabulary). A system using pretraining and a statistical analysis of the data from the pretraining speakers has been reported (Furui 1980), but this uses phoneme templates rather than word templates. Other adaptive recognition systems that have been developed or proposed (Bridle and Ralls 1985, Sugiyama 1986, Niimi and Kobayashi 1986) also rely on having reference patterns (or synthesis parameters: Bridle and Ralls 1985) for phonemic or syllabic units, from which the word reference patterns are built. However, adaptation to word-independent speaker characteristics by a spectral transformation is feasible in a word-based recogniser and has been applied with some success (Choukri, Chollet and Grenier 1986). A speaker adaptation method based on vector quantisation (see below) has also been devised (Shikano, Lee and Reddy 1986).

The second possibility, a speaker-independent recogniser, has been investigated by various researchers (Rabiner 1978, Levinson *et al.* 1979, Rabiner *et al.* 1979, Rabiner and Wilpon 1979a). It is here that multiple reference utterances for each vocabulary word become very important. Because of the variations in pronunciation among speakers, a template derived from an utterance of a word by a single training speaker may not be adequate for recognition of the same word spoken by another speaker. Thus utterances of the same words by several different training speakers must be used in the creation of the templates if input from a range of subsequent speakers is to be recognised accurately.

There are various possible ways to make use of multiple tokens (replications) of a word in making templates for it. One possible approach (Rabiner 1978) is to use each token separately as a template. This, however, results in an excessive amount of computation in the recognition process (unless the number of tokens is fairly small, in which case they may not adequately cover all the possible variations) – since each unknown word must be compared with all the templates for all the reference words. Another possibility is to average the representations for all the tokens to produce a single word template. (The averaging of word representations is not altogether straightforward: they must first be aligned so that corresponding frames match up – using linear alignment, DTW or some other procedure (Niimi 1978, Zelinski and Class 1983, Liu 1984) – before taking the average of the representations in each frame.) However, if the variations among different speakers' pronunciations are large, this will not provide adequately for all the variants. Better speed and accuracy in recognition can be obtained by more sophisticated techniques for choosing or creating templates from the reference tokens. A clustering analysis (Rabiner 1978; Levinson *et al.* 1979; Rabiner *et al.* 1979; Rabiner and Wilpon 1979a; Mokkedem, Hügli and Pellandini 1986) can be carried out, forming clusters of similar tokens, and then a template can be formed for each cluster, by averaging the tokens in the cluster or by some other procedure. This clustering of tokens is done separately for each word in the vocabulary. Typically the recognition performance improves as the number of cluster templates used per word of the vocabulary is increased up to about 10, but this improvement levels off as further templates are added beyond that number (Rabiner *et al.* 1979). Templates obtained by clustering give better recognition accuracy than the same number of templates chosen at random from among the training tokens (Rabiner *et al.* 1979, Rabiner and Wilpon 1979a). (Clustering experiments with a speaker-trained recogniser have also been reported (Rabiner and Wilpon 1979b). The results were qualitatively similar to those for the speaker-independent system, though fewer templates were required for optimal performance. The difficulty in applying speaker-specific clustering in practice is that several training repetitions of the vocabulary are required.)

A further option is progressive adaptation of the templates to the speaker's characteristics by updating during the recognition process (Damper and MacDonald 1984; McInnes, Jack and Laver 1986b). In this case, initial templates are still required, which may be speaker-dependent or speaker-independent, but they are modified by weighted averaging or some other method so as to accord more closely with recognised instances of the words in the input speech. Another form of adaptation, in a speaker-independent system with more than one template per word, relies not on modification of templates but on elimination of those templates which are found to correspond poorly to the current speaker's pronunciations (Hewett, Holmes and Young 1986). This can improve the speed of recognition and can also eliminate errors resulting from the matching of unsuitable templates.

Where the vocabulary contains words which differ only in small portions, recognition errors are liable to occur because of differences between those parts of the templates which represent similar parts of words. For instance, if the vocabulary contains the words *stalactite* and *stalagmite*, and the templates for these words differ in the initial *stala-* section, the input word *stalactite* may well be recognised as *stalagmite*, because the linguistically insignificant difference between the templates in the (longer) *stala-* section outweighs the significant difference between *gm* and *ct*. Such errors may be prevented (Moore, Russell and Tomlinson 1983; Wilcox, Lowerre and Kahn 1983) by combining the similar parts of templates for different words, so that, for example, the frame representations for the initial and final parts of the *stalagmite* template are identical to those for the corresponding parts of the *stalactite* template, and separate data for the two words are stored only for the distinguishing portions.

Experiments show (Dautrich, Rabiner and Martin 1983b; Rollins and Wiesen 1983; Baker and Pinto 1986) that, for recognition in noisy conditions, it is best to conduct the training in the same level of noise in which the unknown input words are to be spoken. If the characteristics of continuous background noise during the recognition session are analysed, this noise can be added to the original templates (obtained in quiet conditions) to improve the performance in the noisy environment (Aktas *et al.* 1986).

*Search Path Constraints*

As mentioned above, constraints of continuity and monotonicity are required for the time registration path, and it may also be desirable to impose restrictions on the steepness of the slope and the sharpness of changes in slope to prevent excessive distortion of the patterns being aligned.

These constraints can conveniently be combined in a specification of which points are allowed as predecessors to a given point. Let the possible sequences of recent preceding points at the point $(m, n)$ be $P(1), \ldots, P(i), \ldots, P(r)$. A condition $C(i)$ may be imposed on the permissibility of $P(i)$. For each permitted sequence

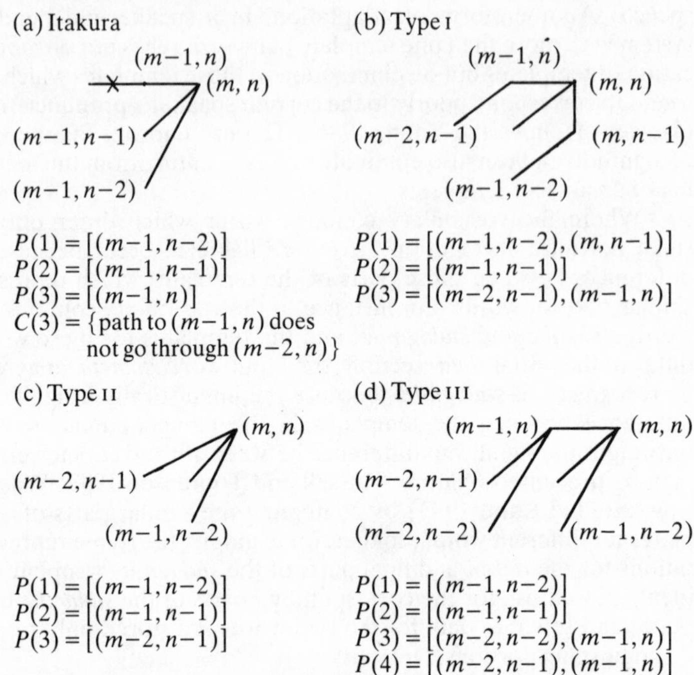


*Figure* 1.4. Local path constraints.

$P(i)$, an accumulated distance $D^i(m, n)$ is calculated. The accumulated distance $D(m, n)$ is then defined by

$$D(m, n) = \min [D^1(m, n), \dots, D^i(m, n), \dots, D^r(m, n)] \quad (1.6)$$

For example, the Itakura constraints (Itakura 1975; Rabiner, Rosenberg and Levinson 1978; Myers, Rabiner and Rosenberg 1980b), as illustrated in figure 1.4(a), have $P(1) = [(m-1, n-2)]$; $P(2) = [(m-1, n-1)]$; and $P(3) = [(m-1, n)]$, permitted on the condition $C(3)$ that the best path to $(m-1, n)$ (already found by the DTW algorithm) does not go through $(m-2, n)$. Note that in this case each of the sequences $P(1)$, $P(2)$ and $P(3)$ consists of a single point. These constraints force the average slope of the path to be no less than 0.5 and no greater than 2.0. Other sets of constraints, denoted by Type I, Type II and Type III (Myers, Rabiner and Rosenberg 1980b), are also shown in figure 1.4, with the sequences of preceding points permitted. In each case, no conditions $C(i)$ are imposed.

When conditions $C(i)$, such as $C(3)$ in the Itakura constraints, are allowed, the optimality principle of dynamic programming is no longer strictly valid. The Itakura constraints give only an approximation to the optimisation of the path which is afforded by the Type III constraints, since the optimal path may be excluded from consideration if it contains a step (1, 0) starting at a point which can be reached by a partial path (which is not a part of the optimal complete path) ending with a step (1, 0). However, comparative recognition tests (Myers, Rabiner and Rosenberg 1980b; Rabiner 1982) have shown that the Itakura constraints give a better rate of correct recognition than the Type III constraints. Very similar results to those using Itakura constraints were obtained (Myers, Rabiner and Rosenberg 1980b) when Types I and II constraints were used.

All these constraints impose the same limits on the average slope of the path. Further examples of local path constraints, permitting different ranges of slopes, are given in Sakoe and Chiba (1978); the best results were obtained when the range of slopes permitted was from 0.5 to 2.0, as in the above instances. When the search area in the input-reference plane is reduced to a narrow band about the linear path (see below), however, a restriction on the range of slopes may not be necessary (Paliwal, Agrawal and Sinha 1982).

The Itakura and Type III constraints are not symmetric with respect to the reference and input directions. They could be modified by interchanging occurrences of $m$ and $n$ in the specifications of the sequences $P(i)$. This would correspond to having the reference frames in the horizontal direction and the input frames in the vertical direction in figures 1.2 and 1.4 (Myers, Rabiner and Rosenberg 1980b). Another modification (Das 1980) is to have the reference or input pattern in the horizontal direction according to which pattern has more frames. Having the longer pattern in the horizontal direction has the advantage that the steps which involve skipping a frame in the vertical direction are likely to occur less often.

The weight given to the frame distance at each point on a time registration path in calculating the accumulated distance, and hence the overall distance for the path, can be made to vary according to the slopes of the segments of the path near that point (Sakoe and Chiba 1978; Myers, Rabiner and Rosenberg 1980b; Bisiani and Waibel 1982), and also according to other variables, such as the positions of the frames in the reference and input patterns (Bridle 1973; Russell, Moore and Tomlinson 1983; McCullough 1983; Nakagawa 1983).

The simplest forms of weighting are those where the weight on each frame distance depends only on the immediately preceding step in the path. The two most usual weighting schemes are an asymmetric one, in which the weight at each step is equal to the distance (number of frames) moved in one (input or reference) direction, and a symmetric one, in which the weight is equal to the sum of the distances moved in both directions. In the asymmetric scheme the sum of the weights for the whole path is $M$ (the number of input frames) or $N$ (the number of reference frames), depending on which direction is chosen. In the symmetric scheme the sum is $M+N$.

One modification (Okochi and Sakai 1982) that can be applied to any weighting scheme involves using the weighting coefficient defined by a step in the path to multiply the distances at both the initial point of that step and its final point (rather than just the distance at the final point). This trapezoidal weighting has a smoothing effect, since the total weight on a frame distance is the average of the weights derived from the steps in the path preceding and following the point.

There are many other possible weighting schemes, such as those designed to penalise very steep or shallow slopes (Gray and Markel 1976), but the simple asymmetric scheme seems to have been the most frequently used. The choice of direction on which to base the weighting can be quite significant when this scheme is used (Das 1980; Myers, Rabiner and Rosenberg 1980b): it is better to base the weighting on the progress of the path through the input pattern rather than the reference pattern (Myers, Rabiner and Rosenberg 1980b), or the longer of the two patterns rather than the shorter one (Das 1980). Such weights based on the input utterance have been found to give a better recognition rate than other options (Myers, Rabiner and Rosenberg 1980b), and weighting in the input direction also has the advantage that no template-specific normalisation of the word distances is required in making a recognition decision, because the total weight depends only on the length of the input word.

It is less usual to make the weighting depend on variables other than the slope of the path. However, it has been suggested (Bridle 1973) that the first and last parts of a word, for instance, may exhibit greater variability than the intermediate section, and in that case, to compensate for this, a smaller weight can be assigned to each frame distance in the corresponding parts of the time registration path than to each one in the rest of the path. (A more sophisticated adjustment of the weighting, which is made to depend on the rate of spectral change in the part of the word being matched, is described in Elenius and Blomberg (1982); but it is probably better to treat this as a modification of the frame distance function. The same applies to the weighted spectral slope metrics described in Olano (1983)). Also, the weighting on points far from the linear path (the straight line joining (1, 1) to ($M$, $N$)) could be made large, in order to penalise excessive variations in the parts of the reference and input words being matched together.

A two-pass recognition procedure (Rabiner and Wilpon 1981) and a subsequent modification of it (Tribolet, Rabiner and Wilpon 1982) use special weights in a second-stage distance calculation, following the ordinary DTW matching, to improve discrimination between similar words. For each pair of words to be distinguished, a weighting function is used which takes larger values in those regions where the words differ most. These procedures require preliminary computation to determine the weighting functions, but have been found to improve recognition rates. They are rather similar in concept to the frame-specific and discriminatory distance functions (Bocchieri and Doddington 1986a,b) mentioned above.

To penalise paths containing steps with slopes other than 1.0, additive penalties on such steps may be imposed, instead of multiplicative weights. In this case, the contribution made to the overall distance by the slope of each part of the path does not depend on the frame distances in that part of the alignment. A method for deriving such penalties on the possible path steps for each part of a word individually, according to the frequency with which these steps tend to occur, has been proposed (Russell, Moore and Tomlinson 1983), and has been observed to improve discrimination between words differing mainly in their time scales, although having no beneficial effect on recognition in other cases. These word-dependent penalties are similar to the word-dependent transition probabilities in the Markov modelling approach described below.

Correctly locating the beginning and end of an utterance is not in general an easy task (Vaissière 1985). Speech sounds must be distinguished from background noise, and in particular from breath noise, clicks etc. made by the speaker at the end of the utterance. If the endpoint detection rule is based on signal amplitude or voicing, some sounds, notably voiceless fricatives such as the *f* in *five*, may be classified as background noise, and thus excluded from the utterance. The risk of this may be reduced by adjusting the thresholds for distinction between speech and non-speech signals, but then there will be an increased tendency for background noise to be included as speech.

To allow for errors in endpoint identification, the endpoint constraints on the time registration path may be relaxed to allow it to start and finish anywhere in specified regions at the beginnings and ends of the input and reference patterns, instead of only at the points (1, 1) and ($M$, $N$) respectively (Rabiner, Rosenberg and Levinson 1978). This can be expected to result in some increase in the amount of computation required, as the region of the input-reference plane where frame distances are calculated will typically be increased.

Having a choice of pairs of endpoints complicates the identification of the optimal path. Firstly, the decision must be made whether to perform a separate time warp from each possible initial point or to include them all as points from which paths may come in a single warp (in which case the procedure will tend to favour paths from later initial points, since these paths will tend to have smaller accumulated distances). Secondly, whichever way that decision is made, at the end of the warping process there will be a number of complete paths to choose from, with different final points, and possibly also different initial points. Simply choosing the path with the smallest overall distance has the disadvantage that the paths are of different lengths and there will be a bias towards the shorter paths, and so it may be desirable to normalise the overall distances for path length before comparing them (Rabiner, Rosenberg and Levinson 1978). The edge-free staggered array DP algorithm (Furui 1986a) avoids these problems, where the weighting is proportional to the sum of the distances moved in the input and reference directions, by allowing paths to start and finish anywhere on selected diagonals of slope $-1$ (where $m+n = 1$ and where $m+n = M+N$, respectively). This requires the inclusion of sections of the reference and input signals beyond the detected word endpoints. An endpoint relaxation technique can correct errors due to bad endpoint detection (Rabiner, Rosenberg and Levinson 1978; Rabiner 1982), but, depending on the characteristics of the vocabulary used, may also introduce errors which outweigh the improvements (Haltsonen 1984).

Another endpoint modification technique (Haltsonen 1984, 1985a; Jouvet and Schwartz 1984) is a procedure in which a silence or noise frame, which can be matched repeatedly with successive frames of the input speech, is appended to each end of each template, and extra frames of the input signal beyond the detected utterance endpoints are used. This reduces the incidence of errors due to exclusion of parts of the input utterance, while allowing any non-speech frames in the extra input regions to be matched to the initial and final silence frames and so generate no errors (since the distance added by the matching of a given input frame to silence will be the same for each template, as the silence frames of all the templates are the same). A modification of this technique has been formulated (Haltsonen 1985b) in which no prior detection of word endpoints is required.

Another possibility, to improve endpoint alignment without calculating so many extra frame distances, is to do some preliminary testing at the beginning and end regions and, in each of these regions, to choose the alignment so that the local match is optimised. This procedure (Das 1982) gives new initial and final points for the time registration path, which is then determined in the usual way with these new points replacing (1, 1) and $(M, N)$. Thus the extra distance calculations involved are only in small initial and final regions. In a comparative test (Haltsonen 1984) this technique was found to increase recognition accuracy slightly, but not as much as the technique with silence frames described above.

When the endpoints of the time registration path are specified exactly, the area of the input-reference plane in which the path may lie is a parallelogram determined by the maximum and minimum slopes allowed by the local path constraints. When the endpoints are variable, it is a significantly larger polygon. It is possible to impose further restrictions on the area to be searched for possible paths; this reduces the number of local and accumulated distance computations, and may also improve the accuracy of the recogniser by preventing excessive distortion. Eliminating certain areas from the search is equivalent to assigning infinite weights to all local distances at points in these areas, and so these area restrictions are technically a special case of the weighting on points far from the linear path mentioned above.

One simple form of area restriction is to exclude points more than a fixed number of reference (or input) frames away from the straight line joining (1, 1) to $(M, N)$ (Das 1980; Paliwal, Agrawal and Sinha 1982). An even simpler method is to use the line of slope 1.0 from (1, 1) instead of the line from (1, 1) to $(M, N)$, so that $(m, n)$ is excluded if $|m-n|$ exceeds some constant value $\epsilon$ (Sakoe and Chiba 1978; Myers, Rabiner and Rosenberg 1980b; Paliwal, Agrawal and Sinha 1982). This has the disadvantage that if the durations of the reference and input words differ by more than the chosen value $\epsilon$ there will be no permissible time registration path.

A more sophisticated area restriction method is the adaptive one used in the UELM (unconstrained endpoints, local minimum) algorithm (Rabiner, Rosenberg and Levinson 1978). Here the values of $n$ considered for a given $m$ are those not more than $\epsilon$ away from $\tilde{n}$, where the value of $\tilde{n}$ is chosen to minimise $D(m-1, \tilde{n})$. Note that in this case no constraint can be imposed on the final point: if the matching is successful, the algorithm will find the final point itself.

*Word Length Normalisation*

There are certain advantages in having the same number of frames in each of the two words being matched together. A simple area restriction (as described above) can sensibly be applied, as can weighting coefficients which penalise steps of slopes other than 1.0 (since the optimal path can be expected to be fairly close to the linear path from (1, 1) to $(M, N)$, which in this case $(M=N)$ has slope 1.0).

If the durations of the words are determined before the frame representations are computed, it is possible to adjust the frame separations so as to have the same number of frames in every word (whether a reference template or an input word: Gupta, Bryan and Gowdy 1978). This, however, introduces a time-lag in the operation of the recogniser, and requires the calculation of frame representations to be adaptable for varying frame separations, although with overlapping windows the latter is not such a significant point since the separation can be changed while keeping the window length constant without leaving gaps of unused data between windows.

Another way to adjust the numbers of frames is a linear interpolation technique (Myers, Rabiner and Rosenberg 1980b). In this method, frame representations are computed with a standard time separation in the usual way, and then once they have all been stored (and the duration of the word is known) they are replaced by a preset number of frame representations at equal intervals throughout the word. These frame representations are calculated from the original representations by linear interpolation. The reference templates are stored in length-normalised form, so that only the input word needs to be normalised before each word recognition. The time warping cannot start until the whole input word has been read in, but the time-lag will be less than with the preceding method provided that the interpolation procedure takes less time than the calculation of frame representations from raw sampled input. Also, for this interpolation procedure only the frame representations, rather than the individual sample values, must be stored until the word has been fully read in, so that it is likely to be more economical in use of memory than the preceding method. Experiments show (Myers, Rabiner and Rosenberg 1980b) that, when word lengths are normalised, imposing a restriction $|m-n| \leqslant \epsilon$ can improve the recognition slightly, rather than degrading it.

Another way to apply the interpolation procedure would be to leave the input frame representations as they are and apply the interpolation to the reference words to normalise each of them to the length of the input word. This would have the advantage of adapting the number of frames used so that more were used for a long input word than for a short one, thus achieving a better combination of recognition performance and economy in frame distance calculations, but would require the interpolation to be done each time for all the reference words, rather than just for the one input word.

A trace segmentation procedure (Kuhn and Tomaschewski 1983; Pieraccini and Billi 1983; Gauvain, Mariani and Lienard 1983) has some similarities to the word length normalisation technique outlined above. It uses linear interpolation to form new frame representations from the original ones, and it results in word representations of a fixed length. However, there is an important difference, in that the new representations are not regularly spaced in time, but are chosen so that the amount of spectral change in the speech signal from one representation to the next is constant over all parts of the word.

Bandpass filter energies are extracted at regular time intervals in the usual way, and extra frames representing silence are appended to the utterance, one at the beginning and one at the end. If the vectors of filter outputs for the successive frames, including the initial and final silence frames, are denoted by $\mathbf{x}(0), \mathbf{x}(1), \ldots, \mathbf{x}(J)$ (where the number of non-silence frames is $J-1$), then

$$T = \sum_{j=1}^{J} d(\mathbf{x}(j-1), \mathbf{x}(j)) \tag{1.7}$$

where $T$ is the total length of the trace in $Q$-dimensional space ($Q$ = number of filters) formed by joining the points defined by consecutive frame vectors. Here $d$ is a distance function for vectors, such as the absolute value distance. The trace is divided into $S$ segments each of length $T/S$, where typically $S$ is about $J/3$. The vectors defining the segment boundaries are computed, and these are used as the new frame vectors (together with the initial and final silence vectors).

This trace segmentation procedure, followed by the usual DTW alignment of the (new) frame vectors, makes up the word matching method called DYPATS (dynamic programming after trace segmentation: Kuhn and Tomaschewski 1983). It not only adjusts each word to a standard number of frames but also gives greater weight to parts of a word where spectral change is occurring, which is probably more sensible than giving equal weight to equal segments of time. Moreover, it can result in considerable savings in computation and data storage, especially when used in conjunction with a strict search area restriction, while maintaining or improving on the level of recognition accuracy obtained with straightforward DTW (Kuhn and Tomaschewski 1983).

Although the formulation above is for a recognition system using bandpass filter representations, trace segmentation can be used with other acoustic representations (Ney 1983; McInnes, Jack and Laver 1986a). The appending of silence frames may not be appropriate in this case. Results suggest that trace segmentation is less effective for cepstral coefficients than for bandpass filter energies.

Another form of trace segmentation (Pieraccini and Billi 1983; Gauvain, Mariani and Lienard 1983) divides the trace into segments of a predetermined constant length, rather than into a predetermined number of segments. This has the advantage that processing can start as soon as the first input frame vectors are calculated, but does not result in normalisation of all words to the same number of frames.

There are various other acoustically based non-linear segmentation techniques (Greer, Lowerre and Wilcox 1982; Brown 1982; Pieraccini and Billi 1983; Chuang and Chan 1983). These rely, like trace segmentation, on measuring distances between nearby frame vectors of a word (but not always successive ones) and applying a threshold to determine where to start a new segment. Other possible methods of deriving a sequence of vectors to represent a segmented word, instead of interpolation at the segment boundaries, are selection of the nearer of the two neighbouring input vectors at each segment boundary (Gauvain, Mariani and Lienard 1983) and averaging of all the vectors in each segment (Ney 1983). A comparison of these methods (McInnes, Jack and Laver 1986a) indicates that selection gives poorer results than interpolation, and that averaging is better than interpolation when the number of segments per word is small (less than about half the average number of frames per word) but interpolation becomes slightly better as the number of segments approaches the number of frames.

*The Recognition Decision*

Once an unknown word has been read in and compared by the DTW procedure with all the reference templates, there will be a list of overall distances (perhaps normalised for template length) from the input word to the various templates. It has been found beneficial (Clotworthy and Smith 1986) to normalise the distance for each word according to the word's variability, as determined from a statistical analysis; but this requires a number of reference utterances for each word of the vocabulary.

The way in which these distances are used will depend on various aspects of the system. If there are several templates for each vocabulary word, it may be decided to assess a reference word by the distances to all or several of its templates rather than just by the distance to the nearest of them. Decision rules for this case are discussed below. But even where there is just one template for each word there are different ways to use the distances.

In a recogniser with one template per word, the most straightforward decision procedure is to recognise the input word as the word whose template gives the smallest distance. A modification is to make a recognition decision like this only if the distance is less than some fixed value (Itakura 1975), or only if it is less than the second-smallest distance by at least some fixed difference or ratio (Rabiner *et al.* 1979), or only on both of these conditions (Levinson, Rosenberg and Flanagan 1978), and to give a reject or no recognition response if the condition is not met. This reduces the rate of wrong recognitions, at the expense of a rejection rate which will generally include some rejections of correct nearest templates. This modification is appropriate for a system in which high reliability of recognised words is desired and the operator does not mind having to repeat some words when they are not clearly recognised the first time.

Another option is for the output from the template matching process to consist not just of a single best-matching word but of an ordered list of the best several candidates (Rabiner *et al.* 1979, Rabiner and Wilpon 1979a). This is particularly useful in a system where there are restrictions on which word sequences may occur (due to the syntax or format of the input: Levinson, Rosenberg and Flanagan 1978; Levinson 1978), since these restrictions can be used to eliminate some of the words listed as possible by the template matching process at each position in the sequence. This procedure can be applied, for example, in a directory assistance system, to the recognition of names spelt out (so that the words to be recognised are letter names), with the restriction on the sequence being that the name formed must be in the directory (Rosenberg and Schmidt 1979, Rabiner and Levinson 1981).

In the case of several templates for each word, there are various decision rules that can be applied to define the best identification for an input word, among them the nearest neighbour (NN), $K$-nearest neighbour (KNN) and majority vote decision rules.

The NN rule is the simplest: the input word is recognised as the word corresponding to the template whose distance from the input pattern is smallest. This is similar in implementation to the procedure described above for the case where each word in the vocabulary is represented by just one template.

The KNN rule (Rabiner *et al.* 1979, Rabiner and Wilpon 1979a) takes the average, for each word in the vocabulary, of the distances from the input word to the $K$ nearest templates of the vocabulary word, and then chooses the vocabulary word for which this average is smallest. With 10–12 speaker-independent templates per word, 2 or 3 is a suitable value for $K$ (Rabiner *et al.* 1979, Rabiner and Wilpon 1979a, Rabiner and Levinson 1981). Notice that the NN rule is a particular instance of the KNN rule, namely that in which $K=1$.

The majority vote rule (sometimes confusingly called KNN: Gupta, Bryan and Gowdy 1978) takes into account the $K$ templates (of whatever words) nearest to the input pattern, and chooses the word represented by the greatest number of these $K$ templates. In the event of a tie among different words, the NN criterion is used to decide among them. Here $K$ will usually be larger for this rule than for the KNN rule: the value $K=7$ has been found to give good results (Gupta, Bryan and Gowdy 1978; Gupta, Lennig and Mermelstein 1984) when there are 12 or 14 templates per word. This too reduces to the NN rule when $K=1$, and indeed also when $K=2$.

Many other decision rules could be devised, such as modifications of the KNN rule using weighted averages, or combinations of majority vote rules with several different values of $K$. A procedure using NN or majority vote, depending on features of the distribution of templates around the input pattern, was found (Gupta, Bryan and Gowdy 1978) to give better recognition performance than either of the two rules used alone.

*Reductions in Computation and Storage Requirements*

DTW algorithms are computationally expensive, especially when the vocabulary is large and each input word has to be compared with every one of the templates. Various means of reducing the amount of computation, and in some cases also the storage requirements (for reference data or for quantities used in the algorithm such as accumulated distances), have been proposed.

One way to eliminate some of the computation is to impose some sort of accumulated distance threshold on each template as it is being matched with the input word, and to abandon the matching process if the threshold is exceeded (Itakura 1975, Rabiner *et al.* 1979, Rabiner and Levinson 1981). Another procedure (Gupta, Bryan and Gowdy 1978) which has a similar effect is to carry out the DTW matching with the input word for all the templates in parallel and, at certain stages, to exclude from consideration prespecified proportions of the templates (choosing those with the largest accumulated distances). Indeed, both of these elimination procedures can be applied together. The amount of computation required can be reduced by such methods by a factor of 5–10 (Itakura 1975; Gupta, Bryan and Gowdy 1978).

A beam search strategy (Bisiani and Waibel 1982; Greer, Lowerre and Wilcox 1982) can be adopted, in which all the templates are matched in parallel, and at each input frame only those templates – or only those partial paths, in whatever template – which have accumulated distances less than a fixed threshold above the best current accumulated distance are retained. This has an advantage of adaptiveness over the fixed absolute threshold method, and, unlike the fixed-proportion exclusion method, it allows the number of templates under consideration at each stage to depend on whether there are many templates with accumulated distances close to the current minimum.

A branch-and-bound or best-first strategy (Bisiani and Waibel 1982) involves continuing, at each stage, the path (in whatever template) which has the least accumulated distance. Thus not all paths under consideration at any stage will necessarily have reached the same input frame. (A search strategy decision must be taken after each accumulated distance calculation.) A pruning technique can then be applied, by which any path which falls behind the longest current partial path (in the input direction) by a set number of frames is abandoned. This method of pruning has the same advantages mentioned above for beam searching. A drawback of the branch-and-bound approach is the large number of accumulated distance comparisons required to identify the paths to be extended.

If some words in the vocabulary are very dissimilar to the input word, it may not be necessary even to begin the DTW matching for these words since they can be eliminated by some simpler procedure before they get to the DTW stage. For instance, a match limiter has been employed (Kaneko and Dixon 1983) which makes an initial comparison of each template with the input, using only duration and three averaged spectra to represent each word. Then only the $K$ words (for some fixed value of $K$) with the best scores, out of a much larger vocabulary, are passed on to the second stage which performs standard DTW matching. Procedures similar in principle to this have been devised using a variety of other comparison methods to obtain the preliminary scores (Guo-tian 1985; Pan *et al.* 1985; Pan, Soong and Rabiner 1985; Bergh, Soong and Rabiner 1985; Aktas *et al.* 1986; Miwa and Kido 1986; McInnes, Jack and Laver 1986b). A threshold on the distance in the preliminary match, possibly depending on the distance obtained for the best-scoring template, can be applied to decide which words are passed on, as an alternative to specifying a fixed number of words (Pan *et al.* 1985; Pan, Soong and Rabiner 1985; McInnes, Jack and Laver 1986b). This two-stage decision procedure can be extended to three or more stages (Pan *et al.* 1985; McInnes, Jack and Laver 1986b) with increasingly accurate and computationally intensive comparisons.

In a multiple-template system, another option is to match initially only a certain number of the templates for each word, and on the basis of the distances obtained to select a few words for matching of all their templates (Nomura and Nakatsu 1986).

Another approach, which can be combined with the techniques already described, concentrates not on eliminating templates but on reducing the amount of computation to be done for each template in aligning it with the input word. Some examples of this sort of reduction have already been examined, namely the restrictions on the area in which paths are allowed. But there are other, more sophisticated techniques, and some of these are described below.

One modification is an ordered graph search (OGS: Brown and Rabiner 1982a,b). For each point $i$, $f(i)$ is defined as the minimal total distance on a path from $(1, 1)$ to $(M, N)$ going through $i$, and is estimated by $f'(i) = g(i) + h'(i)$, where $g(i)$ is the accumulated distance on the path selected from $(1, 1)$ to $i$ and $h'(i)$ is an estimate of the distance $h(i)$ on the optimal path from $i$ to $(M, N)$. Then, at each stage in the process, the point $i$ with the smallest value of $f'(i)$ (among points at the ends of existing partial paths) is expanded, i.e. an accumulated distance $g(j)$ is calculated for each point $j$ which may be the successor of $i$ on a path: $g(j) = g(i) + d(j)$ (if weight on $d(j) = 1$). The process continues until an attempt is made to expand $(M, N)$, when the completion of the path is recognised and the overall distance is taken to be $g(M, N)$. This procedure reduces the number of frame distances calculated, but requires considerably more computation for other parts of its execution than the standard DTW search: thus its usefulness depends on details of the implementation. If $h'(i)$ is taken to be 0 for every point $i$, the OGS technique reduces to the branch-and-bound technique described above, applied to each template separately.

Both the computation of frame distances for each word pair matched and the storage requirement for templates can be reduced by reducing the number of frame representations stored for each word. One way to do this (Tappert and Das 1978) is (in contrast to storing separately the representations for successive frames where these are similar) to store only the first frame representation for a steady section of speech. Representations for succeeding frames are treated as being identical to the first one as long as they differ from it by less than some preset amount. Then, when it comes to calculating frame distances, if the $m$th input frame has been classified as being the same as the $(m-1)$th, for instance, then $d(m, n)$ is not calculated anew if $d(m-1, n)$ has already been calculated. In the system in which this procedure was implemented, it was found that the reference storage requirements could be reduced by a factor of 2, and the frame distance calculations reduced by a factor of 4 (since the reduction was applied to both input and reference frames), without significant loss of recognition performance (Tappert and Das 1978).

Various methods of variable frame rate coding and acoustic segmentation have been devised (Silverman and Dixon 1980; Greer, Lowerre and Wilcox 1982; Brown 1982; Kuhn and Tomaschewski 1983; Pieraccini and Billi 1983; Gauvain, Mariani and Lienard 1983; Chuang and Chan 1983; Scagliola and Sciarra 1986) to reduce the number of (frame or segment) representations per word. Information about the durations of the segments represented may be stored, and used to control the matching of segments (Silverman and Dixon 1980; Greer, Lowerre and Wilcox 1982; Scagliola and Sciarra 1986) or to determine how many times each representation should be repeated in a segment expansion procedure (Brown 1982). The latter case is very similar to the procedure of Tappert and Das (1978) described above. If there is no expansion by repetition of representations, there is a saving in accumulated distance computation, in addition to the reduction in the number of frame distances to be calculated, but in this case the accuracy of the recogniser tends to deteriorate as the number of segments per word is reduced (Brown 1982).

The DYPATS process (Kuhn and Tomaschewski 1983) already described reduces computation and storage – because $S<J$ and so the number of frames in each word is reduced. In addition, the trace segmentation stage provides a large part of the time warping required (in the case of reference and input being the same word) and so the DTW search can be restricted to a narrow band around the

diagonal from (0, 0) to $(S, S)$ without much loss of recognition accuracy.

The technique of vector quantisation (VQ: Sugamura, Shikano and Furui 1983; Niles, Silverman and Dixon 1983; Pieraccini and Billi 1983; Levinson, Rabiner and Sondhi 1983; Kammerer, Küpper and Lagger 1984; Glinski 1985) can be used to reduce the reference storage requirements and the number of frame distances to be calculated, particularly if the number of templates is large. This technique generally results in some degradation of the recognition performance.

Vector quantisation can be one-sided or two-sided (Pieraccini and Billi 1983, Glinski 1985). In the case of one-sided quantisation, a codebook is constructed, consisting of vectors of speech frame parameters, and each frame of each template is represented by one of the codebook vectors, usually the one nearest to the actual frame vector. When an input frame vector is received, it is matched with all the codebook vectors, and the distances thus obtained are used in place of the true input-reference frame distances in the DTW algorithm. The templates are stored as sequences of codebook vector indices, rather than sequences of actual vectors. The codebook vectors themselves are stored separately. In two-sided VQ, both reference and input frames are represented by vectors from the codebook. This has the advantage that the distances for all pairs of vectors in the codebook can be stored as a table, so that they do not have to be calculated afresh when they are required. Some computation will have to be done, however, to identify which codebook vector should be used to represent each new input frame. This need not involve matching the input vector with all the codebook vectors: a faster VQ technique such as binary tree coding (Glinski 1985) can be used – though this may reduce the recognition accuracy. Depending on the size of the codebook, a large amount of memory may be required for the table of distances.

*Dynamic Programming applied to Hidden Markov Models*

One approach to speech recognition is to construct a statistical model (rather than a template or templates) of each word in the vocabulary, and recognise each input word as that word of the vocabulary whose model assigns the greatest likelihood (probability or probability density) to the occurrence of the observed input pattern. One type of statistical model (Bahl and Jelinek 1975; Billi 1982; Brown, Lee and Spohrer 1983; Rabiner, Levinson and Sondhi 1983; Levinson, Rabiner and Sondhi 1983; Sugawara *et al.* 1985; Paul 1985; Juang *et al.* 1985) is the hidden Markov model

(HMM) described below. If such a model is adopted a form of dynamic programming algorithm, called the Viterbi algorithm (Forney 1973), can be applied to calculate, for each word's model, the likelihood for the optimal matching of the input to a sequence of states of the model (Forney 1973; Bahl and Jelinek 1975; Billi 1982; Rabiner, Levinson and Sondhi 1983; Levinson, Rabiner and Sondhi 1983; Juang *et al.* 1985).

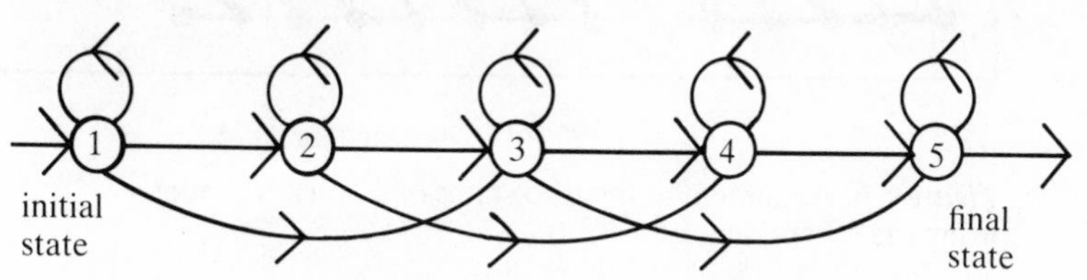


*Figure* 1.5. State transition network for a Markov model.

The essential idea of the HMM approach is that each word in the vocabulary is represented by a set of states, including an initial state and a final state, with probabilities of transitions from state to state, and for each state a probability distribution for the emission of a vector of acoustic parameters. (The states and transitions can be shown in the form of a network, as in figure 1.5.) When the word is spoken, the process is assumed to be in the initial state when the word begins, and then to make state transitions at time intervals equal to the separations between speech frames, in such a way that it is in the final state when the word ends. At each time frame, a vector is emitted whose probability distribution is that associated with the current state. An alternative formulation (Bahl and Jelinek 1975, Sugawara *et al.* 1985) has the emission probabilities associated with the transitions rather than with the states. The states themselves are not assumed to be observable, but only the emitted frame vectors which are probabilistically related to the states – hence the use of the word *hidden*. The name *Markov* is applied because the state transitions form a first-order Markov chain: that is, the probabilities of transitions from a given state depend only on the identity of that state, and not on which states the process has been in at previous times.

To apply the Viterbi dynamic programming algorithm to the matching of a sequence of input frames to a given HMM, the states of the Markov model are arranged along the reference axis, as in figure 1.6, where the possible transitions at each input frame are shown in the form of a trellis (Forney 1973) and the optimal alignment is marked by the heavy line. The path constraints, local

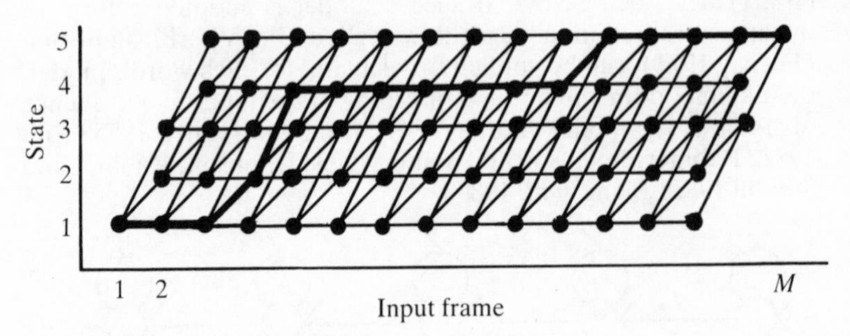


*Figure* 1.6. Alignment of input to states of a Markov model using the Viterbi algorithm.

distances and accumulated distances are defined as follows.

At each step, the path must advance exactly one frame in the input direction, and can move in the reference direction to any state to which there is a possible transition from the preceding state. Thus, for instance, if the allowable transitions to state $n$ are from states $n-2$, $n-1$ and $n$, the possible predecessors of the point $(m, n)$ are $(m-1, n-2)$, $(m-1, n-1)$ and $(m-1, n)$. Note that the possible steps in the path are not necessarily the same at all points (even ignoring endpoint effects): they depend on which transitions exist among the particular states in the part of the Markov model that the path has reached. In the case illustrated, however, the model has a regular structure, and so the possible path steps exhibit a corresponding regularity.

The distance between an input frame vector and a state is the negative of the logarithm of the probability (or probability density) assigned to that vector by the probability distribution associated with the state. An additive quantity (rather than a multiplicative weight as is usual in template matching) is assigned to each step in a path according to which state transition takes place there. The quantity added is the negative of the logarithm of the transition probability. The accumulated distance at any point on a path is the sum of the contributions (local distances and negative log transition probabilities) from the points and path steps up to and including that point. This is the negative of the logarithm of the probability (or probability density) of the sequence of transitions and emissions represented by the path and the input frames so far. All transition and emission probabilities are assumed to be independent of one another, so that the probability of the sequence is the product of the individual probabilities, and hence its negative log probability is the sum of the individual negative log probabilities. The reason for working in the logarithmic domain is the computational convenience of adding instead of multiplying.

The Viterbi algorithm reduces to a simple form of DTW (with $P(i)$ consisting of a single point $[(m-1, n_i)]$ for each $i$, weighting according to movement in the input direction, and an appropriate distance function), if a state is defined for each template frame and all the transition probabilities are equal – apart from an additive quantity, depending only on the input frame number, due to the addition of negative log transition probabilities. The emission probability distribution for each state in the HMM corresponds to a frame vector in the template and the associated frame distance function. In particular, a multivariate Gaussian distribution with a diagonal covariance matrix yields negative log probabilities which are weighted Euclidean distances from the mean vector of the distribution. The relation between HMMs with Gaussian autoregressive probability densities and DTW with an LPC-based distance measure is developed in detail in Juang (1984). Similarly, HMM formulations of other forms of DTW algorithm can be devised. One corresponding to the Type III constraints (Cravero *et al.* 1984), for instance, involves two copies of each state (strictly, two states with the same emission probability distribution but different transitions); the second copy is used when a state is repeated (corresponding to a step (1, 0) in the path). Thus DTW is in principle a special case of the Viterbi algorithm. The peculiarities of DTW are that the number of states (template frames) per word tends to be larger than in other word recognition techniques using the Viterbi algorithm (and indefinite repetition is not usually permitted) and that the variability of the transition and emission probabilities from one state to another is more constrained.

The Viterbi algorithm does not, strictly speaking, obtain the likelihood of the observed input speech given the HMM for the hypothesised word. It obtains the likelihood of the more restricted event that, given the HMM, the observed input vectors are emitted from a specific sequence (the optimal sequence) of states. The overall likelihood of the observed input vectors, given the model, is the sum of the likelihoods with specified sequences of states. The above algorithm can be adapted to calculate this by first transforming to the multiplicative domain (multiplying probabilities, instead of adding their negative logarithms), and then at each point adding the accumulated partial-path probabilities, instead of selecting the maximal one, before multiplying by the local likelihood. This adapted algorithm is related to the forward-backward algorithm described below (Rabiner, Levinson and Sondhi 1983) where the quantity computed at each point is the forward probability. This calculation of the overall likelihood (called Baum-Welch scoring: Rabiner, Levinson and Sondhi 1983) is more expensive computationally than Viterbi scoring, because of the multiplications involved.

The emission probability distribution for each state can be discrete or continuous. In the discrete case (Billi 1982; Rabiner, Levinson and Sondhi 1983; Levinson, Rabiner and Sondhi 1983; Sugawara *et al.* 1985), the output of the model is assumed to be a sequence of symbols from a finite alphabet. To cope with input consisting of acoustic parameter vectors, vector quantisation must be applied; then the HMMS are matched with the sequence of VQ codebook indices derived from the input. In the continuous case, the distribution associated with each state is typically some form of Gaussian distribution, or a mixture of such distributions (Juang *et al.* 1985, Rabiner *et al.* 1985, Juang and Rabiner 1985, 1986) – though a system incorporating mixtures of non-parametric distributions, obtained using a Parzen estimator, has been reported (Soudoplatoff 1986). Continuous distributions have been found to yield better recognition performance than discrete ones (Juang *et al.* 1985, Rabiner *et al.* 1985). For the case where the emission probability distributions are discrete and are associated with transitions rather than states, an alternative formulation (Bahl, Jelinek and Mercer 1975) has separate transitions, with appropriately adjusted probabilities, corresponding to all the possible emitted symbols. In this case the Markov model is explicit, rather than hidden, but essentially the same parameter estimation and recognition algorithms can be applied.

Models for words which are similar in some (initial and final) parts can be combined into a single model, thus economising on storage and computation – rather as the templates for such words are combined in the discriminative technique described above. The problem of computing appropriate transition probabilities for such models is addressed in Kamp (1985). Standardised forms of state transition network have been devised. One such is the Bakis model (Billi 1982; Bourlard, Wellekens and Ney 1984), which allows a state to be repeated indefinitely or to be omitted: thus the local path constraints are those of the simple example illustrated in figure 1.3. (The transition network in figure 1.5 is for a five-state Bakis model.)

A comparison of digit recognition results using speaker-trained HMMS with different structures is given in Russell and Cook (1986).

The best results were obtained when the number of states was large (e.g. 20) and arbitrary left-to-right transitions were permitted. Most other results reported have been for models with fewer states – typically about 5 (Rabiner, Levinson and Sondhi 1983; Levinson, Rabiner and Sondhi 1983; Sugawara *et al.* 1985; Juang *et al.* 1985; Rabiner *et al.* 1985; Kamp 1985; Juang and Rabiner 1985, 1986). Markov models have been used particularly in connected word recognition (Bahl, Jelinek and Mercer 1975; Kaplan, Reddy and Kato 1980; Cravero *et al.* 1984), as described below.

The training procedure for each word of the vocabulary is rather more complex for an HMM-based recogniser than for a template-based one. First, the number of states to be used to represent the word must be decided, and allowable transitions between states must be defined (Rabiner, Levinson and Sondhi 1983; Levinson, Rabiner and Sondhi 1983; Russell and Cook 1986). Then the emission and transition probabilities must be estimated and this can be done in various ways.

One method of estimating the transition probabilities (Jelinek 1976; Billi 1982; Bourlard, Wellekens and Ney 1984; Cravero *et al.* 1984) is to apply the forward-backward algorithm (described below) or the Viterbi algorithm iteratively to the training data for the word being modelled. Before this is done, each state in the word model must be assigned a probability distribution for the emission of frame vectors. This is done by analysis of the distribution of frame vectors from the training data which might (as indicated, perhaps, by a clustering analysis) correspond to that state. Then the transition probabilities are estimated by starting with (for example) equal probabilities for all transitions from each state, applying the forward-backward or Viterbi algorithm to each training token of the word and adjusting the probabilities to accord with the frequencies of occurrence of the transitions. The emission probability distributions can be re-estimated at the same time by accumulating statistics on the vectors in the training data that are associated by the algorithm with each state. This procedure can be iterated several times. The forward-backward algorithm (Bahl, Jelinek and Mercer 1975; Jelinek 1976; Juang 1984) is a recursion procedure which accumulates statistics from all possible alignments of the data with state sequences of the model, rather than only the alignment with the optimal state sequence which is found by the Viterbi algorithm. The iterative training algorithm which uses these statistics is called the Baum-Welch algorithm (Rabiner, Levinson and Sondhi 1983; Levinson, Rabiner and Sondhi 1983). Statistics on the vectors occurring with each state are assigned weights in the training procedure which correspond to the probabilities with which the state-vector combinations occur. Similarly, the re-estimated probabilities for transitions from a state are proportional to the weighted sums of occurrences of these transitions over all possible alignments – where the weights are again the probabilities assigned to the occurrences by the model with its existing parameters. To compute these weighted sums efficiently, they are formulated in terms of forward and backward probabilities, which can be calculated by the forward-backward algorithm.

There is experimental evidence to suggest that the Viterbi algorithm, though theoretically inferior, since it does not provide maximum-likelihood estimates (when applied to training) or the overall likelihood of a word given a model (when applied to recognition), is as good as the forward-backward algorithm in practice, at least for some applications (Rabiner, Levinson and Sondhi 1983; Levinson, Rabiner and Sondhi 1983; Russell and Cook 1986). A Bayesian estimation procedure for estimating emission probabilities, assuming the availability of transition probability estimates, is described in Brown, Lee and Spohrer (1983). This also involves iterative application of the Viterbi algorithm to training data, but uses speaker-independent, prior probability distribution estimates for the model parameters, whose inclusion was found to improve the recognition performance of the models generated. It could be modified to incorporate progressive estimation of transition probabilities. A maximum mutual information estimation procedure, differing from the usual maximum likelihood estimation in that all the models' parameters are estimated together to optimise the discrimination between different words, has been observed to yield improved recognition performance (Bahl *et al.* 1986).

There are other algorithms, including some based on Lagrangian techniques, which can be used to estimate both emission and transition probabilities (Rabiner, Levinson and Sondhi 1983; Levinson, Rabiner and Sondhi 1983).

A problem with many HMM training techniques is that they find locally optimal values in the space of model parameters which are not guaranteed to be globally optimal. Improvements in this respect have been obtained (at considerable computational expense) using a simulated annealing process (Paul 1985), in which random perturbations are allowed to shake the parameter values out of a local optimum so that the globally optimal values can be found. Another approach (Russell and Cook 1986) is to improve the chance of reaching the global optimum by deriving a good set of initial parameter values from training data by a word segmentation procedure.

The advantages of the HMM approach are that more of the information contained in several training tokens can be incorporated in a single word model than could be incorporated in a single template, and that probabilistic information from frequency-of-occurrence statistics can be conveniently represented. However, to make use of these advantages several training tokens of each word are required. One HMM per word has been found to give fairly good results in speaker-independent word recognition, though not always as good as those obtained (with considerably more computation) using multiple templates (Rabiner, Levinson and Sondhi 1983; Levinson, Rabiner and Sondhi 1983). Improvements have been found to result from using two HMMS per word (one for male and one for female speakers, or by a clustering procedure), instead of pooling all the training data to make a single model for each word (Juang *et al.* 1985, Rabiner *et al.* 1985, Poritz and Richter 1986).

An adaptation procedure for HMMS, based on the forward-backward algorithm, has been implemented (Sugawara, Nishimura and Kuroda 1986); this allows models to be improved as more examples of the words become available, by a weighted averaging of the original training statistics and those derived from the new input, and can be used to adapt the models to a new speaker.

A shortcoming of Markov models for representing spoken words is the assumption that the state transition and emission probabilities at each frame of the word are independent of the state transitions that have occurred earlier in the utterance and the frame vectors that have been emitted at earlier times. In particular, the probability that the process will remain in a given state at frame $m$, given that it is in that state at frame $m-1$, is taken to be independent of the number of frames preceding frame $m-1$ during which the process has been in that state. In the common and computationally convenient case where each state has a transition to itself to allow indefinite repetition, this gives rise to an exponential probability distribution for the duration of each state, which does not correspond well to the characteristics of real speech (Russell and Moore 1985, Juang *et al.* 1985, Russell and Cook 1987). Modifications of HMM-based recognition have been proposed to overcome this problem, by incorporating durational information into the dynamic programming procedure (Russell and Moore 1985, Juang *et al.* 1985, Rabiner *et al.* 1985, Russell and Cook 1987), or by adjusting the word distances after alignment to take account of duration probabilities (Juang *et al.* 1985, Rabiner *et al.* 1985). Hidden semi-Markov models (Russell 1985, Russell and Cook 1987), which incorporate a specification of the probability distribution of the duration of each state, can be trained by an extension of the forward-backward algorithm (Russell 1985). Improvements in recognition accuracy have been observed to result in some cases (Russell and Moore 1985, Juang *et al.* 1985, Rabiner *et al.* 1985). It is also possible to improve state duration modelling, while retaining the computational simplicity afforded by the first-order Markov property, by using multiple copies of each state (i.e. states with identical emission probabilities) in some suitably devised configuration (Cook and Russell 1986, Russell and Cook 1987).

*Word Spotting*

Usually in a speech recognition system the aim is to recognise all the words spoken. The word spotting task is an exception to this. Here the aim is to find all occurrences of designated key words in a sample of connected speech, and there is no attempt to identify the other words in the utterance. This has applications in selection of those parts of a large corpus of speech (consisting for instance of intercepted radio messages) which are of interest for some specific purpose.

A method of matching a key word template to appropriate portions of a sample of connected speech has been described (Bridle 1973). Like the isolated word recognition procedures already described, this involves time warping and frame distances; but in other respects there are differences due to the different nature of the problem. Instead of being given input with a specified word-beginning frame or region, the procedure must investigate each input frame to see whether an instance of the key word begins there. When a likely beginning of a key word is found, the end of the word is still undetermined, and so no final input frame can be specified to constrain the warping path. The decision required as output from the template matching process is not which word or sequence of words best matches the input pattern, but rather whether any parts of the input pattern are instances of the key word, and, if so, which parts. The method described depends on defining a local similarity function $F$ which is a weighted sum of frame similarity values at recent points along a partial path, with the weights decaying exponentially away from the current point. The procedure does not however give an overall distance or similarity measure for a completed path.

The search for a region matching the key word is illustrated in figure 1.7. From each input frame, a path is started, and is extended as long as the values taken by $F$ continue to be above a set threshold. A path completed subject to this condition, so that $F[P(m, n)]$

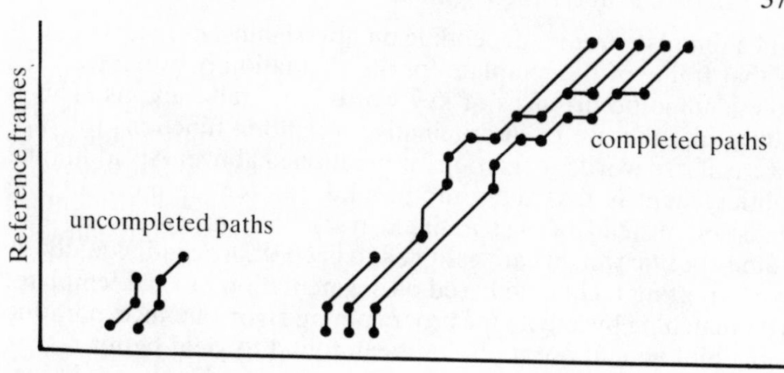


*Figure* 1.7. Application of DTW to word spotting.

exceeds the threshold for every initial part $P(m, n)$ of the path, is taken as indicating an occurrence of the key word. There may well be several completed paths very close together, perhaps with points in common. In such a case a rule can be applied to ensure that the word is recognised as occurring only once there.

A modified version (Christiansen and Rushforth 1977) of this word spotting method involves the use of multiple templates to form a composite template during the matching procedure. For each key word, several templates are stored, perhaps from different speakers. Then, for each point $(m, n)$ on a warping path, the frame similarity used is the maximum of the similarities between the $m$th input frame and the $n$th frames of all the templates. Thus the template effectively used for a path has its $n$th frame chosen from the $n$th frames of the available templates so as to maximise the similarity at the appropriate point on the path.

The threshold on $F$ must be chosen to achieve a satisfactory rate of spotting of genuine occurrences of key words without introducing too many false alarms. It is best to choose the threshold value for each speaker and each key word individually. One problem with this is that the input speakers may not be available for training, and so some method of automatic adjustment to the characteristics of speakers is desirable (Wohlford, Smith and Sambur 1980).

A secondary testing procedure can be applied (McCullough 1983) for identification of the better putative hits, using template-specific linear combinations of various matching statistics, to add to the information obtained in the primary matching process. A weight is assigned to each combination of a frame matching statistic and a template frame, depending on how significant that statistic is at that frame of the template for discrimination between true hits (on genuine occurrences of key words) and false alarms. This is rather like the use of discriminative weighting functions for two-pass isolated word recognition, as mentioned above. An additional subtlety here is that there are several different frame matching statistics, instead of just a single frame distance measure, and frame-specific weights are assigned to each statistic individually. A word spotting technique based on segmentation of each template, with matching by a dynamic programming algorithm incorporating limits on segment duration, has been found to yield better results than the whole-word matching method with standard DTW (Kawabata and Kohda 1986).

*Connected Word Recognition*

In a connected word recognition system, the input consists of words spoken without gaps between them. (Connected speech is not necessarily fluent speech: the latter has more coarticulation between words, making the application of a template matching system rather more difficult.) The whole utterance is typically a sentence or a string of digits. Thus the recognition involves not only deciding what each word is but also finding where words begin and end, and, usually, deciding how many words the utterance contains. (A rather similar problem occurs in isolated word recognition if the gaps between words are not long enough to be distinguished, by their duration, from stop consonant silences within words. The Quiktalk algorithm (Welch and Oxenberg 1980) is designed to cope with this problem, and is similar in principle to the two-level algorithm for connected word recognition described below.) A DTW system with word templates can make all these decisions together. Various DTW algorithms for connected word recognition have been devised; these are described in the paragraphs below.

A two-level algorithm has been proposed (Sakoe 1979) in which, in the first stage, each word template is matched against every part of the input, and then, in the second stage, these matchings are combined to give a recognition of a whole string of words. In the first part of the two-level algorithm, word level matching, a template is chosen for each permissible input section (permissible in the sense that it is of such a length that at least one of the templates can be matched to it) to minimise the word distance obtained by matching against that part of the input. The output of the word level matching consists of a template index and a corresponding word distance for each permissible pair of starting and ending frames. The second part of the algorithm is phrase level matching. This uses the information given by the word level matching and builds up the word reference string by a dynamic programming search in the $(l, m)$ plane, where $l$ is the number of the word (counting from the start of the utterance) and $m$ is the input frame number. The accumulated distance $G(l, m)$ at each point is obtained by minimising over all possible beginning frames for an $l$th word ending at frame $m$. The path ends at $(L, M)$, where $L$ is the number of words in the utterance and $M$ is the number of input frames. If the number of words in the utterance has been specified, $L$ is fixed and so the endpoint $(L, M)$ is automatically determined. If not, several paths may be completed, with different values of $L$ at their endpoints, corresponding to matchings of concatenations of different numbers of reference words to the input. In this latter case, $L$ is chosen so as to minimise the total distance $G(L, M)$.

Two implementations of this two-level matching procedure have been given (Sakoe 1979): one suitable for computer simulation, in which tables of frame distances for all possible input-reference frame pairs are computed and referred to in the course of the word level matching, and the phrase level matching is not begun until the word level matching has been completed; and one suitable for real-time recognition, in which, as each input frame is read in, the word level and phrase level matching procedures are advanced to incorporate words and partial phrases ending at that input frame.

An algorithm similar to the two-level algorithm but allowing an overlap or a gap between successive words has been formulated and tested (Tajima, Komura and Sato 1986). The best results (with templates formed from connected speech as well as from isolated words) were obtained when no overlap was allowed but a gap of up to 15 or 20 frames (150 or 200 ms) was permitted.

A sampling algorithm which reduces the computation involved by matching templates only to selected parts of the input has been described (Myers, Rabiner and Rosenberg 1980a, 1981; Rabiner and Schmidt 1980a,b). This builds up strings of words from left to right, and at the $l$th stage keeps a list of the best few candidate strings of identifications of the first $l$ words, with their ending frames and accumulated distances. To extend a partial recognition string, templates are matched to the portion of input speech following its ending frame; the UELM search area constraints are used. An overlap of up to $\Delta$ frames between successive words is allowed and, within the range defined by this parameter, every $I$th input frame is tried as a starting frame. A string is considered complete when its last word ends within a specified number of frames of the end of the input pattern. Once all strings of words have been either completed or abandoned, the one with the smallest overall distance can be taken as the recognition of the utterance, or else a post-error correction procedure can be employed. In the post-error correction, a whole-utterance template is constructed for each of the best few complete strings already found, by concatenating the templates in the string, and is matched against the input utterance by a large-scale constrained-endpoint DTW using either the usual frame distance measure or some simpler measure such as one distinguishing only among voiced, unvoiced and silent frames of speech. The final decision as to which string best matches the input can then be based either purely on the distance obtained in this way or on the average of this and the original distance derived in building up the string. A modified version of the sampling algorithm, with reduced weighting near word boundaries where coarticulation occurs, is described in Gauvain and Mariani (1982). Unlike the other connected word DTW algorithms described here, the sampling algorithm does not guarantee optimality of the sequence of words found.

The level-building algorithm (Myers and Rabiner 1981a) is proposed as a more computationally efficient implementation of the basic idea of the two-level algorithm. The fundamental difference between the two is that, whereas the two-level algorithm minimises accumulated distance at each possible partial phrase ending frame $m$, firstly over template index $v$ (separately for each possible beginning frame $m'$ of a word ending at $m$) and then over word beginning frame $m'$, the level-building algorithm minimises it firstly over $m'$ (separately for each $v$) and then over $v$. (Again this is all done separately for each possible value of $l$ corresponding to the ending frame $m$.) The meaning of *level* in this algorithm is different from that in the two-level algorithm. Here there is no division into word level and phrase level matching, and a level is a value of $l$. The improvement in efficiency (over Sakoe's second-version algorithm (Sakoe 1979)) is because duplication of frame distance calculations occurs only due to matching the same word at different levels in the same region of the input pattern. There is no duplication for different beginning frames of the same word at the same level, since all the matching for a given word and level is carried out in a single application of the basic DTW procedure. The operation of the level-building algorithm is illustrated in figure 1.8. The computation proceeds level by level, and, at each level, template by template. At the beginning of each level, there are various possible previous frames, each with its accumulated distance de-

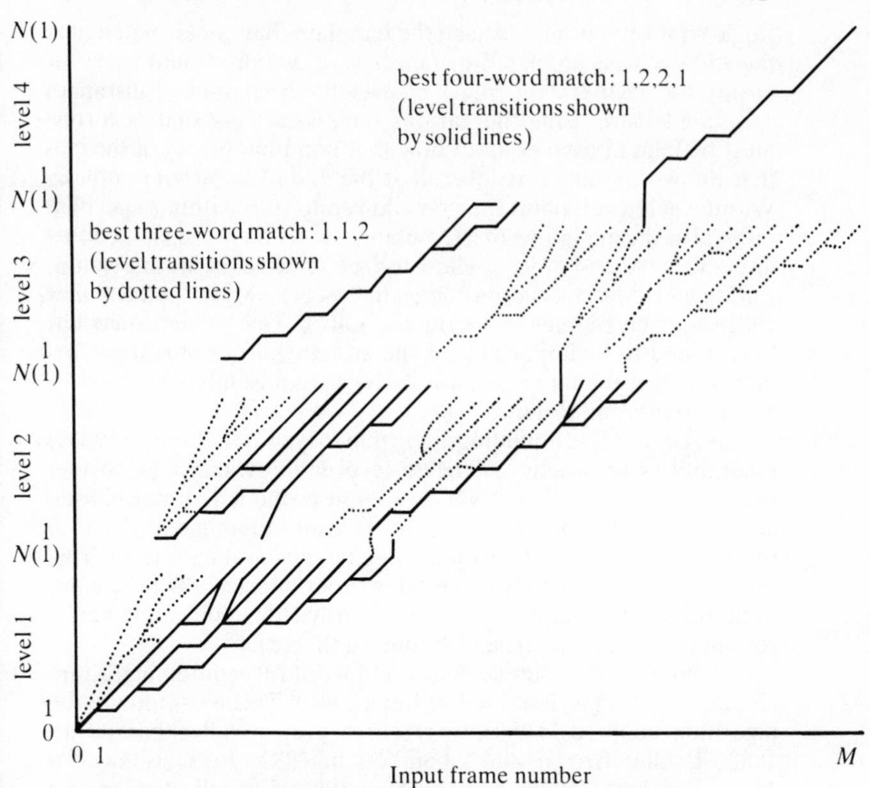


*Figure* 1.8. Level-building algorithm with a two-word vocabulary and the restriction $L<5$. Within each level, solid lines show matching of template 1; dotted lines, of template 2. $N(1) = N(2) + 1$.

rived from the previous levels. In the case of the first level, there is only one previous frame, namely input frame 0 (added for convenience of notation), with accumulated distance 0. In the notation used above for the two-level algorithm, the accumulated distance to input frame $m$ at the beginning of level $l$ is $G(l-1, m)$. For each template, a DTW matching is carried out starting from these previous-level ending frames and their accumulated distances, which gives a set of ending frames for that template on the current level, with corresponding accumulated distances. The optimal starting frame of the word does not have to be found separately for each ending frame since paths to all the ending frames are found by the single DTW operation. Once all the templates have been matched at the $l$th level, accumulated distances $G(l, m)$ are found by minimising for each ending frame $m$ over the accumulated distances calculated there using the various templates. As usual, a record must be kept at each point of how that point has been reached, so that the words can be recovered at the end of the whole process. Within each level, paths may be allowed to start within a specified interval at the beginning of a template, and to end within a specified interval at the end of it, to allow for the reduced forms of pronunciation which often occur in connected speech (Myers and Rabiner 1981a,b,c; Gagnoulet and Couvrat 1982). This modification has been found to be important for the attainment of optimal performance, when the templates are derived from isolated-word utterances (Myers and Rabiner 1981c).

A reduced level-building algorithm (Myers and Rabiner 1981a) incorporates a restriction of the $l$th-level ending frame $M(l)$ to a set range around the value $m$ which minimises the length-normalised accumulated distance $G(l, m)/m$, and an accumulated distance threshold to eliminate badly matching templates at each level. This reduced algorithm involves the same order of frame distance computations as the sampling algorithm, but yields significantly better recognition rates (Myers and Rabiner 1981b,c).

A conceptually simple connected word recognition algorithm (Vintsyuk 1971) is described in Ney (1984). This is essentially the algorithm adopted in various research projects (Peckham *et al.* 1982; Bridle, Brown and Chamberlain 1982, 1983; Nakagawa 1983; Bourlard, Wellekens and Ney 1984; Landell, Naylor and Wohlford 1984; Cravero *et al.* 1984; Jouvet and Schwartz 1984). It has the advantage over the algorithms described above (apart from the first version of the two-level algorithm) that each frame distance is calculated just once. It also requires only a fairly small amount of storage for data obtained during its execution. The principle of its operation is illustrated in figure 1.9. For the purpose of the dynamic programming procedure, the word templates are made into a compound template, with special rules for going from the end of one word template to the beginning of another. A frame in this compound template has a word template index $v$, and a frame number $n$ within the word template. A point in the input-reference plane can be specified by three coordinates $(m, n, v)$, where as usual $m$ is the input frame. Where $n>1$ (or 2, if path constraints such as those in figure 1.4 are used), the permitted predecessors of $(m, n, v)$ are as specified by the usual path constraints, except for the extra coordinate $v$. Where $n=1$ (or 2, depending on the path con-

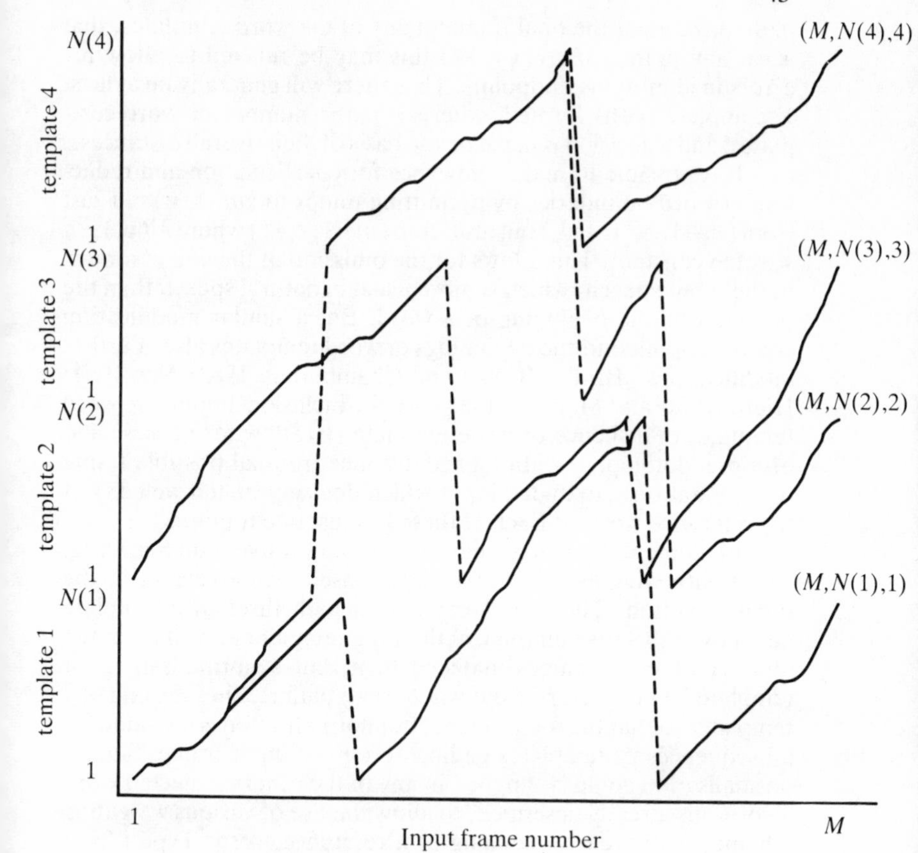


*Figure* 1.9. One-stage algorithm. Note the variation in number of words matched from one path to another: word strings matched are 1,4,4; 1,4,3,2; 1,1,2,1; and 2,3,2,2,3.

straints), the preceding point may be $(m-1, N(v'), v')$ for any template index $v'$ (or it may be any preceding point in template $v$ permitted by the ordinary path constraints, such as $(m-1, 1, v)$). (Here $N(v)$ is the number of frames in the $v$th word template $R(v)$.) Apart from this modification to the local path constraints, the DTW proceeds as for the isolated word case, one input frame at a time. One or more columns of distances, depending on the path constraints, need to be stored at each stage in the procedure. As is usual in a connected word system, a record must also be kept of the word templates used on the path to each point reached. A complete path may end at the final frame of any of the word templates, that is, at any point $(M, N(v), v)$; this may be relaxed to allow for errors in identifying endpoints. Thus there will generally be at least $V$ complete paths formed, where $V$ is the number of word templates, and a decision is made on the basis of their overall distances.

It is possible to make allowance for coarticulation and reduction at word boundaries by permitting jumps to $(m, 1, v)$ not just from $(m-1, N(v'), v')$ but also from $(m-1, n, v')$ where $N(v')-n$ $<$ some constant. This allows for the omission of the end of a word in the input speech, which is more usual in natural speech than the omission of the beginning of a word. But a similar modification could be applied to the beginnings of word templates also. Further modifications (Bridle, Brown and Chamberlain 1982; Ney 1984; Hieronymus and Majurski 1985) are the inclusion among the word templates of a silence or noise template (to allow for pauses) and of a pseudotemplate with a fixed distance from all possible frame representations (to match input which does not match well any of the reference words). Each of these has just one frame.

Implementations have been described (Jouvet and Schwartz, 1984) with weights other than those based on movement in the input direction. The total weight on a path through a template depends in this case on the length of the template as well as on the number of input frames matched to it, but a normalisation for template length is carried out whenever a path reaches the end of a template, so that the total weight after normalisation is the same for all sequences of templates ending at a given input frame. Similar normalisation could be applied in any of the other connected word algorithms already described, to allow the use of various weighting schemes within each matching of a reference word. Type I local path constraints and a symmetric weighting scheme were found to be better than Itakura constraints and the asymmetric scheme. A modification has also been described (Nakagawa 1983) in which different weights are assigned to different frames of a template (cf. the frame-specific weighting suggested in Bridle 1973), with a similar normalisation for total weight at the end of each word. A connected word recognition algorithm for use with variable frame rate analysis, incorporating weighting according to the durations of the frames being matched together, has been implemented, and was found to give better results than without use of duration information. The variable frame rate analysis technique (similar to trace segmentation with averaging) improved the recognition performance while reducing the amount of computation (Scagliola and Sciarra 1986).

One disadvantage of the one-stage algorithm in its basic form (relative to the two-level and level-building algorithms) is that it includes no specification of the number of words in the utterance. Such a specification could, however, be incorporated by means of syntactic constraints (Bridle, Brown and Chamberlain 1982), as described below. Implementations of the one-stage algorithm designed to reduce the number of memory access operations required are described in Watari (1986).

In a connected word recogniser with a large vocabulary, it is desirable to be able to make use of any syntax known to apply to the input utterance, in order to reduce the number of words that have to be tried at each point, and to increase the recognition accuracy by ruling out likely-sounding but impossible word sequences. The use of syntactic information has been found to result in greatly improved recognition accuracy in a system with input of sentences spoken as isolated words (Levinson, Rosenberg and Flanagan 1978; Levinson 1978; Rabiner and Levinson 1981). All the above algorithms can be modified fairly easily to incorporate syntactic constraints.

In the sampling algorithm, the matching of templates to extend partial recognition strings can be restricted so that only syntactically permissible words are tried. In the level-building algorithm, levels can be replaced by states in a loop-free transition network representing a simple grammar, and adding 1 to the level number $l$ to get to the next level when passing on accumulated distances can be replaced by making a transition to another state (Myers and Rabiner 1981a). The record kept of the path taken to each point will have to include information about which state transitions have been made. This adaptation of the level-building algorithm is described in more detail in Myers and Levinson (1982). By arranging the computation so that all levels are extended in parallel along the input, it is possible to dispense with the requirement that the syntax be loop-free, since in this case each transition need not always be from a preceding level.

The two-level algorithm can be modified (Myers and Rabiner 1981c) to allow the application of syntax, specified again by a transition network (not necessarily loop-free). In this case, the phrase level matching would have syntactic states in place of word numbers, and for each input section the optimal template and distance would have to be determined (in the word level matching) not just from the whole vocabulary but from each subset of the vocabulary corresponding to a syntactic state transition. An advantage of the two-level algorithm (Myers and Rabiner 1981c) is that it can generate the $K$ best candidate strings, for any specified value of $K$ – which can be useful in an application where the direct implementation of word sequence constraints is unwieldy (as in the recognition of spellings of names from a directory: Rosenberg and Schmidt 1979). This is done by recording the $K$ best words and distances for each input section (or all the word distances if $K>V$), and retaining the $K$ best partial strings of words at each point in the phrase level matching.

In the one-stage algorithm, each reference word can be given a 'from predecessor' set, which is the set of all words which may precede it; then the preceding points allowed for $(m, 1, v)$ will include $(m-1, N(v'), v')$ only for those template indices $v'$ for which $R(v')$ is in $R(v)$'s from predecessor set (Ney 1984). The points at which warping paths may start will be $(1, 1, v)$ only for those values of $v$ for which $R(v)$ is designated as a word which may begin an utterance. Similarly, paths will be allowed to end at points $(M, N(v), v)$ only where the words $R(v)$ are permissible final words. If the words which may precede $R(v)$ are different according to its position in the sentence, several copies of $R(v)$ can be included in the compound template, each with a different from predecessor set, and will be treated as different words during the DTW procedure, except that the frame distances can be computed just once (for each input frame) and used for all the copies. The syntactic network need not be loop-free. A particular case, where in fact the network is loop-free, is where the syntax specifies only the number, $L$, of words in the utterance. Then there are $L$ copies of each template, the from predecessor set of the $l$th copy of a template consists of the $(l-1)$th copies of all the templates, the permissible initial words are the first copies of all templates and the final words are the $L$th copies. The algorithm with this syntax is identical to the level-building algorithm. Details of the implementation of the syntactically constrained one-stage algorithm are given in Bridle, Brown and Chamberlain (1982) and Peckham *et al.* (1982).

If a Markov modelling approach is adopted, an integrated syntactic and acoustic network can be constructed (Kaplan, Reddy and Kato 1980; Cravero *et al.* 1984), in which each word in the syntactic network is replaced by the transition network for the Markov model of the word (with possible insertion of silence states at word boundaries). Then the Viterbi algorithm can be applied to this integrated network just as it would be applied to the transition network for a single word. This connected word recognition procedure corresponds to the one-stage algorithm with syntactic constraints in the same way that the isolated word Viterbi algorithm corresponds to ordinary isolated word DTW. Other connected word recognition procedures using HMMS include one using stack decoding (Jelinek, Bahl and Mercer 1975; Bahl, Jelinek and Mercer 1975; Jelinek 1976), and one using level building with Viterbi matching for each individual word (Rabiner and Levinson 1985). The first of these, unlike procedures using the Viterbi algorithm, computes an overall likelihood for the recognised word sequence, by storing and adding the probabilities for all alignments of the sequence of word models with the input. Probabilistic information on the occurrence of words and word sequences can conveniently be incorporated into a Markov-model-based recogniser (Bahl, Jelinek and Mercer 1975).

Because words often have different characteristics when spoken connectedly (due to coarticulation with preceding and following words), training on isolated words may not be satisfactory for a connected word recognition system. To improve performance in the presence of interword coarticulation, a technique of training on words from connected speech may be used. Training tokens extracted from connected speech can be used to construct templates corresponding to different speaking rates and degrees of coarticulation, which can then be used alongside the templates obtained from isolated word tokens (Rabiner, Bergh and Wilpon 1982; Rabiner, Wilpon and Terrace 1984). In an HMM-based recogniser, strings of words may be used for training, instead of isolated words (Brown, Lee and Spohrer 1983). It is also possible in a template-based recogniser to improve templates trained initially on isolated or manually extracted words, by matching concatenated templates to known connected strings of words to find the word boundaries, and so extracting instances of the words for further training (Rabiner, Wilpon and Juang 1986; Tajima, Komura and Sato 1986).

*Concluding Remarks*

Published results (White and Neely 1976, Sakoe and Chiba 1978, Kuhn and Tomaschewski 1983, Chuang and Chan 1983, Taylor 1986) show that dynamic time warping is a powerful method of performing time alignment of speech patterns, and provides a substantial improvement over linear time registration as measured by recognition rates. Some methods of word matching without nonlinear time alignment have, however, yielded good results. These include methods using segment-specific VQ codebooks (without fine temporal information: Burton and Shore 1985; Burton, Shore and Buck 1985), or histograms of occurrences of VQ labels (Watanuki and Kaneko 1986), in equal-duration segments, for which linear alignment is adequate because the temporal variations are represented implicitly in the histograms or codebooks. These methods require several training utterances for each word, and are thus most applicable to speaker-independent recognition.

The major disadvantage of DTW, especially for large vocabularies, is the large amount of computation required, particularly for the numerous frame distance calculations. Other methods of non-linear time alignment have been devised (Nara *et al.* 1982, Chuang and Chan 1983) which require less computation, but the recognition accuracies of these methods are inferior to those obtained with DTW. Therefore a good deal of effort has been devoted to finding ways to reduce the computational load without losing recognition accuracy.

Among the most promising reduction techniques are preliminary trace segmentation (Kuhn and Tomaschewski 1983; Pieraccini and Billi 1983; Gauvain, Mariani and Lienard 1983) (combined with search area restrictions: Kuhn and Tomaschewski 1983; Gauvain, Mariani and Lienard 1983); accumulated distance thresholds (Itakura 1975; Rabiner *et al.* 1979), progressive rejection procedures (Gupta, Bryan and Gowdy 1978) and beam searching (Bisiani and Waibel 1982; Greer, Lowerre and Wilcox 1982), for rejection of templates (and, in the case of beam searching, individual paths) before completion of the matching; and, for large vocabularies in particular, match limiting by an initial simple comparison (Kaneko and Dixon 1983; Guo-tian 1985; Pan *et al.* 1985; Pan, Soong and Rabiner 1985; Bergh, Soong and Rabiner 1985; Aktas *et al.* 1986; Miwa and Kido 1986). (Note, however, that the first and last of these may not allow the DTW matching to start until the input word has been completed.) These three techniques could be combined quite effectively in an isolated word recogniser. Beam searching is also applicable to connected word recognition using the one-stage algorithm or Markov models, and has frequently been so applied (Bahl, Jelinek and Mercer 1975; Kaplan, Reddy and Kato 1980; Bridle, Brown and Chamberlain 1982; Bridle, Brown and Chamberlain 1983; Ney 1983; Cravero *et al.* 1984).

The variable-length trace segmentation described in Pieraccini and Billi (1983) and Gauvain, Mariani and Lienard (1983) has some of the merits of the fixed-length trace segmentation procedure, and allows the matching to begin during the speaking of the input word; but it does not result in normalisation to a constant number of frames per word (and hence the severe search area restrictions of Kuhn and Tomaschewski (1983) cannot be applied).

This technique, unlike standard trace segmentation, can be applied to connected word recognition with an unknown number of words in the input utterance (Gauvain and Mariani 1982, Ney 1983, Gauvain and Mariani 1984). Numerous acoustic segmentation techniques have been devised (Tappert and Das 1978; Greer, Lowerre and Wilcox 1982; Brown 1982; Gauvain and Mariani 1982, 1984; Ney 1983; Kuhn and Tomaschewski 1983; Pieraccini and Billi 1983; Gauvain, Mariani and Lienard 1983; Chuang and Chan 1983; Lienard and Soong 1984), varying in the criteria for defining segments, the derivation of a representation for each segment (or each segment boundary) and the use, if any, made of information on segment durations.

Vector quantisation (Billi 1982; Sugamura, Shikano and Furui 1983; Pieraccini and Billi 1983; Rabiner, Levinson and Sondhi 1983; Levinson, Rabiner and Sondhi 1983; Kammerer, Küpper and Lagger 1984; Bourlard, Wellekens and Ney 1984; Landell, Naylor and Wohlford 1984; Cravero *et al.* 1984; Jouvet and Schwartz 1984) can reduce the amount of computation to be done in evaluating frame distances. However, there tends to be a loss of accuracy in recognition; this can be counteracted by using a large quantisation codebook, but then significant computational savings are achieved only when the vocabulary is fairly large.

One motive for applying the above computation reduction techniques is the desirability (for most applications) of real-time or near-real-time recognition. Another approach altogether (though the two can in some cases be combined) is to use special hardware on which DTW can be performed rapidly. Descriptions of devices designed specifically for DTW computation are given in Ackland, Weste and Burr (1981), Peckham *et al.* (1982), Ishizuka *et al.* (1983), MacAllister (1983), Brown *et al.* (1984), Kavaler *et al.* (1984), Ackenhausen (1984), Kitazume, Ohira and Endo (1985), Yoder and Jamieson (1985), Mann and Rhodes (1986) and Charot, Frison and Quinton (1986). DTW algorithms lend themselves to parallel processing; several parallel processing schemes for DTW are described and compared in Yoder and Siegel (1982), and the arrays of Ackland, Weste and Burr (1981), MacAllister (1983) and Yoder and Jamieson (1985) implement two of these schemes. Systolic array architectures for connected word recognition, using algorithms similar to the two-level algorithm, are described in Charot, Frison and Quinton (1986). A pipelined structure for beam searching is described in Anantharaman and Bisiani (1985).

The choice of local path constraints and weighting functions in the basic DTW algorithm tends not to make very much difference to the quality of recognition obtained; however, the results of Myers, Rabiner and Rosenberg (1980a, 1981) and Rabiner (1982) suggest that Itakura or Type II constraints (in the notation of Myers, Rabiner and Rosenberg (1980b), as defined above), with weighting according to movement in the input direction, give a slight advantage in recognition accuracy over other combinations while being quite economical computationally. Where paths are confined to a narrow band from the initial to the final point of the warp, the ban on repetition of the step (1, 0) in the Itakura constraints has been found to be unhelpful (Paliwal, Agrawal and Sinha 1982).

The superiority of Itakura constraints to Type III (Rabiner 1982) might seem surprising, since the former constraints yield only an approximation to the optimal alignment given by the true dynamic programming formulation of the latter. Presumably the explanation is that the word distance computed using the Itakura constraints tends to be more suboptimal for incorrect-word matches than for correct-word ones. This can be understood by considering the nature of the cases where an optimal path, obtained with Type III constraints, is liable to be excluded by the condition for horizontal steps in the Itakura constraints. Such cases include those where one frame of the template (assumed to be on the vertical axis) is similar to several successive input frames but the adjacent frames of the template are dissimilar. Horizontal steps will tend to occur at that template frame during the search for the alignment path, and so a horizontal step in the minimal-distance path may well be forbidden because of a previously chosen horizontal step, one input frame earlier, at the same frame in the template. Such instances of isolated well-matching template frames seem more likely to occur in the alignment of incorrect-word templates than with correct-word templates.

Results of experiments on endpoint relaxation indicate that in some cases it can be helpful (Rabiner, Rosenberg and Levinson 1978; Das 1980; Rabiner 1982) but in others it increases the error rate (Dautrich, Rabiner and Martin 1983b; Haltsonen 1984). Methods of adjusting the endpoints by preliminary testing (Das 1980, Haltsonen 1984) or methods using extended input and initial and final silence frames appended to the templates (Haltsonen 1984) (cf. the noise template method of Jouvet and Schwartz (1984)) appear to be better than simple endpoint relaxation. The appropriateness of any endpoint modification technique will depend on the characteristics of the method used to locate the endpoints in the first place, and on the conditions in which the recogniser is being used.

Any speech recogniser is only as good as its reference patterns. For speaker-dependent recognition a robust training procedure (Dautrich, Rabiner and Martin 1983b,c; McInnes, Jack and Laver 1986b), which obtains two sufficiently similar repetitions of each word and averages them, can help to ensure that the templates formed are free from mispronunciations and extraneous noise. For speaker-independent recognition, a clustering analysis of training data from a representative set of speakers is an appropriate means of deriving templates (Rabiner 1978; Levinson *et al.* 1979; Rabiner *et al.* 1979; Rabiner and Wilpon 1979a; Mokkedem, Hügli and Pellandini 1986).

A development of the idea of training to the speaker is progressive adaptation of templates during the input of speech to be recognised. Whenever a word is recognised and the recognition is acceptable to the user (shown by the absence of an attempt to correct it), the template for that word is adapted to take account of information contained in the newly recognised instance of the word; or, if it is sufficiently different from the existing template or templates, an additional template can be created. Some work has already been done on incorporating template adaptation in a word recognition system (Damper and MacDonald 1984; Landell, Wohlford and Bahler 1986; McInnes, Jack and Laver 1986b; Meddis 1986), but there is probably considerable scope for further research in this area. A further possible development is adaptation of the best-matching template away from the input word when this word is known to have been recognised incorrectly, with the aim of reducing the likelihood of a recurrence of the same misrecognition (McInnes, Jack and Laver 1986b). A training procedure incorporating such adjustment away from incorrectly recognised words has been reported, using weighted averaging with a negative weight (Niimi 1978).

Recognisers using whole word templates tend to perform badly on vocabularies containing words which differ only in short sections. Refinements to the basic template matching method, such as discriminative networks (Moore, Russell and Tomlinson 1983; Wilcox, Lowerre and Kahn 1983) to eliminate the effects of linguistically insignificant differences, and two-pass decision procedures using weighting to emphasise distinguishing features (Rabiner and Wilpon 1981; Tribolet, Rabiner and Wilpon 1982), result in improved performance on such vocabularies. In the special case where durational information is important for distinguishing words, the techniques of Moore (1982) and Russell, Moore and Tomlinson (1983) can be helpful. In each case, DTW is still used, with appropriate modifications. The maximum mutual information training procedure for HMMS (Bahl *et al.* 1986) is similar to these techniques, especially the discriminative weighting: it takes into account characteristics of the vocabulary as a whole, rather than only of each word individually.

For recognition of connected speech, the one-stage algorithm (Bridle, Brown and Chamberlain 1982, 1983; Nakagawa 1983; Ney 1984) has certain advantages over the other algorithms described: it calculates each frame distance only once, and it can easily be adapted to incorporate syntactic constraints. If the number of words in each input utterance is fixed, however, the level building algorithm (Myers and Rabiner 1981a,b) may be better (Ackenhausen 1984), since it always finds the best-matching string of $L$ words for a specified value of $L$. The one-stage algorithm can be converted into an implementation of the level building algorithm by imposing appropriate constraints.

The use of templates extracted from connected speech can significantly improve the performance of a connected word recogniser (Rabiner, Bergh and Wilpon 1982; Rabiner, Wilpon and Juang 1986). It is also helpful to permit the omission of the end (or beginning) of a word, which commonly occurs in connected speech (Myers and Rabiner 1981c). However, the word template matching approach has the fundamental limitation that, because the basic unit is the word (rather than something smaller), it is difficult to make full allowance for coarticulation effects by incorporating detailed phonological rules. Thus it may be better to use a phoneme-based (or phonetic-segment-based) strategy where natural continuous speech is to be recognised (Vaissière 1985). The phonemes used in a recognition system will not necessarily be what would be called phonemes in standard linguistics and it may well be found necessary (Furui 1980) to include several realisations of each (strictly so-called) phoneme.

Another disadvantage of the word template matching approach is the amount of training required for each new speaker (in a speaker-trained system) or for each new word added to the vocabulary (in a speaker-independent system using clustering analysis for template creation). This too is less of a problem in a system based on units smaller than words, since there are fewer reference patterns to be determined, and also partial-vocabulary training (Furui 1980) is more feasible.

Even in a system where the primary recognition strategy is one of segmentation into phoneme-sized (or smaller) units, or of matching with phoneme templates, DTW matching of word templates may have a part to play. As suggested in Cook (1976), it can be used to verify the recognition hypotheses produced by the primary analysis. The verifier described in Cook (1976) used templates generated by a synthesis-by-rule program, but it would be quite possible to use natural word templates instead. For verification in a continuous speech recogniser, an algorithm such as ZIP (Chamberlain and Bridle 1983) could be used, with either concatenated word templates or synthesised speech on the reference axis. The post-error correction procedure of Rabiner and Schmidt (1980) is a verification procedure using concatenated word templates.

Dynamic programming has been successfully applied to speech recognition using hidden Markov nodels (Kaplan, Reddy and Kato 1980; Billi 1982; Brown, Lee and Spohrer 1983; Rabiner, Levinson and Sondhi 1983; Levinson, Rabiner and Sondhi 1983; Bourlard, Wellekens and Ney 1984; Cravero *et al.* 1984). While template matching using DTW is technically a form of hidden Markov model matching by the Viterbi algorithm, in practice there have tended to be two distinguishable approaches. The Markov modelling approach has the advantage that more information about the variability of the pronunciation of a word can be incorporated into a Markov model than into a template. The use of Markov models instead of templates has been found to lead to improved accuracy in recognition (Bourlard, Wellekens and Ney 1984), or to similar accuracy with reduced computation (Rabiner, Levinson and Sondhi 1983; Levinson, Rabiner and Sondhi 1983; Cravero *et al.* 1984). However, a large amount of training is required for a Markov modelling system (Rabiner, Levinson and Sondhi 1983; Levinson, Rabiner and Sondhi 1983), which may make it unsuitable for some applications. The case for progressive adaptation during recognition sessions is probably even stronger for Markov model systems than for template-based systems, since the asymptotic optimal performance is reached more slowly as the amount of training data increases. A method which could be applied to accomplish such adaptation has been described, and has been found to allow improvement of models (Sugawara, Nishimura and Kuroda 1986).

Word recognition systems using dynamic programming techniques have been produced commercially by various manufacturers. Some of the available recognisers are designed for isolated word recognition only, but others, such as the NEC DP-100 (Kaplan, Reddy and Kato 1980) and DP-200 and the Logos system (Peckham *et al.* 1982) incorporate connected word algorithms. A few commercially produced recognisers, available or under development, use Markov models (Jelinek *et al.* 1985). Comparative evaluations of speech recognisers in terms of accuracy and other features have been published (Doddington and Schalk 1981, Lea 1983, Bell and Becker 1983, Woodard and Lea 1984), and comparisons of human and machine speech recognition have been reported (Vickroy, Silverman and Dixon 1982). Standards for evaluating recognition systems are in the process of development (Rubinchek 1983).

An important aspect of many potential applications of isolated and connected word recognition is that the possible sequences of words are strongly constrained by the syntax, semantics and pragmatics of the task (Rabiner and Levinson 1981, Levinson 1985). The use of syntactic constraints to restrict the output of the recognition system can greatly improve its accuracy, whether the input consists of a sequence of isolated words (Levinson, Rosenberg and Flanagan 1978; Levinson 1978; Rabiner and Levinson 1981) or of a connected utterance (Levinson 1985, Rabiner and Levinson 1985). In a dialogue system, the use of a semantic model to exploit the relations between successive sentences can further improve the performance by allowing acoustic recognition errors to be corrected automatically, or detected and corrected by querying the user (Levinson 1985, Levinson and Shipley 1980, Rabiner and Levinson 1981); and where understanding rather than transcription is the goal some word errors may not affect the outcome (Levinson and Shipley 1980). Thus the task may be accomplished with much greater reliability than would be possible using the word-level acoustic pattern-matching alone.

The above discussion indicates that, although so much research has already been done on dynamic programming algorithms for speech pattern alignment and on speech recognition systems using whole-word pattern matching, there are still areas where further investigations could be worthwhile. These include the formation of templates, networks or HMMS (in an initial training process or by adaptation), and the derivation of effective matching procedures and decision rules to be used with them, to provide optimal discrimination among the words of a specific vocabulary (Rabiner and Wilpon 1981; Tribolet, Rabiner and Wilpon 1982; Moore, Russell and Tomlinson 1983; Wilcox, Lowerre and Kahn 1983; Bahl *et al.* 1986; Hewett, Holmes and Young 1986; McInnes, Jack and Laver 1986b; Meddis 1986); and the design of whole speech systems, incorporating use of the available information at all levels (Rabiner and Levinson 1981) and appropriate interaction between the machine and the user (Rabiner and Levinson 1981; Waterworth 1984; McInnes, Jack and Laver 1986b). It should also be noted that the dynamic programming principle has applications to many other pattern matching problems as well as to word matching – both within the field of speech recognition and in other fields (such as recognition of handwriting). Other applications in speech recognition are in matching of templates or models for subword units (Billi 1982; Sauter 1985; Mergel and Ney 1985; Glassman 1985; Bourlard, Kamp and Wellekens 1985; Brassard 1985; Colla, Scagliola and Sciarra 1985); in comparing a phoneme lattice derived from unknown input speech with reference lattices or models (Kobayashi and Niimi 1985); in automatic training of a recogniser on connected words or subword units (Ney 1985, Haltsonen and Ruusunen 1985); and in dynamic frequency warping, to improve the usual spectral distance function (Hunt 1985, Blomberg and Elenius 1986) or to compare frames represented by sets of formant frequencies (Hunt 1985). Research into a particular application may yield results useful in other applications.

## REFERENCES


Ackenhausen, J. G. (1984) The CDTWP: a programmable processor for connected word recognition. *Proc. IEEE Int. Conf. Acoust., Speech, and Signal Process.*, paper 35.9.

Ackland, B., N. Weste & D. J. Burr (1981) An integrated multiprocessing array for time warp pattern matching. *Eighth Annual Symposium on Computer Architecture* (published in *Sigarch Newsletter* 9, no.3, 197-215).

Aktas, A., B. Kammerer, W. Küpper & H. Lagger (1986) Large-vocabulary isolated word recognition with fast coarse time alignment. *Proc. IEEE-IECEJ-ASJ Int. Conf. Acoust., Speech, and Signal Process.*, 709-12.

Anantharaman, T. S. & R. Bisiani (1985) Custom data-flow machines for speech recognition. *Proc. IEEE Int. Conf. Acoust., Speech, and Signal Process.*, 1847-50.

Bahl, L. R., P. F. Brown, P. V. de Souza & R. L. Mercer (1986) Maximum mutual information estimation of hidden Markov model parameters for speech recognition. *Proc. IEEE-IECEJ-ASJ Int. Conf. Acoust., Speech, and Signal Process.*, 49-52.

Bahl, L. R. & F. Jelinek (1975) Decoding for channels with insertions, deletions, and substitutions with applications to speech recognition. *IEEE Trans. Information Theory IT-21*, 404-11.

Bahl, L. R., F. Jelinek & R. L. Mercer (1975) A maximum likelihood approach to continuous speech recognition. *IEEE Trans. Pattern Anal. and Machine Intelligence PAMI-5*, 179-90.

Baker, J. M. & D. F. Pinto (1986) Optimal and suboptimal training strategies for automatic speech recognition in noise, and the effects of adaptation on performance. *Proc. IEEE-IECEJ-ASJ Int. Conf. Acoust., Speech, and Signal Process.*, 745-8.

Bell, D. W. & R. W. Becker (1983) Designing experiments to evaluate speech I/O devices and applications, *Speech Technology 1,* no.4, 70-9.

Bellman, R. (1957) *Dynamic Programming,* Princeton University Press.

Bergh, A. F., F. K. Soong & L. R. Rabiner (1985) Incorporation of temporal structure into a vector-quantization-based preprocessor for speaker-independent, isolated-word recognition. *AT&T Tech. J. 64,* 1047-63.

Billi, R. (1982) Vector quantization and Markov source models applied to speech recognition. *Proc. IEEE Int. Conf. Acoust., Speech and Signal Process.,* 574-7.

Bisiani, R. & A. Waibel (1982) Performance trade-offs in search techniques for isolated word speech recognition. *Proc. IEEE Int. Conf. Acoust., Speech and Signal Process.,* 570-3.

Blomberg, M., R. Carlson, K. Elenius & B. Granstrom (1984) Auditory models in isolated word recognition. *Proc. IEEE Int. Conf. Acoust., Speech, and Signal Process.,* paper 17.9.

Blomberg, M. & K. Elenius (1986) Nonlinear frequency warp for speech recognition. *Proc. IEEE-IECEJ-ASJ Int. Conf. Acoust., Speech, and Signal Process.,* 2631-4.

Bocchieri, E. L. & G. R. Doddington (1986a) Speaker independent digit recognition with reference frame-specific distance measures. *Proc. IEEE-IECEJ-ASJ Int. Conf. Acoust., Speech, and Signal Process.,* 2699-702.

—— (1986b) Frame-specific statistical features for speaker independent speech recognition. *IEEE Trans. Acoust., Speech, and Signal Process. ASSP-34,* 755-64.

Bourlard, H., Y. Kamp & C. J. Wellekens (1985) Speaker dependent connected speech recognition via phonemic Markov models. *Proc. IEEE Int. Conf. Acoust., Speech, and Signal Process.,* 1213-16.

Bourlard, H., C. J. Wellekens & H. Ney (1984) Connected digit recognition using vector quantisation. *IEEE Int. Conf. Acoust., Speech, and Signal Process.,* paper 26.10.

Brassard, J. P. (1985) Integration of segmenting and nonsegmenting approaches in continuous speech recognition. *Proc. IEEE Int. Conf. Acoust., Speech, and Signal Process.,* 1217-20.

Bridle, J. S. (1973) An efficient elastic-template method for detecting given words in running speech. *British Acoust. Soc. Meeting,* 1-4.

Bridle, J. S., M. D. Brown & R. M. Chamberlain (1982) An algorithm for connected word recognition. *Proc. IEEE Int. Conf. Acoust., Speech and Signal Process.,* 899-902.

—— (1983) Continuous connected word recognition using whole word templates. *The Radio and Electronic Engineer 53,* 167-S5.

Bridle, J. S., K. M. Ponting, M. D. Brown & A. W. Borrett (1984) A noise compensating spectrum distance measure applied to automatic speech recognition. *Proc. Inst. Acoust. 6,* part 4, 307-14.

Bridle, J. S. & M. P. Ralls (1985) An approach to speech recognition using synthesis-by-rule, in *Computer Speech Processing* (eds F. Fallside &

W. A. Woods) pp.277-92. Prentice Hall International.

Brown, M. K. & L. R. Rabiner (1982a) Dynamic time warping for isolated word recognition based on ordered graph searching techniques. *Proc. IEEE Int. Conf. Acoust., Speech and Signal Process.*, 1255-8.

—— (1982b) An adaptive, ordered, graph search technique for dynamic time warping for isolated word recognition. *IEEE Trans. Acoust., Speech, and Signal Process. ASSP-30*, 535-44.

Brown, M. K., R. Thorkildsen, Y. H. Oh & S. S. Ali (1984) The DTWP: an LPC based dynamic time warping processor for isolated word recognition. *Proc. IEEE Int. Conf. Acoust., Speech, and Signal Process.*, paper 25B.5.

Brown, P. F., C. H. Lee & J. C. Spohrer (1983) Bayesian adaptation in speech recognition. *Proc. IEEE Int. Conf. Acoust., Speech and Signal Process.*, 761-4.

Brown, R. W. (1982) Segmentation for data reduction in isolated word recognition. *Proc. IEEE Int. Conf. Acoust., Speech and Signal Process.*, 1262-5.

Burton, D. K. & J. E. Shore (1985) Speaker-dependent isolated word recognition using speaker-independent vector quantization codebooks augmented with speaker-specific data. *IEEE Trans. Acoust., Speech, and Signal Process. ASSP-33*, 440-3.

Burton, D. K., J. E. Shore & J. T. Buck (1985) Isolated word speech recognition using multisection vector quantization codebooks. *IEEE Trans. Acoust., Speech, and Signal Process. ASSP-33*, 837-49.

Chamberlain, R. M. & J. S. Bridle (1983) ZIP: a dynamic programming algorithm for time-aligning two indefinitely long utterances. *Proc. IEEE Int. Conf. Acoust., Speech and Signal Process.*, 816-19.

Charot, F., P. Frison & P. Quinton (1986) Systolic architectures for connected speech recognition. *IEEE Trans. Acoust., Speech, and Signal Process. ASSP-34*, 765-79.

Chollet, G. F. & C. Gagnoulet (1982) On the evaluation of speech recognisers and data bases using a reference system. *Proc. IEEE Int. Conf. Acoust., Speech and Signal Process.*, 2026-9.

Choukri, K., G. Chollet & Y. Grenier (1986) Spectral transformations through canonical correlation analysis for speaker adaptation in ASR. *Proc. IEEE-IECEJ-ASJ Int. Conf. Acoust., Speech, and Signal Process.*, 2659-62.

Christiansen, R. W. & C. K. Rushforth (1977) Detecting and locating key words in continuous speech using linear predictive coding. *IEEE Trans. Acoust., Speech, and Signal Process. ASSP-25*, 361-7.

Chuang, C. K. & S. W. Chan (1983) Speech recognition using variable frame rate coding. *Proc. IEEE Int. Conf. Acoust., Speech and Signal Process.*, 1033-6.

Clotworthy, C. J. & F. J. Smith (1986) Spoken word variation and probability estimation. *IEE Conf. Pub. 258 (Speech Input/Output; Techniques and Applications)*, 27-30.

Colla, A. M., C. Scagliola & D. Sciarra (1985) A connected speech recognition system using a diphone-based language model. *Proc. IEEE Int. Conf. Acoust., Speech, and Signal Process.*, 1229-32.

Cook, A. E. & M. J. Russell (1986) Improved duration modelling in hidden Markov models using series-parallel configurations of states. *Proc. Inst. of Acoust. 8,* part 7, 299-306.

Cook, C. (1976) Word verification in a speech understanding system. *1976 IEEE Int. Conf. Record on Acoust., Speech, and Signal Process.* 553-6.

Cravero, M., L. Fissore, R. Pieraccini & C. Scagliola (1984) Syntax driven recognition of connected words by Markov models. *IEEE Int. Conf. Acoust., Speech, and Signal Process.*, paper 35.5.

Damper, R. I. & S. L. MacDonald (1984) Template adaptation in speech recognition. *Proc. Inst. Acoust. 6,* part 4, 293-9.

Das, S. K. (1980) Some experiments in discrete utterance recognition. *Proc. IEEE Int. Conf. Acoust., Speech and Signal Process.*, 178-81 (later edition published in *IEEE Trans. Acoust., Speech, and Signal Process. ASSP-30,* 535-44).

Dautrich, B. A., L. R. Rabiner & T. B. Martin (1983a) On the use of filter banks for isolated word recognition. *Proc. IEEE Int. Conf. Acoust., Speech and Signal Process.*, 1057-60.

—— (1983b) The effects of selected signal processing techniques on the performance of a filter-bank-based isolated word recognizer. *Bell System Tech. J. 62,* 1311-36.

—— (1983c) On the effects of varying filter bank parameters on isolated word recognition. *IEEE Trans. Acoust., Speech, and Signal Process. ASSP-31,* 793-806.

Davis, S. B. & P. Mermelstein (1980) Comparison of parametric representations for monosyllabic word recognition in continuously spoken sentences. *IEEE Trans. Acoust., Speech, and Signal Process ASSP-28,* 357-66.

Dologlou, Y. & J. M. Dolmazon (1984) Comparison of a model of the peripheral auditory system and L.P.C. analysis in a speech recognition system. *Proc. IEEE Int. Conf. Acoust., Speech, and Signal Process.*, paper 17.10.

Doddington, G. R. & T. B. Schalk (1981) Speech recognition : turning theory to practice. *IEEE Spectrum 18,* no.9, 26-32.

Elenius, K. & M. Blomberg (1982) Effects of emphasizing transitional or stationary parts of the speech signal in a discrete utterance recognition system. *Proc. IEEE Int. Conf. Acoust., Speech and Signal Process.*, 535-8.

Forney, G. D. (1973) The Viterbi algorithm. *Proc. IEEE 61,* 268-78.

Furui, S. (1980) A training procedure for isolated word recognition systems. *IEEE Trans. Acoust., Speech, and Signal Process. ASSP-28,* 129-36.

—— (1986a) Speaker-independent isolated word recognition using dynamic features of speech spectrum. *IEEE Trans. Acoust., Speech,*

*and Signal Process. ASSP-34*, 52-9.
—— (1986b) Speaker-independent isolated word recognition based on emphasized spectral dynamics. *Proc. IEEE-IECEJ-ASJ Int. Conf. Acoust., Speech, and Signal Process.*, 1991-4.
Gagnoulet, C. & M. Couvrat (1982) Seraphine: a connected word speech recognition system. *Proc. IEEE Int. Conf. Acoust., Speech and Signal Process.*, 887-90.
Gauvain, J. L. & J. Mariani (1982) A method for connected word recognition and word spotting on a microprocessor. *Proc. IEEE Int. Conf. Acoust., Speech and Signal Process.*, 891-4.
—— (1984) Evaluation of time compression for connected word recognition. *IEEE Int. Conf. Acoust., Speech, and Signal Process.*, paper 35.10.
Gauvain, J. L., J. Mariani & J. S. Lienard (1983) On the use of time compression for word-based recognition. *Proc. IEEE Int. Conf. Acoust., Speech and Signal Process.*, 1029-32.
Glassman, M. S. (1985) Hierarchical DP for word recognition. *Proc. IEEE Int. Conf. Acoust., Speech, and Signal Process.*, 886-9.
Glinski, S. C. (1985) On the use of vector quantization for connected-digit recognition. *AT&T Tech. J. 64*, 1033-45
Gray, A. H. & J. D. Markel (1976) Distance measures for speech processing. *IEEE Trans. Acoust., Speech, and Signal Process. ASSP-24*, 380-91.
Greer, K., B. Lowerre & L. Wilcox (1982) Acoustic pattern matching and beam searching. *Proc. IEEE Int. Conf. Acoust., Speech and Signal Process.*, 1251-54.
Guo-tian, Z. (1985) On associative recognition of isolated Chinese word. *Proc. IEEE Int. Conf. Acoust., Speech, and Signal Process.*, 37-40.
Gupta, V. N., J. K. Bryan & J. N. Gowdy (1978) A speaker-independent speech-recognition system based on linear prediction. *IEEE Trans. Acoust., Speech, and Signal Process. ASSP-26*, 27-33.
Gupta, V. N., M. Lennig & P. Mermelstein (1984) Decision rules for speaker-independent isolated word recognition. *Proc. IEEE Int. Conf. Acoust., Speech, and Signal Process.*, paper 9.2.
Haltsonen, S. (1984) An endpoint relaxation method for dynamic time warping algorithms. *Proc. IEEE Int. Conf. Acoust., Speech, and Signal Process.*, paper 9.8.
—— (1985a) Improved dynamic time warping methods for discrete utterance recognition. *IEEE Trans. Acoust., Speech, and Signal Process. ASSP-33*, 449-50.
—— (1985b) Recognition of isolated-word sentences from a large vocabulary using dynamic time warping methods. *IEEE Trans. Acoust., Speech, and Signal Process. ASSP-33*, 1026-7.
Haltsonen, S. & P. Ruusunen (1985) Collection of phoneme samples using time alignment and spectral stationarity of speech signals. *Proc. IEEE Int. Conf. Acoust., Speech, and Signal Process.*, 1561-4.

Hewett, A. J., G. Holmes & S. J. Young (1986) Dynamic speaker adaptation in speaker-independent word recognition. *Proc. Inst. of Acoust. 8,* part 7, 275-82.

Hieronymus, J. L. & W. J. Majurski (1985) A reference speech recognition algorithm for benchmarking and speech data base analysis. *Proc. IEEE Int. Conf. Acoust., Speech, and Signal Process.*, 1573-6.

Hunt, M. J. (1985) A robust formant-based speech spectrum comparison measure. *Proc. IEEE Int. Conf. Acoust., Speech, and Signal Process.*, 1117-20.

Hunt, M. J. & C. Lefebvre (1986) Speech recognition using a cochlear model. *Proc. IEEE-IECEJ-ASJ Int. Conf. Acoust., Speech, and Signal Process.*, 1979-82.

Ishizuka, H., M. Watari, H. Sakoe, S. Chiba, T. Iwata, T. Matsuki & Y. Kawakami (1983) A microprocessor for speech recognition. *Proc. IEEE Int. Conf. Acoust., Speech and Signal Process.*, 503-6.

Itakura, F. (1975) Minimum prediction residual principle applied to speech recognition. *IEEE Trans. Acoust., Speech, and Signal Process. ASSP-23,* 67-72.

Jelinek, F. (1976) Continuous speech recognition by statistical methods. *Proc. IEEE 64,* 532-56.

Jelinek, F., L. R. Bahl & R. L. Mercer (1975) Design of a linguistic statistical decoder for the recognition of continuous speech. *IEEE Trans. Information Theory IT-21,* 250-6.

Jelinek, F. *et al.* (1985) A real-time, isolated-word, speech recognition system for dictation transcription. *Proc. IEEE Int. Conf. Acoust., Speech, and Signal Process.*, 858-61.

Jouvet, D. & R. Schwartz (1984) One-pass syntax-directed connected-word recognition in a time-sharing environment. *IEEE Int. Conf. Acoust., Speech, and Signal Process.*, paper 35.8.

Juang, B. H. (1984) On the hidden Markov model and dynamic time warping for speech recognition – a unified view. *AT&T Bell Laboratories Tech. J. 63,* 1213-43.

Juang, B. H. & L. R. Rabiner (1985) Mixture autoregressive hidden Markov models for speech signals. *IEEE Trans. Acoust., Speech, and Signal Process. ASSP-33,* 1404-13.

—— (1986) Mixture autoregressive hidden Markov models for speaker independent isolated word recognition. *Proc. IEEE-IECEJ-ASJ Int. Conf. Acoust., Speech, and Signal Process.*, 41-4.

Juang, B. H., L. R. Rabiner & J. G. Wilpon (1986) On the use of bandpass liftering in speech recognition. *Proc. IEEE-IECEJ-ASJ Int. Conf. Acoust., Speech, and Signal Process.*, 765-8.

Juang, B. H., L. R. Rabiner, S. E. Levinson & M. M. Sondhi (1985) Recent developments in the application of hidden Markov models to speaker-independent isolated word recognition. *Proc. IEEE Int. Conf. Acoust., Speech, and Signal Process.*, 9-12.

Kammerer, B., W. Küpper & H. Lagger (1984) Special feature vector coding and appropriate distance definition developed for a speech recognition system. *Proc. IEEE Int. Conf. Acoust., Speech, and Signal Process.*, paper 17.4.

Kamp, Y. (1985) State reduction in hidden Markov chains used for speech recognition. *IEEE Trans. Acoust., Speech, and Signal Process. ASSP-33,* 1138-45.

Kaneko, T. & N. R. Dixon (1983) A hierarchical decision approach to large-vocabulary discrete utterance recognition. *IEEE Trans. Acoust., Speech, and Signal Process. ASSP-31,* 1061-6.

Kaplan, G., R. Reddy & Y. Kato (1980) Words into action. *IEEE Spectrum 17,* no.6, 22-9.

Kavaler, R., R. W. Brodersen, T. G. Noll, M. Lowy and H. Murveit (1984) A dynamic time warp IC for a one thousand word recognition system. *Proc. IEEE Int. Conf. Acoust., Speech, and Signal Process.*, paper 25B.6.

Kawabata, T. & M. Kohda (1986) Word spotting taking account of duration change characteristics for stable and transient parts of speech. *Proc. IEEE-IECEJ-ASJ Int. Conf. Acoust., Speech, and Signal Process.*, 2307-10.

Kitazume, Y., E. Ohira & T. Endo (1985) LSI implementation of a pattern matching algorithm for speech recognition. *IEEE Trans. Acoust., Speech, and Signal Process. ASSP-33,* 1-4.

Kobayashi, Y. & Y. Niimi (1985) Matching algorithms between a phonetic lattice and two types of templates – lattice and graph. *Proc. IEEE Int. Conf. Acoust., Speech, and Signal Process.*, 1597-1600.

Kuhn, M. H. & H. H. Tomaschewski (1983) Improvements in isolated word recognition. *IEEE Trans. Acoust., Speech, and Signal Process. ASSP-31,* 157-67.

Landell, B. P., J. A. Naylor & R. E. Wohlford (1984) Effect of vector quantisation on a continuous speech recognition system. *Proc. IEEE Int. Conf. Acoust., Speech, and Signal Process.*, paper 26.11.

Landell, B. P., R. E. Wohlford & L. G. Bahler (1986) Improved speech recognition in noise. *Proc. IEEE-IECEJ-ASJ Int. Conf. Acoust., Speech, and Signal Process.*, 749-51.

Lea, W. A. (1983) Selecting the best speech recognizer for the job. *Speech Technology 1,* no.4, 10-29.

Levinson, S. E. (1978) The effects of syntactic analysis on word recognition accuracy. *Bell System Tech. J. 57,* 1627-44.

—— (1985) A unified theory of composite pattern analysis for automatic speech recognition, in *Computer Speech Processing* (eds F. Fallside & W. A. Woods) pp.243-75. Prentice Hall International.

Levinson, S. E., L. R. Rabiner & M. M. Sondhi (1983) Speaker independent isolated digit recognition using hidden Markov models. *Proc. IEEE Int. Conf. Acoust., Speech and Signal Process.*, 1049-52.

Levinson, S. E., A. E. Rosenberg & J. L. Flanagan (1978) Evaluation of a word recognition system using syntax analysis. *Bell System Tech. J. 57,* 1619-26.

Levinson, S. E. & K. L. Shipley (1980) A conversational-mode airline information and reservation system using speech input and output. *Bell System Tech. J. 59,* 119-37.

Levinson, S. E., L. R. Rabiner, A. E. Rosenberg & J. G. Wilpon (1979) Interactive clustering techniques for selecting speaker-independent reference templates for isolated word recognition. *IEEE Trans. Acoust., Speech, and Signal Process. ASSP-27,* 134-41.

Lienard, J. S. & F. K. Soong (1984) On the use of transient information in speech recognition. *Proc. IEEE Int. Conf. Acoust., Speech, and Signal Process.,* paper 17.3.

Liu, Y. J. (1984) On creating averaging templates. *Proc. IEEE Int. Conf. Acoust., Speech, and Signal Process.,* paper 9.1.

MacAllister, J. (1983) Systolic arrays for dynamic programming in speech recognition systems. *Proc. IEEE Int. Conf. Acoust., Speech and Signal Process.,* 507-10.

McCullough, D. P. (1983) Secondary testing techniques for word recognition in continuous speech. *Proc. IEEE Int. Conf. Acoust., Speech and Signal Process.,* 300-3.

McInnes, F. R., M. A. Jack & J. Laver (1986a) Comparative study of time segmentation and segment representation techniques in a DTW-based word recogniser. *IEE Conf. Pub. 258 (Speech Input/Output; Techniques and Applications),* 21-6.

—— (1986b) An isolated word recognition system with progressive adaptation of templates. *Proc. Inst. of Acoust. 8,* part 7, 283-90.

Mann, J. R. & F. M. Rhodes (1986) A wafer scale DTW multiprocessor. *Proc. IEEE-IECEJ-ASJ Int. Conf. Acoust., Speech, and Signal Process.,* 1557-60.

Meddis, R. (1986) Towards an auditory primal sketch. *Proc. Inst. of Acoust. 8,* part 7, 589-96.

Mergel, D. & H. Ney (1985) Phonetically guided clustering for isolated word recognition. *Proc. IEEE Int. Conf. Acoust., Speech, and Signal Process.,* 854-7.

Miwa, J. & K. Kido (1986) Speaker-independent word recognition for large vocabulary using pre-selection and non-linear spectral matching. *Proc. IEEE-IECEJ-ASJ Int. Conf. Acoust., Speech, and Signal Process.,* 2695-8.

Mokeddem, A., H. Hügli & F. Pellandini (1986) New clustering algorithms applied to speaker independent isolated word recognition. *Proc. IEEE-IECEJ-ASJ Int. Conf. Acoust., Speech, and Signal Process.,* 2691-4.

Moore, R. K. (1982) Locally constrained dynamic programming in automatic speech recognition. *Proc. IEEE Int. Conf. Acoust., Speech and Signal Process.,* 1270-3.

Moore, R. K., M. J. Russell & M. J. Tomlinson (1983) The discriminative network: a mechanism for focusing recognition in whole-word pattern matching. *Proc. IEEE Int. Conf. Acoust., Speech and Signal Process.,* 1041-4.

Myers, C. S. & S. E. Levinson (1982) Speaker independent connected word recognition using a syntax-directed dynamic programming procedure. *IEEE Trans. Acoust., Speech, and Signal Process. ASSP-30*, 561-5.

Myers, C. S. & L. R. Rabiner (1981a) A level building dynamic time warping algorithm for connected word recognition. *IEEE Trans. Acoust., Speech, and Signal Process. ASSP-29*, 284-97.

—— (1981b) Connected digit recognition using a level-building DTW algorithm. *IEEE Trans. Acoust., Speech, and Signal Process. ASSP-29*, 351-63.

—— (1981c) A comparative study of several dynamic time-warping algorithms for connected-word recognition. *Bell System Tech. J. 60*, 1389-409.

Myers, C. S., L. R. Rabiner & A. E. Rosenberg (1980a) An investigation of the use of dynamic time warping for word spotting and connected speech recognition. *Proc. IEEE Int. Conf. Acoust., Speech and Signal Process.*, 173-7.

—— (1980b) Performance tradeoffs in dynamic time warping algorithms for isolated word recognition. *IEEE Trans. Acoust., Speech, and Signal Process. ASSP-28*, 623-35.

—— (1981) On the use of dynamic time warping for word spotting and connected word recognition. *Bell System Tech. J. 60*, 303-25.

Nakagawa, S. (1983) A connected spoken word recognition method by O(n) dynamic programming pattern matching algorithm. *Proc. IEEE Int. Conf. Acoust., Speech and Signal Process.*, 296-9.

Nara, Y., K. Iwata, Y. Kijima, S. Kimura, S. Sasaki & J. Tanahashi (1982) Large-vocabulary spoken word recognition using simplified time-warping patterns. *Proc. IEEE Int. Conf. Acoust., Speech and Signal Process.*, 1266-9.

Neben, G., R. J. McAulay & C. J. Weinstein (1983) Experiments in isolated word recognition using noisy speech. *Proc. IEEE Int. Conf. Acoust., Speech and Signal Process.*, 1156-9.

Ney, H. (1983) Experiments in connected word recognition. *Proc. IEEE Int. Conf. Acoust., Speech and Signal Process.*, 288-91.

—— (1984) The use of a one-stage dynamic programming algorithm for connected word recognition. *IEEE Trans. Acoust., Speech, and Signal Process. ASSP-32*, 263-71.

—— (1985) A script-guided algorithm for the automatic segmentation of continuous speech. *Proc. IEEE Int. Conf. Acoust., Speech, and Signal Process.*, 1209-12.

Niimi, Y. (1978) A method for forming universal reference patterns in an isolated word recognition system. *Proc. 4th Int. Joint Conf. Pattern Recognition.*

Niimi, Y. & Y. Kobayashi (1986) Synthesis of speaker-adaptive word templates by concatenation of the monosyllabic sounds. *Proc. IEEE-IECEJ-ASJ Int. Conf. Acoust., Speech, and Signal Process.*, 2651-4.

Niles, L., H. F. Silverman & N. R. Dixon (1983) A comparison of three feature vector clustering procedures in a speech recognition paradigm. *Proc. IEEE Int. Conf. Acoust., Speech and Signal Process.*, 765-8.

Nocerino, N., F. K. Soong, L. R. Rabiner & D. H. Klatt (1985) Comparative study of several distortion measures for speech recognition. *Proc. IEEE Int. Conf. Acoust., Speech, and Signal Process.*, 25-8.

Nomura, T. & R. Nakatsu (1986) Speaker-independent isolated word recognition for telephone voice using phoneme-like templates. *Proc. IEEE-IECEJ-ASJ Int. Conf. Acoust., Speech, and Signal Process.*, 2687-90.

Okochi, M. & T. Sakai (1982) Trapezoidal DP matching with time reversibility. *Proc. IEEE Int. Conf. Acoust., Speech and Signal Process.*, 1239-42.

Olano, C. A. (1983) An investigation of spectral match statistics using a phonemically marked data base. *Proc. IEEE Int. Conf. Acoust., Speech and Signal Process.*, 773-6.

Paliwal, K. K., A. Agrawal & S. S. Sinha (1982) A modification over Sakoe and Chiba's dynamic time warping algorithm for isolated word recognition. *Proc. IEEE Int. Conf. Acoust., Speech and Signal Process.*, 1259-61.

Pan, K. C., F. K. Soong & L. R. Rabiner (1985) A vector-quantization-based preprocessor for speaker-independent isolated word recognition. *IEEE Trans. Acoust., Speech, and Signal Process. ASSP-33*, 546-60.

Pan, K. C., F. K. Soong, L. R. Rabiner & A. F. Bergh (1985) An efficient vector-quantization preprocessor for speaker independent isolated word recognition. *Proc. IEEE Int. Conf. Acoust., Speech, and Signal Process.*, 874-7.

Paul, D. B. (1985) Training of HMM recognisers by simulated annealing. *Proc. IEEE Int. Conf. Acoust., Speech, and Signal Process.*, 13-16.

Peckham, J., J. Green, J. Canning & P. Stephens (1982) Logos – a real time hardware continuous speech recognition system. *Proc. IEEE Int. Conf. Acoust., Speech and Signal Process.*, 863-6.

Pieraccini, R. & R. Billi (1983) Experimental comparison among data compression techniques in isolated word recognition. *Proc. IEEE Int. Conf. Acoust., Speech and Signal Process.*, 1025-8.

Poritz, A. B. & A. G. Richter (1986) On hidden Markov models in isolated word recognition. *Proc. IEEE-IECEJ-ASJ Int. Conf. Acoust., Speech, and Signal Process.*, 705-8.

Rabiner, L. R. (1978) On creating reference templates for speaker independent recognition of isolated words. *IEEE Trans. Acoust., Speech, and Signal Process. ASSP-26*, 34-42.

—— (1982) Notes on some factors affecting performance of dynamic time warping algorithms for isolated word recognition. *Bell System Tech. J. 61*, 363-73.

Rabiner, L. R., A. Bergh & J. G. Wilpon (1982) An embedded training procedure for connected digit recognition. *Proc. IEEE Int. Conf. Acoust., Speech and Signal Process.*, 1621-4.

Rabiner, L. R. & S. E. Levinson (1981) Isolated and connected word recognition – theory and selected applications. *IEEE Trans. Communications COM-29*, 621-59.

—— (1985) A speaker-independent, syntax-directed, connected word recognition system based on hidden Markov models and level building. *IEEE Trans. Acoust., Speech, and Signal Process. ASSP-33*, 561-73.

Rabiner, L. R., S. E. Levinson & M. M. Sondhi (1983) On the application of vector quantisation and hidden Markov models to speaker-independent, isolated word recognition. *Bell System Tech. J. 62*, 1075-1105.

Rabiner, L. R., A. E. Rosenberg & S. E. Levinson (1978) Considerations in dynamic time warping algorithms for discrete word recognition. *IEEE Trans. Acoust., Speech, and Signal Process. ASSP-26*, 575-82.

Rabiner, L. R. & C. E. Schmidt (1980a) A connected digit recognizer based on dynamic time warping and isolated digit templates. *Proc. IEEE Int. Conf. Acoust., Speech and Signal Process.*, 194-8.

—— (1980b) Application of dynamic time warping to connected digit recognition. *IEEE Trans. Acoust., Speech, and Signal Process. ASSP-28*, 377-88.

Rabiner, L. R. & J. G. Wilpon (1979a) Speaker-independent isolated word recognition for a moderate size (54 word) vocabulary. *IEEE Trans. Acoust., Speech, and Signal Process. ASSP-27*, 583-7.

—— (1979b) Application of clustering techniques to speaker-trained isolated word recognition. *Bell System Tech. J. 58*, 2217-33.

—— (1981) A two-pass pattern-recognition approach to isolated word recognition. *Bell System Tech. J. 60*, 739-66.

Rabiner, L. R., J. G. Wilpon & B. H. Juang (1986) A continuous training procedure for connected digit recognition. *Proc. IEEE-IECEJ-ASJ Int. Conf. Acoust., Speech, and Signal Process.*, 1065-8.

Rabiner, L. R., J. G. Wilpon & S. G. Terrace (1984) A directory retrieval system based on connected letter recognition. *IEEE Int. Conf. Acoust., Speech, and Signal Process.*, paper 35.4.

Rabiner, L. R., B. H. Juang, S. E. Levinson & M. M. Sondhi (1985) Recognition of isolated digits using hidden Markov models with continuous mixture densities. *AT&T Tech. J. 64*, 1211-34.

Rabiner, L. R., S. E. Levinson, A. E. Rosenberg & J. G. Wilpon (1979) Speaker-independent recognition of isolated words using clustering techniques. *IEEE Trans. Acoust., Speech, and Signal Process. ASSP-27*, 336-49.

Rollins, A. & J. Wiesen (1983) Speech recognition and noise. *Proc. IEEE Int. Conf. Acoust., Speech and Signal Process.*, 523-6.

Rosenberg, A. E. & C. E. Schmidt (1979) Automatic recognition of spoken spelled names for obtaining directory listings. *Bell System Tech. J. 58*, 1797-1823.

Rubinchek, B. (1983) Towards standards for speech I/O systems. *Speech Technology 1,* no.4, 40-2.

Russell, M. J. (1985) Maximum likelihood hidden semi-Markov model parameter estimation for automatic speech recognition. *RSRE Memorandum 3837.*

Russell, M. J. & A. E. Cook (1986) Experiments in speaker-dependent isolated digit recognition using hidden Markov models. *Proc. Inst. of Acoust. 8,* part 7, 291-8.

—— (1987) Experimental evaluation of duration modelling techniques for automatic speech recognition. *Proc. IEEE Int. Conf. Acoust., Speech, and Signal Process.*, 2376-9.

Russell, M. J. & R. K. Moore (1985) Explicit modelling of state occupancy in hidden Markov models for automatic speech recognition. *Proc. IEEE Int. Conf. Acoust., Speech, and Signal Process.*, 5-8.

Russell, M. J., R. K. Moore & M. J. Tomlinson (1983) Some techniques for incorporating local timescale variability information into a dynamic time-warping algorithm for automatic speech recognition. *Proc. IEEE Int. Conf. Acoust., Speech and Signal Process.*, 1037-40.

Sakoe, H. (1979) Two-level DP-matching – a dynamic programming-based pattern matching algorithm for connected word recognition. *IEEE Trans. Acoust., Speech, and Signal Process. ASSP-27,* 588-95.

Sakoe, H. & S. Chiba (1978) Dynamic programming algorithm optimization for spoken word recognition. *IEEE Trans. Acoust., Speech, and Signal Process. ASSP-26,* 43-9.

Sauter, L. C. (1985) Isolated word recognition using a segmental approach. *Proc. IEEE Int. Conf. Acoust., Speech, and Signal Process.*, 850-3.

Scagliola, C. & D. Sciarra (1986) Two novel algorithms for variable frame analysis and word matching for connected word recognition. *Proc. IEEE-IECEJ-ASJ Int. Conf. Acoust., Speech, and Signal Process.*, 1105-8.

Shikano, K., K. F. Lee & R. Reddy (1986) Speaker adaptation through vector quantization. *Proc. IEEE-IECEJ-ASJ Int. Conf. Acoust., Speech, and Signal Process.*, 2643-6.

Silverman, H. F. & N. Rex Dixon (1980) State constrained dynamic programming (SCDP) for discrete utterance recognition. *Proc. IEEE Int. Conf. Acoust., Speech and Signal Process.*, 169-72.

Soudoplatoff, S. (1986) Markov modeling of continuous parameters in speech recognition. *Proc. IEEE-IECEJ-ASJ Int. Conf. Acoust., Speech, and Signal Process.*, 45-8.

de Souza, P. & P. J. Thomson (1982) LPC distance measures and statistical tests with particular reference to the likelihood ratio. *IEEE Trans. Acoust., Speech, and Signal Process. ASSP-30,* 304-15.

Sugamura, N., K. Shikano & S. Furui (1983) Isolated word recognition using phoneme-like templates. *Proc. IEEE Int. Conf. Acoust., Speech and Signal Process.*, 723-36.

Sugawara, K., M. Nishimura & A. Kuroda (1986) Speaker adaptation for a hidden Markov model. *Proc. IEEE-IECEJ-ASJ Int. Conf. Acoust., Speech, and Signal Process.*, 2667-70.

Sugawara, K., M. Nishimura, K. Toshioka, M. Okochi and T. Kaneko (1985) Isolated word recognition using hidden Markov models. *Proc. IEEE Int. Conf. Acoust., Speech, and Signal Process.*, 1-4.

Sugiyama, M. (1986) Unsupervised speaker adaptation methods for vowel templates. *Proc. IEEE-IECEJ-ASJ Int. Conf. Acoust., Speech, and Signal Process.*, 2635-8.

Tajima, K., M. Komura & Y. Sato (1986) Connected word recognition by overlap and split of reference patterns and its performance evaluation tests. *Proc. IEEE-IECEJ-ASJ Int. Conf. Acoust., Speech, and Signal Process.*, 1101-4.

Tappert, C. C. & S. K. Das (1978) Memory and time improvements in a dynamic programming algorithm for matching speech patterns. *IEEE Trans. Acoust., Speech, and Signal Process. ASSP-26,* 583-6.

Taylor, M. R. (1986) Comparative isolated word recognition experiments. *Proc. Inst. of Acoust. 8,* part 7, 265-73.

Tohkura, Y. (1986) A weighted cepstral distance measure for speech recognition. *Proc. IEEE-IECEJ-ASJ Int. Conf. Acoust., Speech, and Signal Process.*, 761-4.

Tribolet, J. M., L. R. Rabiner & M. M. Sondhi (1979) Statistical properties of an LPC distance measure. *IEEE Trans. Acoust., Speech, and Signal Process. ASSP-27,* 550-8.

Tribolet, J. M., L. R. Rabiner & J. G. Wilpon (1982) An improved model for isolated word recognition. *Bell System Tech. J. 61,* 2289-312.

Vaissière, J. (1985) Speech recognition: a tutorial, in *Computer Speech Processing* (eds F. Fallside & W. A. Woods) pp.191-242. Prentice Hall International.

Vickroy, C. A., H. F. Silverman & N. R. Dixon (1982) Study of human and machine discrete utterance recognition (DUR). *Proc. IEEE Int. Conf. Acoust., Speech and Signal Process.*, 2022-5.

Vintsyuk, T. K. (1968) Speech discrimination by dynamic programming. *Kibernetika (Cybernetics) 4,* 81-8.

—— (1971) Element-wise recognition of continuous speech consisting of words of a given vocabulary. *Kibernetika (Cybernetics) 7,* no.2, 361-72.

Watanuki, O. & T. Kaneko (1986) Speaker-independent isolated word recognition using label histograms. *Proc. IEEE-IECEJ-ASJ Int. Conf. Acoust., Speech, and Signal Process.*, 2679-82.

Watari, M. (1986) New DP matching algorithms for connected word recognition. *Proc. IEEE-IECEJ-ASJ Int. Conf. Acoust., Speech, and Signal Process.*, 1113-16.

Waterworth, J. A. (1984) Interaction with machines by voice – human factors issues. *British Telecom Technol. J. 2,* no.4, 56-63.

Welch, J. R. & S. C. Oxenberg (1980) Reduction of minimum word-boundary gap lengths in isolated word recognition. *Proc. IEEE Int. Conf. Acoust., Speech and Signal Process.*, 190-3.

White, G. M. & R. B. Neely (1976) Speech recognition experiments with linear prediction, bandpass filtering, and dynamic programming. *IEEE Trans. Acoust., Speech, and Signal Process. ASSP-24,* 183-8.

Wilcox, L., B. Lowerre & M. Kahn (1983) Use of a priori knowledge of vocabulary for real time discrete utterance recognition. *Proc. IEEE Int. Conf. Acoust., Speech and Signal Process.*, 1045-8.

Wohlford, R. E., A. R. Smith & M. R. Sambur (1980) The enhancement of wordspotting techniques. *Proc. IEEE Int. Conf. Acoust., Speech and Signal Process.*, 209-12.

Woodard, J. P. & W. A. Lea (1984) New measures of performance for speech recognition systems. *Proc. IEEE Int. Conf. Acoust., Speech, and Signal Process.*, paper 9.6.

Yoder, M. A. & L. H. Jamieson (1985) Simulation of a highly parallel system for word recognition. *Proc. IEEE Int. Conf. Acoust., Speech, and Signal Process.*, 1449-52.

Yoder, M. A. & L. J. Siegel (1982) Dynamic time warping algorithms for SIMD machines and VLSI processor arrays. *Proc. IEEE Int. Conf. Acoust., Speech and Signal Process.*, 1274-7.

Zelinski, R. & F. Class (1983) A learning procedure for speaker-dependent word recognition systems based on sequential processing of input tokens. *Proc. IEEE Int. Conf. Acoust., Speech and Signal Process.*, 1053-6.

TWO : J. HARRINGTON

# ACOUSTIC CUES FOR AUTOMATIC RECOGNITION OF ENGLISH CONSONANTS

The identification of a set of acoustic features for the automatic extraction of phonetic segments from the acoustic waveform is at the core of many of the speech understanding systems of the ARPA project (Lea 1980) and many of the continuous speech recognition systems being developed at various laboratories worldwide.[1] In such systems, the complexity of the task is broken down into two stages whereby a segmentation of the acoustic waveform is undertaken into broad categories such as *stop, fricative* and *sonorant* and these categories are progressively refined to a level of phonetic detail that would ultimately enable a lexical access component to hypothesise the correct string of words. In this chapter, a preliminary exploration is made of identifying the acoustic features which would enable the refinement of such broad categories to be made. The basis for this exploration is not empirical, but an analysis of three different kinds of studies in acoustic phonetics over the last 30–40 years: predictions of the acoustic characteristics of phonetic segments based on calculations derived from a vocal tract model (Fant 1970, 1980); the identification of acoustic characteristics of phonetic segments from spectrograms (Fischer-Jørgensen 1954, Lehiste and Peterson 1961) and, more recently, from digital signal processing techniques (Kewley-Port 1982, Zue 1976); and the specification of the minimal cues which listeners need to identify phonemes based on synthesis and labelling experiments (Cooper, Delattre, Liberman, Borst and Gerstman 1952). The first two types of study are probably most relevant here, although the discussion will be enriched by the results of experiments in the third type of study when these are considered relevant.

It must be stressed from the outset that only those acoustic features which enable a progressive *phonetic* (as opposed to phonemic) refinement of the broad categories stop, fricative and sonorant are considered here. The discontinuity of an acoustic to phonemic mapping is readily apparent in considering the problem of relating English /l/ to the acoustic waveform. /l/ belongs to the class of consonants which are often labelled approximant (Catford 1977) or sonorant (Chomsky and Halle 1968). We might therefore assume that the automatic extraction of /l/ depends on the identification of the broad category sonorant from the acoustic waveform and the subsequent progressive differentiation of /l/ from consonants such as /m/, /n/, /ŋ/, /j/, /w/, /r/ and all vowels and diphthongs which might also be categorised as sonorant. However, the automatic extraction of /l/ in *plea* is likely to encounter serious difficulties using this procedure, for the reason that /l/ in this word is realised predominantly as a voiceless [l̥]: since a turbulent airstream underlies the production of [l̥], it is more likely that the acoustic waveform of *plea* will be resolved into the categories (stop + *fricative* + sonorant), rather than (stop + *sonorant* + sonorant). The lesson from this example is clear enough: it cannot be assumed that the features which are created using phonemic criteria are isomorphic with the features that are established on acoustic criteria. Instead, the relationship between acoustic waveform and phonemes is a two-stage process of specifying both the allophones of each phoneme and the distribution of the allophones across whatever broad categories we choose to extract from the acoustic waveform (thus, /l/ has at least three allophones, [l], [ɫ] and [l̥]; an acoustic analysis of the first two allophones will result in their broad classification as sonorant, while [l̥] will be classified as fricative). Having hypothesised *a priori* the distribution of allophones across these broad categories, a subsequent task is to define the set of features for their within-category differentiation: it will be necessary, therefore, to define a set of features which would enable the separation of [l̥] from [f], [h], [θ], [s], [ʃ] and all the other allophones which are hypothesised to fall into the class *fricative*.

In setting up acoustic features for the progressive phonetic refinement of the broad categories *stop, fricative* and *sonorant,* reference will be made, therefore, at various stages to the relationship of these categories to the major allophones of phonemes. The problem of speaker normalisation is not treated in this paper and the acoustic features are designed primarily for the extraction of phonetic segments of British English, Received Pronunciation (RP). A summary is presented at the end of each major section of the acoustic features in terms of a discrimination tree.

## Oral Stops: Place of Articulation

From an acoustic point of view, an oral stop can be decomposed into five acoustic segments (Fant 1968): *occlusion, transient, frication, aspiration* and *transition*. The *occlusion* corresponds to the interval of complete vocal tract closure and is characterised acoustically by the absence of energy; stops that are voiced during the occlusion are characterised by (low frequency) energy in the 0–500 Hz range. The *transient* corresponds to the release of the closure following a sharp rise in intra-oral air-pressure. On spectrograms, the transient is often detectable as an intense spike of duration around 10 ms (Fant 1973). The combination of high intra-oral pressure and a narrow channel at the point of release results in frication which, in general, should be spectrally similar to a homorganic fricative produced in isolation (Fant 1970). With an increase in the vocal tract opening, frication may give way to *aspiration*, a noise source caused by turbulence at the glottis. While frication may bear some spectral similarity to the homorganic fricative, a formant-like structure, continuous with F2, F3 and F4 of the following vowel, is often detectable during the aspiration stage (Fant 1973). The *transition* is the interval from the time at which the formants are first detectable in the aspiration stage to the *acoustic vowel target* in CV(C) syllables; the acoustic vowel target in this case is considered to be the point in the syllable at which the formant frequency values correspond most closely to those of the same vowel produced in isolation. In many cases, the rate of change of the formants with respect to time is closest to zero at the acoustic vowel target in CV(C) syllables. Another term which is sometimes used for the transition is the *acoustic vowel onglide* (Lehiste and Peterson 1961).

An acoustic analysis of stops shows that it is often not possible to differentiate the transient from frication; the undifferentiated transient and frication stages are often known as the *burst*. The aspiration stage can usually be differentiated from the burst by the presence of a formant-like structure. In voiced stops, which usually lack an aspiration stage in English, the acoustic vowel onglide can begin at the burst.

The following sections report on acoustic cues which can be used to resolve oral stops into place of articulation classes (bilabial, alveolar and velar) and include an analysis of acoustic cues for differentiating voiced stops [b d g] from their voiceless counterparts [p t k].

*Burst*

Fant's (1970) analysis of the relationship between articulation and speech waveform, the spectrographic study of Fischer-Jørgensen (1954) and the LPC-based studies of Zue (1976) and Edwards (1981) all suggest that bilabials lack any main resonances in the 0–10 kHz range. Alveolar stops, on the other hand are characterised by an abrupt rise of the spectral level above 4 kHz which is accentuated by an anti-resonance around 3.5 kHz (Fant 1970). Fant's (1970) study is consistent with the pattern playback experiments at the Haskins Laboratories which show that when listeners are presented with synthetic CV syllables, high frequency bursts were heard as /t/ for all vowels that followed (Cooper, Delattre, Liberman, Borst and Gerstman 1952). Fischer-Jørgensen (1954) reports that the lowest resonance of alveolar bursts is 1.95 kHz with higher resonances at 3, 3.8 and 4.8 kHz. Zue (1976) describes the burst of alveolars as broadband and characterised by energy above 2 kHz while Edwards (1981) reports that the energy of alveolar bursts is distributed throughout the spectrum above 2.5 kHz.

A characteristic spectral feature of the burst of velar stops is a concentration of energy in the middle of the spectrum. This compact concentration of energy (*compact* designates that the energy is distributed over a small frequency range) derives primarily from the cavity in front of the tongue constriction; since the tongue constriction varies in place of articulation from pre-velar for the realisation of /k g/ before front vowels (e.g. *key, geese*) to post-velar for /k g/ realised before back vowels (*caught, Gaul*), the energy concentration of bursts is accordingly context-dependent. For the front allophones of /k g/, the compact energy is distributed around a centre frequency of 3.2 kHz (Fant 1973) or 2.72 kHz (Zue 1976); the compact energy in the bursts of the back allophones of /k g/ can occur around 1.2 kHz (Fant 1973) or 1.77 kHz (Zue, 1976). For most bursts of velar allophones, the compact energy seems to vary with F2 and F3, and possibly even F4 of the following vowels (Fant 1973; Fischer-Jørgensen 1954; Zue 1976). Fischer-Jørgensen (1954) also reports that the bursts of velar allophones are characterised by a high frequency resonance around 5 kHz which does not vary substantially in the context of different vowels; the presence of such a high frequency resonance is also reported in Zue (1976).

A by-product of the finding that the burst of bilabials lacks any main resonances (due to the absence of a front cavity of any appreciable size) is that its intensity should be less than that of alveolars or velars. Zue (1976) finds that the overall RMS amplitude for bursts of bilabial allophones in CV syllables is around 12 dB less than the RMS amplitude of the bursts of alveolar and velar allophones in the same context. The average values are: −27.6 dB for [p]; −16.6 dB for [t] and −17.2 dB for [k]; −28 dB for [b], −15.8 dB for [d] and −16.6 dB for [g]. Where '>' denotes 'the intensity of the burst is greater than', Zue also found that in /s STOP V/ clusters k>t>p; in /s STOP r V/ clusters, t>k>p; in /s STOP V/ clusters, k>t>p; in /STOP r V/ clusters (voiceless stops), t>k>p; in /STOP r V/ clusters (voiced stops), g>d>b; in /STOP l V/ clusters k>p; in /STOP w V/ clusters, t>k; and in /STOP w V/ clusters, g>d. A calculation of the intensity of the burst may therefore enable the separation of bilabial stops on the one hand from alveolar and velar stops on the other. There is a tendency for the intensity of the burst of alveolar stop allophones to be greater than that of velars, but this is not valid for every context. Furthermore, the difference in intensity between bilabials and alveolars/velars is considerably greater than the difference in burst intensity between velars and alveolars. Edwards (1981) has replicated Zue's (1976) finding that alveolar stops have the most intense bursts and bilabials the weakest bursts; like Zue, Edwards finds little difference in intensity between the bursts of velars and alveolars.

Thus the bursts of bilabials, unlike those of alveolars and velars, lack any main resonances and are low in intensity; the prime characteristic of alveolar bursts is a broad distribution of spectral energy above 2 kHz. Velar bursts are characterised by compact spectral energy in the middle of the frequency range which varies with F2 and F3 the following vowel.

In the next section, some features are reported which are designed to distinguish between place of articulation within oral stops based on an analysis of the acoustic properties of the burst. Almost all of these features implement, in one form or another, the distinguishing acoustic properties of the burst identified so far.

*Features.* In an acoustic study of English monosyllables produced by three subjects, Halle, Hughes and Radley (1957) isolated the burst and measured (*a*) its intensity in the 0.7–10 kHz range and (*b*) the intensity in the 2.7–10 kHz range and the difference, (*a* − *b*), was calculated. For the bursts of [t], [d] and those of the front allophones of /k/ and /g/ (henceforth [$k_+$] and [$g_+$] respectively), (*a* − *b*) was 5 dB, or less, for voiceless stops and 8 dB, or less, for voiced stops; for bursts of [p], [b] and the bursts of the back allophones of /k/ and /g/ (henceforth [$k_-$] and [$g_-$] respectively), the difference in intensity was greater than these values. Halle *et al.* found that ($a-b$) successfully separated *acute* ([t], [d], [k+], [g+]) from *grave* ([p], [b], [k-], [g-]) with a 95% success rate. This feature seems to be a successful implementation of the findings, referred to in the preceding section, that bilabials, unlike alveolars, have an even distribution of energy throughout the frequency range. In order to distinguish alveolars from velars within the acute class, a measurement was made of the average level of the spectrum between 300 Hz and 10 kHz ($c$) and this was compared with the average level of the spectrum in the 2–4 kHz region ($d$). The measure ($d-c$) was greater than 5 dB and 8 dB for voiceless and voiced velar bursts respectively; for alveolar bursts, ($d-c$) was less than these values. The calculation of ($d-c$) is based on the evidence discussed above that velar bursts, unlike alveolar bursts, are characterised by a compact concentration of energy in the middle of the spectrum. Halle *et al.* found that ($d-c$) successfully separated 87% of the members of the acute class. With regard to grave, a measurement of the difference in intensity between the two peaks of greatest intensity was made. Subsequently, this difference (dB) was plotted as a function of the frequency (Hz) of the spectral peak of greatest intensity. This feature enabled a distinction to be made between the back allophones of velars and bilabials in 85% of the cases.

There have been several studies published in which an attempt is made to identify invariant acoustic properties of stops. In Stevens and Blumstein (1978) and Blumstein and Stevens (1979, 1980), for example, the gross spectral shape of the speech signal that includes and follows the burst is considered to exhibit invariant acoustic attributes. Based on Fant's (1970) acoustic theory of speech production, the authors claim that the gross shape of the spectrum sampled at the onset of the burst with a window width of 25.6 ms will be *diffuse-falling* for bilabial stops, *diffuse-rising* for alveolars and *compact* for velars. *Diffuse* in this context means that there will be peaks distributed throughout the frequency range as opposed to *compact* in which the peaks are concentrated in the middle of the range; *falling* refers to the fact that the slope of the spectrum is negative for increasing frequency, while *rising* means it is positive (see figure 2.1). Blumstein and Stevens (1980) found that LPC spectra sampled at the burst derived from natural CV and VC utterances were classified correctly 85% of the time by the templates (six subjects, four male and two female, produced CV and VC syllables where C = [p t k b d g], V = [i e a o u] for CV stimuli, and V = [i e a ʌ u] for VC stimuli; five repetitions were made by each subject of each syllable resulting in a total of 900 CV syllables

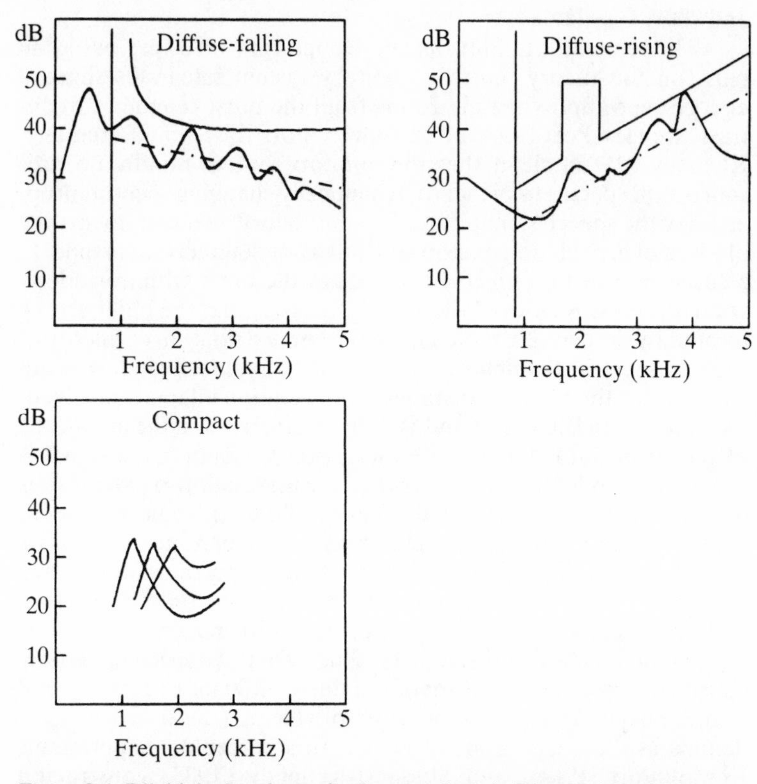


*Figure* 2.1. Bilabial, alveolar and velar templates in Blumstein and Stevens (1979). For a spectrum to be classified as diffuse-falling, a spectral peak is fitted to the top reference line such that all other peaks lie below the line. If, after fitting the spectrum in this way, at least two peaks, one below 2.4 kHz, the other in the 2.4-3.6 kHz range, can be fitted, the spectrum is classified as diffuse-falling (there is no condition on the amplitude of spectral peaks below 1.2 kHz, which is why the top reference line tends to infinity). For a spectrum to be classified as diffuse-rising, a spectral peak greater than 2.2 kHz is identified such that the peak touches the upper reference line and such that all other peaks greater than this frequency fall below the top reference line. When these peaks have been fitted, at least two other peaks must fall *within* the reference lines. The 'hump' in the top reference line of diffuse-rising is to allow for spectra that have a strong concentration at 1.8 kHz (the *locus frequency* for alveolars). The compact template consists of a set of overlapping reference lines from 1.2-3.5 kHz (only the first three are shown here). For a spectrum to be classified as compact, a mid-frequency peak is adjusted to touch one of the peaks that is closest to it in frequency. No other peak must then protrude through the reference line.

and 900 VC syllables).

The templates in Stevens and Blumstein (1980) are based, in part, on the theory that the auditory system integrates spectral energy over approximately 20 ms from the burst release. By contrast, Kewley-Port (1983) and Kewley-Port, Pisoni and Studdert-Kennedy (1983) claim that the auditory system rapidly updates short-term spectra to preserve dynamically changing spectral properties of the speech signal. Based on this theory, acoustic invariance of place of articulation of stop consonants is defined with respect to a succession of LPC spectra taken from the burst with a window-width of 5 ms. Kewley-Port (1983) proposes that the burst can be located by detecting a rapid change in the distribution of energy of the LPC spectra; the first LPC spectrum at the burst (window width 5 ms) forms the basis for distinguishing between bilabial and alveolar stops. As in Blumstein and Stevens' framework, if the amplitude of the spectrum is falling with increasing frequency, the signal is classified as bilabial; if it is rising, a classification of alveolar is made. A spectrum is identified as velar if there is a single prominent peak between 1 kHz and 3.5 kHz which occurs for 3 (not necessarily consecutive) frames from the burst (there are therefore strong similarities to the compact template in Blumstein and Stevens). In addition, since voice onset time has been shown to be greater for velars than bilabials and alveolars (Zue 1976), the spectrum is only classified as a velar stop if there is a delay of 20 ms or more (i.e. 4 frames) from the burst to the onset of voicing (onset of voicing is defined as the onset of an F1 peak). In a perceptual experiment, Kewley-Port, Pisoni and Studdert-Kennedy (1983) constructed two sets of synthetic stimuli: the first preserved the static spectral properties proposed by Blumstein and Stevens (1979, 1980), while the second was based on the dynamic templates proposed by Kewley-Port (1983). The results showed that listeners identified place of articulation significantly better from stimuli based on dynamic properties than from those based on static onset spectra.

Two further perceptual experiments cast doubt on a definition of acoustic invariance in terms of Blumstein and Stevens' static onset spectra. Blumstein, Isaacs and Mertus (1982) synthesised CV stimuli in which the gross shape of the spectrum at stimulus onset was appropriate for one place of articulation, but in which the onset frequencies of the vowel formants cued a different place of articulation: in such a synthetic stimulus, then, the gross shape of the spectrum at burst onset might be appropriate for [b] (i.e. diffuse-falling), but the onset frequencies of the formants appropriate for [d]. If, as Blumstein and Stevens (1980, 660) have proposed, the static onset spectra are the primary cues to which listeners respond in identifying place of articulation, they should override the conflicting cues of the vowel's onset frequencies (in the above example, therefore, listeners should identify /b/, not /d/). The results showed that the majority of correct responses were cued by the onset frequencies, not by the shape of the spectrum. A similar result was obtained by Walley and Carrell (1983); as in Blumstein, Isaacs and Mertus (1982), subjects (adults and children) labelled stimuli in which the gross spectral shape and the onset formant frequencies specified conflicting places of articulation. When subjects were required to identify place of articulation in such stimuli, Walley and Carrell found their responses were determined almost entirely by the formant transitions.

As a result of some of the negative findings discussed above, Lahiri, Gewirth and Blumstein (1984) propose a modification to Blumstein and Stevens' templates which is based on the idea that, for alveolar stops, the change in high frequency energy from burst to voicing onset is less than the change in low frequency energy over the same interval. For bilabial stops, on the other hand, the change in high frequency energy from burst to voicing onset was about the same, or greater than the change in low frequency energy. The template Lahiri *et al.* constructed for the classification of stops consonants in English, Malayalam and French is shown in figure 2.2. Two LPC spectra are taken, one at the burst and the other averaged over the first three glottal pulses after the burst. Connecting lines are drawn between F2 and F4 and the ratio of the difference in energy at high frequencies to the difference in energy at low frequencies is calculated. In figure 2.2, this ratio corresponds to $(d-b)/(c-a)$. A positive ratio of less than 0.5 or a negative ratio with the numerator negative is characteristic of dentals and alveolars; for labials, the ratio is either greater than 0.5 or negative, with the denominator having the negative value. When this metric was applied to 493 natural utterances in English, French and Malayalam, 91% of the stops were correctly classified as either bilabial or alveolar (velar stops were not included in the analysis).

The analysis of Lahiri, Gewirth and Blumstein (1984) is consistent with the results of a study by Ohde and Stevens (1983). Like Lahiri *et al.*, they claim that the spectrum sampled at the burst is more intense in the high frequency region for alveolars than bilabials. The results of a labelling experiment in which subjects classified synthetic stimuli as either bilabial or alveolar is consistent with this view. When the spectrum amplitude of the burst in the F4-F5 region was less than that of the vowel, the responses tended

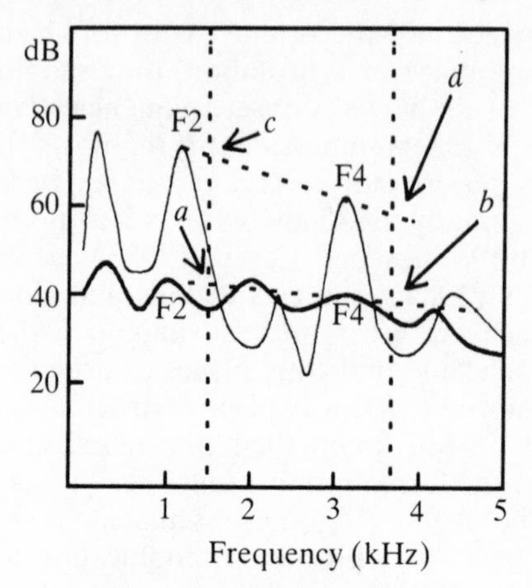


*Figure* 2.2. The metric in Lahiri, Gewirth and Blumstein (1984) for distinguishing between bilabial and alveolar stops in CV syllables. The bold line represents a spectrum taken at the burst, the thin line the spectrum at the vowel onset. The distinction between bilabial and alveolar stops is based on the ratio $(d-b)/(c-a)$.

to be bilabial whereas when the burst amplitude was greater, responses tended to be alveolar; this effect was more pronounced for voiceless, than voiced, stops.

*Formant Locus and Transitions*

The concept of an invariant locus frequency associated with place of articulation of stop consonants emerged both from the work of Potter, Kopp and Green (1966) and experiments with the pattern playback equipment at the Haskins Laboratories (Delattre, Liberman and Cooper 1955). Delattre *et al.* suggested that the place of articulatory closure for each of the labial, alveolar and velar consonants was relatively fixed preceding different vowels; furthermore, such articulatory invariance, they argued, had its acoustic correlate in a relatively fixed starting frequency position of the second formant.

In the experiment by Delattre, Liberman and Cooper (1955), horizontal bars were painted which correspond to the steady-state formants of a given vowel. Subsequently, the slopes of the first part of these bars, which corresponds to the F1 and F2 transitions from the stop consonant to the vowel, were manipulated in a variety of ways until acceptable bilabial, alveolar and velar stops were synthe-

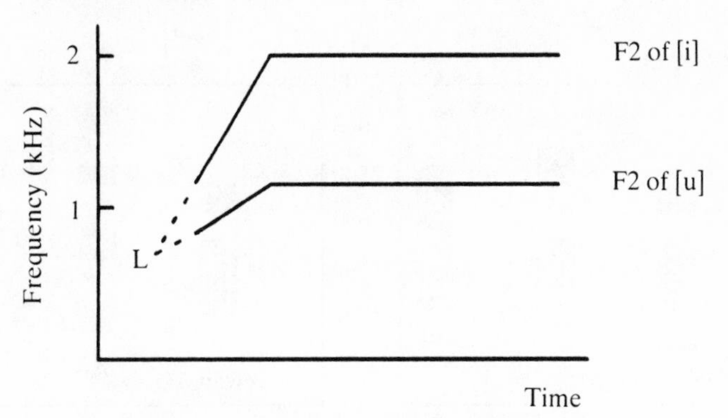


*Figure* 2.3. Relationship between locus frequency, transitions and vowel targets. The formant transitions in the synthesis of [bi] and [bu] extend from 720 Hz (the locus frequency of bilabials, marked *L*) to the steady-state formant frequencies of [i] and [u]; the first 50 ms of the transitions (the dashed lines) are then erased.

sized. Delattre *et al.* found that if the slope of F1 was held constant and rising, the most acceptable [b] and [d] stimuli were synthesized when the F2 transition pointed to locus frequencies of 720 Hz and 1800 Hz respectively. [g] was considered to have a locus at 3 kHz only when the adjoining vowel's second formant frequency was greater than 1.2 kHz. For vowels whose F2 was less than this value, no locus could be found. Figure 2.3 illustrates the relationship between locus frequency, onset of formant transitions and the steady-state formant frequencies in the (hypothetical) synthesis of two syllables, [bi] and [bu].

Given the locus frequency and the frequency values of the second formant of the following vowel, it is possible to predict whether the slope of the second formant transition is rising, falling or level. Since the locus for [b] is low at 720 Hz and since most vowels have F2 above this value, the F2 transition for [bV] syllables should be rising. Similarly, F2 should rise slightly in [di] syllables and should be rising, or level, in [de]; for other [dV] syllables in which F2 of [V] is less than 1.8 kHz, F2 should be falling.

The development of the locus theory was based on the results of synthesis and labelling experiments in which the role of formant frequencies above F2, as well as other fine details of acoustic structure, could not be taken into account. It is perhaps not surprising, therefore, that the identification of loci associated with place of articulation from acoustic analyses of real speech has remained

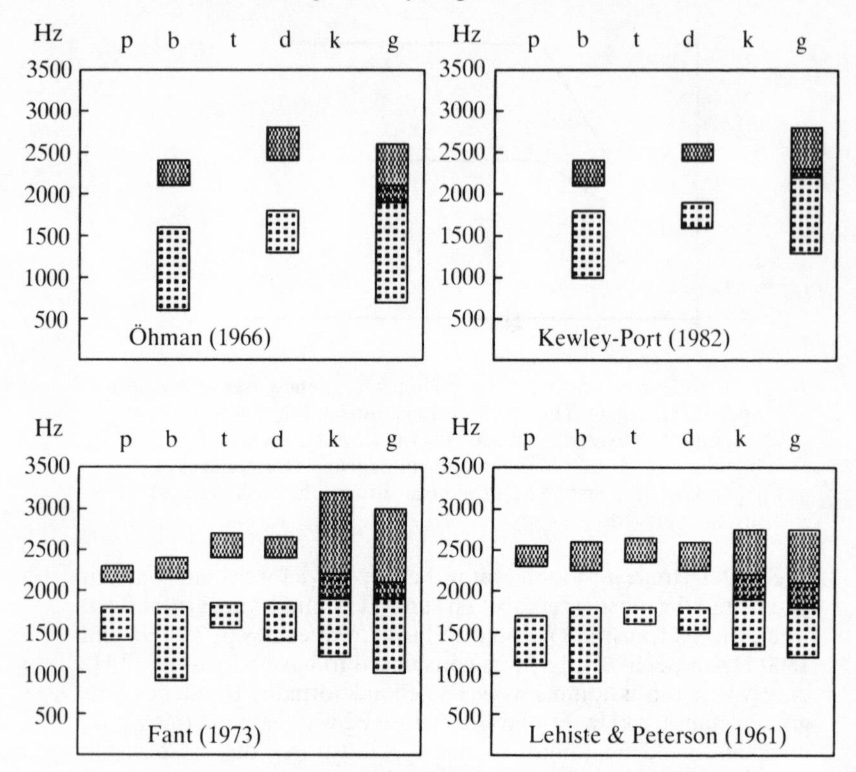


*Figure* 2.4. The range of F2 onset (dotted rectangles) and F3 onset (shaded rectangles) in four studies. The plots of the data in Fant and in Lehiste and Peterson are adapted from a similar display in Fant (1973, 123-4).

elusive. Second, as Fant (1973) points out, the theory of the relationship between invariant place of articulatory closure and F2 loci overlooks the complexities of formant transitions which, in the bilabial series for example, are the product of both lip passage opening and tongue body movement. Thus, although an increase in lip section area at the release of the bilabial cannot result in the downward shift of formants, a superimposed tongue body movement can produce a transition of opposite sign to that induced by the lip passage opening (Fant 1973, 126); the outcome, contrary to the predictions of the locus theory, can be a falling F2 when [p] precedes back vowels.

Figure 2.4 summarises the range of onset frequencies of F2 and

Kewley-Port

| | | i | ɪ | e | ɛ | æ | ɑ | | | o | u |
|---|---|---|---|---|---|---|---|---|---|---|---|
| b | F3 | ↑ | ↑ | ↑ | ↑ | ↑ | ↑ | | | ↑ | ↓ |
| b | F2 | ↑ | ↑ | ↑ | ↑ | ↑ | L | | | ↓ | ↓ |

Fant

| | | i | | e | ɛ | | ɑ | ʉ | | o | u | y | ø | |
|---|---|---|---|---|---|---|---|---|---|---|---|---|---|---|
| p | F3 | ↑ | | ↑ | ↑ | | ↑ | ↓ | | | ↑ | L | L | |
| p | F2 | ↑ | | ↑ | ↑ | | ↓ | L | | ↓ | ↓ | ↓ | ↓ | |

Öhman

| | | | | | | | ɑ | | | o | u | y | ø | |
|---|---|---|---|---|---|---|---|---|---|---|---|---|---|---|
| b | F3 | | | | | | ↑ | | | ↑ | ↑ | ↑ | ↑ | |
| b | F2 | | | | | | | | | | | ↑ | ↑ | |

Fischer-Jørgensen

| | | i | | e | ɛ | a | | | ɔ | o | u | y | ø | œ |
|---|---|---|---|---|---|---|---|---|---|---|---|---|---|---|
| p,b | F3 | ↑ | | ↑ | ↑ | ↑ | | | L | L | L | L | L | L |
| p,b | F2 | | | ↑ | ↑ | ↑ | | | ↓ | ↓ | ↓ | ↑ | ↑ | ↑ |

*Table* 2.1. Bilabial F2 and F3 transitions in four studies.
↑ F2/F3 transition of the acoustic vowel onglide is rising;
↓ F2/F3 transition is falling; L F2/F3 transition is level;
L↑ F2/F3 transition is level or rising; L↓ F2/F3 transition is level or falling. When a cell of a particular segment is empty, the transitions could not be reliably identified or data on transitions is not given in the original study. The arrows are based on a linear interpolation between the 'onset frequency' and the 'vowel steady-state' in Kewley-Port, and between 'instant of release' and 'instant of voice onset' in Fant.

F3 for vowels following oral stop consonants at the three places of articulation taken from four detailed acoustic studies of formant transitions. The data in Fant (1973) were derived from a spectrographic study of the six Swedish stops [p], [t], [k], [b], [d], [g] combined with each of the nine long Swedish vowels [uː], [oː], [ɑː], [ɛ], [eː], [iː], [yː], [uː], [øː] produced by one subject. In Lehiste and Peterson (1961), spectrographic measurements were made both of 1263 CVC monosyllables (where V is a vowel or diphthong in American English) produced by one speaker and a control set of 70 words produced by five speakers of the same

Kewley-Port

| | | i | ɪ | e | ɛ | æ | ɑ | | | o | u |
|---|---|---|---|---|---|---|---|---|---|---|---|
| d | F3 | ↑ | L↓ | L↑ | ↓ | ↓ | ↓ | | | ↓ | ↓ |
| d | F2 | ↑ | ↑ | L↑ | L↓ | ↓ | ↓ | | | ↓ | ↓ |

Fant

| | | i | | e | ɛ | | ɑ | ʉ | | o | u | y | ø | |
|---|---|---|---|---|---|---|---|---|---|---|---|---|---|---|
| t | F3 | ↑ | | L | ↓ | | ↓ | ↓ | | ↓ | ↓ | ↓ | ↓ | |
| t | F2 | ↑ | | ↑ | ↑ | | ↓ | ↓ | | ↓ | ↓ | ↑ | ↓ | |

Öhman

| | | | | | | | ɑ | | | o | u | y | ø | |
|---|---|---|---|---|---|---|---|---|---|---|---|---|---|---|
| d | F3 | | | | | | | | | L↓ | L↓ | ↓ | ↓ | |
| d | F2 | | | | | | ↓ | | | ↓ | ↓ | ↑ | L | |

Fischer-Jørgensen

| | | i | | e | ɛ | a | | | ɔ | o | u | y | ø | œ |
|---|---|---|---|---|---|---|---|---|---|---|---|---|---|---|
| t,d | F3 | ↑ | | ↑ | L↓ | L↓ | | | ↑ | ↑ | ↑ | L↓ | L↓ | L↓ |
| t,d | F2 | | | ↑ | ↑ | | | | ↓ | ↓ | ↓ | L | L | L |

*Table* 2.2. Alveolar F2 and F3 transitions in four studies. See caption to table 2.1 for explanation of symbols.

dialect. The study in Öhman (1966) was based on spectrographic measurements of VCV syllables produced by one speaker in which the consonants were the Swedish voiced stops [b d g] and the vowels [y ø ɑ o u]. Finally, Kewley-Port (1982) used a linear prediction algorithm (Markel and Gray 1976) to analyse the formant frequencies of the voiced stops [b d g] followed by each of the vowels [i ɪ e ɛ æ ɑ o u]. Figure 2.4 shows that only alveolar stops can be considered to have a unique F2 locus whose average value is around 1.8 kHz, as predicted by the Haskins Laboratory experiments. The Haskins Laboratory experiments suggested that velars do not have a unique F2 locus and this is confirmed by the display in figure 2.4 which shows that the F2 onset can vary over a range of roughly 1.5 kHz which overlaps with the F3 range. Bilabials, which also have a wide range of F2 onset frequencies, cannot be considered to have an invariant locus. Kewley-Port (1982) notes that bilabials may have two loci, one at 1645 Hz averaged over the front vowels [i ɪ e ɛ æ] and another at 1090 Hz averaged over the three

Kewley-Port

| | | i | ɪ | e | ɛ | æ | ɑ | | | o | u |
|---|---|---|---|---|---|---|---|---|---|---|---|
| g | F3 | L↓ | ↓ | L↓ | ↓ | L↑ | ↑ | | | L↓ | ↓ |
| g | F2 | ↓ | ↓ | ↓ | ↓ | ↓ | ↓ | | | ↓ | L↓ |

Fant

| | | i | | e | ɛ | | ɑ | ʉ | | o | u | y | ø | |
|---|---|---|---|---|---|---|---|---|---|---|---|---|---|---|
| k | F3 | ↓ | | ↓ | ↓ | | ↑ | ↓ | | ↑ | ↑ | ↓ | L | |
| k | F2 | L | | L | ↓ | | ↓ | ↑ | | ↓ | ↓ | L | ↓ | |

Öhman

| | | | | | | | ɑ | | | o | u | y | ø | |
|---|---|---|---|---|---|---|---|---|---|---|---|---|---|---|
| g | F3 | | | | | | | | | ↑ | ↑ | ↑ | ↑ | |
| g | F2 | | | | | | | | | | | | | |

Fischer-Jørgensen

| | | i | | e | ɛ | a | | | ɔ | o | u | y | ø | œ |
|---|---|---|---|---|---|---|---|---|---|---|---|---|---|---|
| k,g | F3 | ↑ | | ↑ | ↑ | ↑ | | | ↑ | ↑ | ↑ | | | |
| k,g | F2 | L | | L | L | L | | | ↓ | ↓ | ↓ | L↑ | L↑ | L↑ |

*Table* 2.3. Velar F2 and F3 transitions in four studies.
See caption to table 2.1 for explanation of symbols.

back vowels in her study.

Tables 2.1 to 2.3 summarise details on the slope of F2 and F3 of the acoustic vowel onglide transitions in four studies. With regard to bilabials, all studies confirm the Haskins predictions that the F2 and F3 transitions rise preceding front unrounded vowels. These four studies are also consistent with the results of a study by Edwards (1981) which showed 88 % of stop initial syllables which had a rising F2 and a rising F3 in the acoustic vowel onglide were bilabial; only 15 % and 12 % were alveolar and velar respectively. Before back vowels, Kewley-Port, Fant and Fischer-Jørgensen have shown that the F2 transition of bilabials is falling. Table 2.2 shows that there is a tendency for F2 to be rising when alveolar stops precede front, unrounded vowels, and for F2 to be falling when alveolars precede non-front vowels. With regard to velar stops (table 2.3), it seems difficult to make any firm predictions beyond the tendency for F2 to be level, or falling, before vowels that are not both front and round.

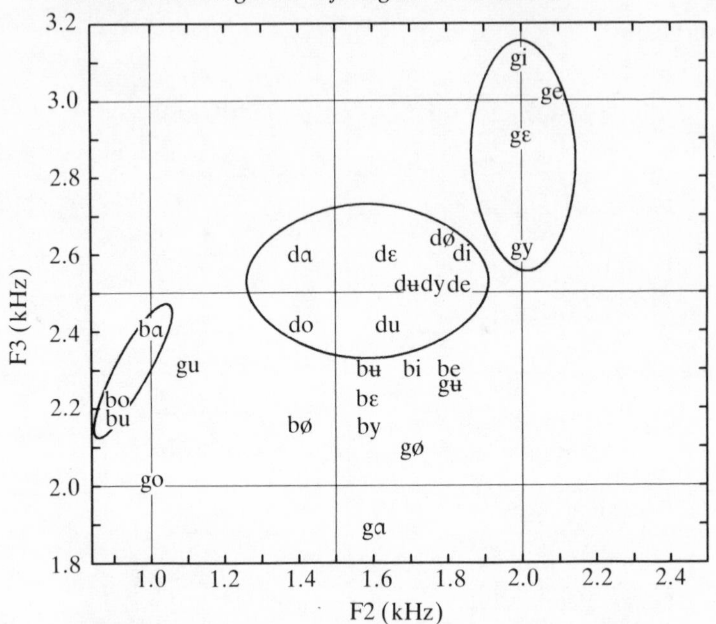


*Figure* 2.5. F2 and F3 onset frequencies in the F2/F3 plane of data in Fant (1973).

As Fant (1973) has suggested, a plot of the onset frequencies (figure 2.5) in the F2-F3 plane shows a tendency towards clustering into three zones that correspond to bilabial, alveolar and velar. Figure 2.5, taken from Fant (1973), is a plot of F2 and F3 at plosion for voiced stops preceding long Swedish vowels; for the purpose of comparison, the onset frequencies of F2 and F3 for the voiced stops in various vowel contexts from Kewley-Port's (1982) data are plotted in figure 2.6. Both figures show that velar stops preceding front vowels cluster in the right of the display while the alveolar stops cluster in the centre. There is a tendency for bilabial stops preceding back vowels to cluster in the far left of the pattern; however, both displays confirm that the other bilabial stops are confused with velar stops preceding back vowels (and [du] in Kewley-Port's data).

Some of this confusion may be resolved by plotting the *change in F2* against the *change in F3.* Change in F2 is defined as the difference between a formant's onset frequency following the release and the formant frequency value of the acoustic vowel target. Such a plot of Kewley-Port's data is shown in figure 2.7. The display

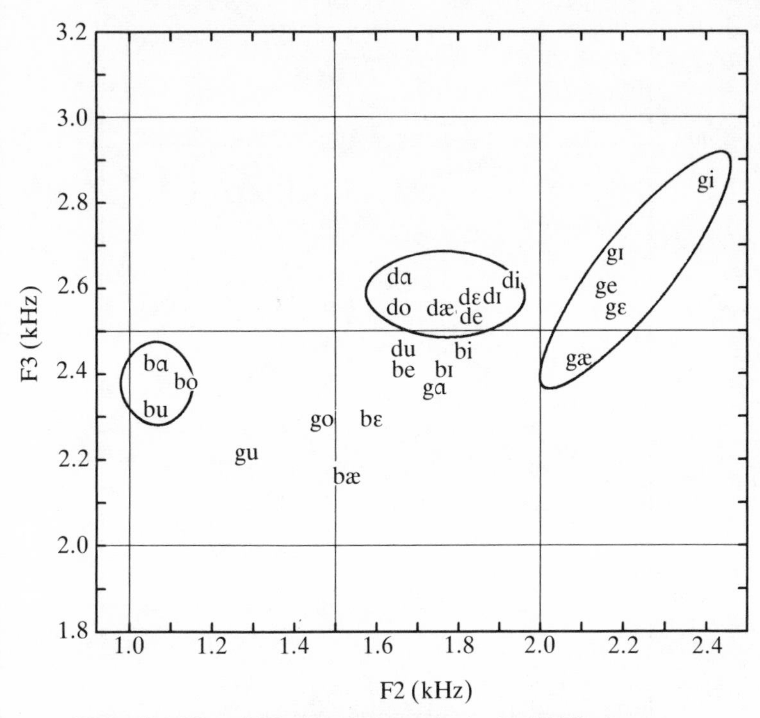


*Figure* 2.6. F2 and F3 onset frequencies in the F2/F3 plane of data in Kewley-Port (1982).

shows that the bilabial stops that are confused in figure 2.6 are well separated from all other stop-vowel pairs in a plot of the *change of F2* against *change in F3.* The reason for this separation is that, with the exception of [ di ], only bilabial stops have both a rising F2 and a rising F3 ( tables 2.1–2.3 ). It is true that [ bo ] and [ bu ] are confused in this display with other CV sequences; however, [ bo ] and [ bu ] are well separated from alveolar and velar stops in the F2-F3 plane of onset frequencies (figure 2.6). It is possible, therefore, that a combination of the plots in figures 2.6 and 2.7 may form the basis for the within-class identification of place of articulation of oral stops.

*Voice Onset Time*

Within the /b d g/ and /p t k/ sets, some studies have shown that velars have the longest, and bilabials the shortest, *voice onset time* (VOT). There is, therefore, some support for the idea that VOT

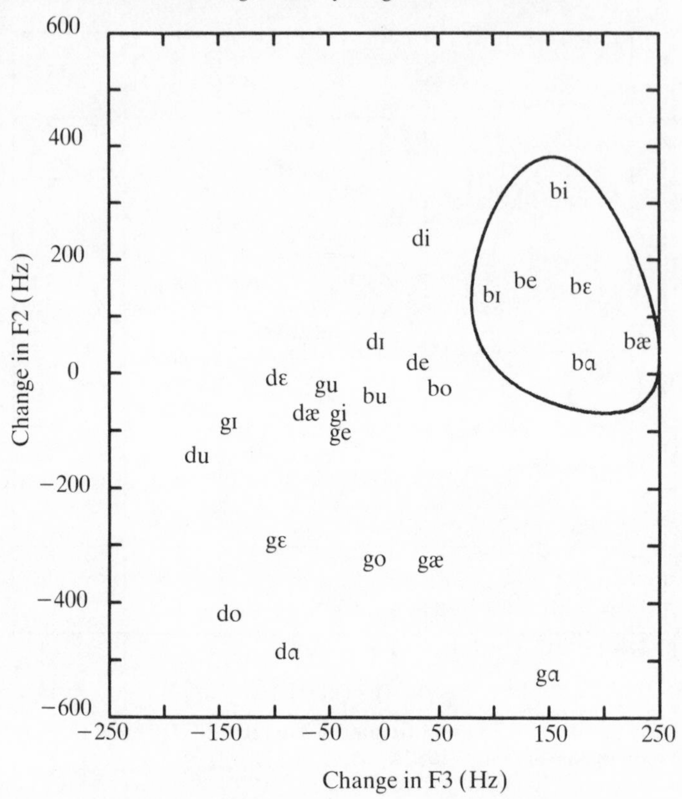


*Figure* 2.7. *Change in F2* against *change in F3* of data in Kewley-Port (1982). *Change* denotes the difference in formant frequency values between the vowel onset frequency and the vowel steady-state.

increases as place of articulation moves from the lips to the back of the oral cavity – Lisker and Abramson 1964, Zue 1976 – however, this hypothesis is not consistent with data in Peterson and Lehiste (1960) which shows that VOT is greater for the voiceless pre-velar stop [k+] than the post-velar stop [k-]). Zue (1976) finds a mean VOT of 13 ms for [b], 19 ms for [d] and 30 ms for [g]. VOT is also greatest for voiceless velars and shortest for voiceless bilabials: the mean durations in Zue's data are 74.5 ms for [k], just over 70 ms for [t] and 58.5 ms for [p]. Edwards (1981) also finds that velar stops have the longest VOTs and labial stops the shortest. Fant (1973) finds the same trend, with mean VOT values of 130 ms for

[k], 120 ms for [t] and 115 ms for [p]. In Peterson and Lehiste (1960), the average duration of aspiration of an initial [p] was 58 ms (averaged over 81 tokens), 69 ms for an initial [t] (73 tokens) and 75 ms for [k] (83 tokens). In Lehiste and Peterson (1961), the average duration of the acoustic vowel onglide was calculated for 1263 words. Average durations of the acoustic vowel onglide were 67 ms for [p], 79 ms for [t] and 88 ms for [k]. For the voiced series [b], [d], [g], values of 51, 68 and 78 ms respectively were found. Kewley-Port (1982) finds VOT is the most effective acoustic cue for the identification of place of articulation (in this study, VOT was defined as the interval from the burst to the onset of a prominent F1 peak accompanied by an abrupt rise in RMS energy). Mean values of VOT were 18 ms for [b], 27 ms for [d] and 31 ms for [g].

### Oral Stops: Voiced/Voiceless Distinction

Voiced and voiceless oral stops in English exhibit differences in the coordination of supralaryngeal activity and the onset of vocal fold vibration which is manifested acoustically as a difference in voice onset time (VOT). VOT can be defined as the duration from the burst to the onset of periodicity (Lisker and Abramson 1967). Table 2.4 summarises the results of VOT for voiced and voiceless stops in the study by Lisker and Abramson (1967) and in the study by Zue (1976) referred to earlier. Lisker and Abramson's study shows that the VOT separation between voiced and voiceless stops decreases for stops produced in sentences compared with initial stops in words produced in isolation. The study also shows there is a tendency for stressed [p t k] to be produced with greater VOTs than unstressed [p t k]. Zue's results for voiceless stops in /həCVC/ nonsense words are in close agreement with the VOT measurements for initial stops in isolated words in the Lisker and Abramson study, although VOT values for voiced stops in Zue are slightly greater. Zue's study also shows that there is a tendency for VOT to increase when stops occur in $C_1C_2V$, where $C_2$ is /l r w j/; VOT for unaspirated stops in /sC/ clusters (*spy, sty, sky*) is close to VOT values for voiced stops in /CV/ syllables.

A problem with the definition of VOT as the duration from the release of the closure to the onset of periodicity is discussed in Fischer-Jørgensen and Hutters (1981). These authors show that, for some speakers, periodic excitation is evident in the lower harmonics before it spreads to higher frequencies. Using photoelectric glottograms, Fischer-Jørgensen and Hutters show that the periodic onset of energy in the higher formants seems to correlate with the time, following stop release, at which the vocal folds first close. In

Lisker and Abramson (1967)

| | b | | p | d | | t | g | | k |
|---|---|---|---|---|---|---|---|---|---|
| Stop initial isolated words (4 speakers) | −101 | 1 | 58 | −102 | 5 | 70 | −88 | 21 | 80 |
| Prestressed stops in sentences (10 spkrs) | −62 | 10 | 35 | −49 | 10 | 48 | −77 | 20 | 55 |
| Unstressed stops in sentences (10 spkrs) | −50 | 10 | 34 | −55 | 15 | 40 | −6 | 21 | 45 |

Zue (1976)

| | b | d | g | CC +V | p | t | k | CC −V | sC | sCC |
|---|---|---|---|---|---|---|---|---|---|---|
| /həCVC/ (3 speakers) | 13 | 19 | 30 | 26 | 59 | 70 | 75 | 86 | 23 | 30 |

*Table* 2.4. Voice onset time in two studies, showing average VOT values in ms (to the nearest ms). For [b d g] in Lisker and Abramson's study there are two averages: the negative values are averages of tokens with voicing lead (i.e. voicing begins during the closure), and the positive values are averages of VOT for all tokens in which voicing begins after the burst. In Zue's study the categories CC+V and CC−V are average VOT values for initial voiced and voiceless stops respectively in which the second consonant is one of the approximants /l r w j/.
sC average VOT values for [sp], [st] and [sk] initial clusters;
sCC average VOT values for [spl], [spɹ], [stɹ] and [skl] initial clusters.

a further experiment, they show that if the start of periodicity in the higher formants is chosen as the point which marks the end of aspiration, observed differences in vowel duration after Danish /p t k/ become much more regular than if the onset of the first harmonic is chosen as the acoustic vowel onset.

The implications of the findings of Fischer-Jørgensen and Hutters for the perception of voicing in stops are considered in a series of experiments in Darwin and Seton (1983). They showed that in two synthetic stimuli, the first of which has periodic energy in F1 beginning at the same time as periodic energy in F2 and F3 and the second of which has periodic energy in F1 that precedes the onset of periodicity in F2 and F3, the probability of identifying the synthetic stops as voiced is greater for the first, than the second, stimulus for the same VOT. These experiments suggest that the voicing cue does not simply depend on the onset of periodic energy in F1 irrespective of the onset of energy in other parts of the spectrum. In a related acoustic study by Edwards (1981), a measurement was made of the interval from the release of the stop closure to the onset of the fundamental, and a second measurement was made from the release to the point at which there was an abrupt rise in the signal amplitude coterminous with the onset of periodic energy in the higher formants. The results of this study, in fact, showed very little difference between the two measurements. The first measurement gave an average VOT of 15 ms for voiced, and 66.6 ms for voiceless stops; the second measurement gave an average of 16.7 ms for voiced, and 68.6 ms for voiceless stops.

*First Formant Characteristics*

Some studies have shown that the *slope of the first formant transition,* the *degree of first formant cutback* and the *onset frequency of the first formant* are also important acoustic cues that underlie the distinction between voiced and voiceless stops. The basis for these cues can be derived, in part, from production theory which has shown that the first formant is very low during complete articulatory obstruction (Delattre, 1951, Stevens and House 1956, Fant 1970) and that if there is voicing during the closure, F1 is again very low, and close to, the fundamental (Fant 1973). In the production of voiced stops, therefore, the first formant will rise rapidly from the time of the burst release to the F1 value of the following vowel target (the rise will be steepest in open vowels whose F1 is high, and flattest in close vowels, such as [i], whose F1 is low). Since periodicity begins around 20 ms after the burst release, the rising first formant will be predominantly pulse excited.

In voiceless stops, on the other hand, the onset of periodicity occurs some 30 ms later than in voiced stops. Therefore, far less of the F1 transition will be pulse excited – indeed, by the time pulse excitation begins, the F1 transition has almost been completed (figure 2.8). On spectrograms, therefore, voiced stops are usually characterised by a voiced, rising F1 transition which is absent in the case of voiceless stops. The absence of a clear F1 transition for voiceless stops is not only attributable to the fact that pulse excitation begins much later in the transition, but also because of the presence of aspiration which occurs in the interval between the burst and onset of voicing: aspiration can, of course, only be produced when the glottis is open and the coupling of the large resonance chamber below the glottis to the supralaryngeal tract is thought to attenuate considerably the first formant (Fant 1970, 1980).

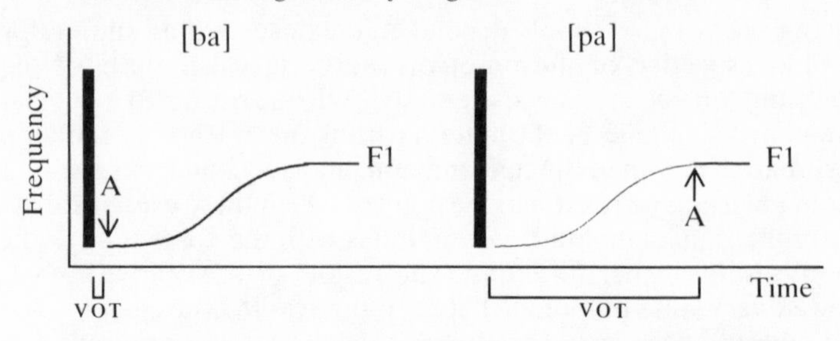


*Figure* 2.8. Schematic outline showing the relationship between VOT, slope of voiced F1 transition and onset frequency of voiced F1 (onset of voiced F1 is marked A, and the vertical black bar designates the burst). In [pa], most of the F1 transition has been completed before voicing begins; consequently, the onset frequency of voiced F1 is higher in [pa] than in [ba].

Some of the earliest work at the Haskins Laboratories (Delattre *et al.* 1952; Liberman, Harris, Kinney and Lane 1961) has confirmed that a *rising F1 transition* is an important cue for the perception of voiced stops. In the experiment by Liberman *et al.*, the onset of the voiced F1 was synchronous with the onset of the higher voiced formants; perception of voiceless stops was then obtained by progressively deleting the transition of the voiced F1 from its onset with respect to the onset time of higher voiced formants. As Fischer-Jørgensen (1969) points out, however, natural speech never has higher voiced formants which begin earlier than the voiced first formant; the cue which elicited voiceless percepts in this experiment was, therefore, probably the absence of a voiced F1 transition, and not a delay of voiced F1 with respect to higher voiced formants. There is also acoustic data which shows that the presence of a rising F1 is a cue for voiced stops: Fischer-Jørgensen (1954, 140) reports that for the voiceless stops, F1 transitions were level, slightly rising and rising for 19, 5 and 5 tokens respectively; the corresponding values for voiced stops were 2, 6 and 20 tokens. These results provide some support, therefore, for the Haskins experiments which suggest that a rising F1 is a cue for a voiced stop. However, Kewley-Port (1982) found that the slope of F1 was either flat, or falling, for [bi di gi bu gu] which suggests that a rising F1 is not a necessary cue of voiced stops.

A study by Lisker (1975) points to the importance of the *onset frequency* of the voiced first formant as a basis for distinguishing voiced from voiceless stops. As figure 2.8 shows, if the onset of voicing in [p t k] occurs late relative to the onset of the F1 transition, the onset frequency of voiced F1 will be higher than in [b d g]. A study by Summerfield and Haggard (1977) is consistent with Lisker's (1975) analysis. In the experiments by Summerfield and Haggard, the onset frequency of F1 and the duration of the transition were varied. The results of labelling experiments showed that VOT and F1 onset frequency may be traded for each other: the lower the onset frequency of F1, the larger the separation interval required between burst release and onset of periodicity to elicit a voiceless percept. As Darwin and Seton (1983) observe, the quantitative effect of F1 onset frequency can be estimated from the data in Summerfield and Haggard (1977). Before open vowels, a 1 Hz change in F1 onset frequency is approximately equivalent to a 0.11 ms change in aspiration duration.

The formant transitions and offset frequency in VC syllables should show a mirror image of those in CV syllables: F1 should therefore fall sharply into the closure when the final stop consonant is voiced and the offset frequency of the (level, or slightly falling) voiced F1 should be higher for voiceless compared with voiced stops. An acoustic study by Wolf (1978) of [ab], [ad], [ag], [at], [ak] showed that F1 averaged over both the first 50 ms and the last 30 ms of the voiced formant was significantly lower for [b d g] than [p t k], but no comment is made on the offset slope of the F1 transition. In another acoustic study of [dap], [dat], [dak], [dab], [dad], [dag], Revoile, Pickett and Holden (1982) report a more sharply falling F1 into the closure and a lower F1 offset frequency for [dab] and [dad] than [dap] and [dat] respectively, but generalisations for the velars were not possible.

*Preceding Vowel Duration*

Several studies have shown for American English that the duration of the vowel before a voiceless oral stop in prepausal syllables is about 60% shorter than before voiced stops; in reviewing a variety of studies which have demonstrated this effect (House 1961, House and Fairbanks 1953, Klatt 1973, Malécot 1970, Peterson and Lehiste 1960, Raphael 1970, Wang 1959, Zimmerman 1958), Wardrip-Fruin (1980) finds a range of 52–69% shorter vowel duration before voiceless than voiced stops. In an analysis of the perceptual cues that distinguish *rapid* and *rabid,* Port (1979) finds a trading relationship between the duration of the vowel and the duration of the closure: an increase in the duration of the vowel (within limits) has to be compensated for by a corresponding increase in the duration of the silent gap for listeners to perceive *rapid*.

Raphael, Dorman and Liberman (1980) have shown that the vowel duration cue that underlies the voiced/voiceless stop distinction is affected by the presence of a prevocalic stop that was added to a synthetic /VC/ syllable: the final consonant is heard more frequently as /d/ (as opposed to /t/) when a consonant precedes the vowel than when the vowel is syllable initial. Fowler's (1984) explanation for this phenomenon is that the vowel in /CV/ and /V/ syllables is *perceived* to be the same duration despite the fact the measured acoustic duration of the vowel shortens when a consonant is added to a /V/ syllable to derive a /CV/ syllable (Lindblom and Rapp 1973). Therefore, if the *acoustic* durations of the vowels in /V/ and /CV/ syllables are equalised, the *perceived* duration of the vowel in the /CV/ syllable will increase compared with that of the /V/ syllable. From this perspective, if a synthesised post-vocalic alveolar stop is added to /CV/ and /V/ syllables for which the vowels have the same *acoustic* duration, the final consonant is more likely to be *perceived* as voiced in the resulting /CVC/ syllable, since a greater perceived vowel duration is a cue for a following voiced stop.

Some studies have shown that preceding vowel duration is not a necessary cue to elicit the perception of a voiced stop (Hogan and Rozsypal 1981, Parker 1974, Wardrip-Fruin 1980, Wolf 1978). In an experiment by Wardrip-Fruin (1980), 52 monosyllabic words produced by two speakers were recorded. Various parts of the syllable were deleted, compressed or expanded and presented to twelve adult listeners who judged whether the syllables ended in a voiced or voiceless stop. Wardrip-Fruin (1980) found that syllable duration was not an adequate cue for the voiced/voiceless distinction since syllables without final transition and final segment information were not heard as better than 60% voiced at any syllable duration. Similarly, when Hogan and Rozsypal (1981) decreased the duration of a vowel before a voiced stop and increased the duration of a vowel before a voiceless stop, they found that expanding the vowel nucleus alone cannot force a change in perception of the final consonant from voiceless to voiced.

For preceding unstressed vowels, vowel duration is a less reliable cue for the identification of stop consonant voicing; Edwards (1981) reports that the mean durations of [ə] in /həCVC/ syllables are 73 ms and 65 ms preceding voiced and voiceless stops respectively. The mean difference of 8 ms agrees well with the results of House and Fairbanks (1953) who made similar measurements in a /huCVC/ context. Edwards finds that the overlap between the distributions of the vowel durations before voiced and voiceless stops is so great that the cue is of limited use for differentiating /p t k/ from /b d g/.

*Other Cues*

Lisker (1957), Port (1976) and Slis and Cohen (1969) find that the *duration of the intervocalic closure* is greater in voiceless stops than voiced stops. Lisker (1978) reports that when the duration of the (silent) closure of a naturally produced token of *rapid* is reduced to less than 100 ms, perception changes to *rabid*. Since, however, there is a large overlap in the ranges of possible intervocalic closure durations of [p] and [b] in *rapid* and *rabid* produced in isolation by one speaker, Lisker (1978) questions whether the duration of the closure could be successfully implemented to differentiate between voiced and voiceless stops by a speech recognition device.

Lisker (1978) has reported that intervocalic [b d g] often exhibit *voicing during the closure* of the stop (referred to as *buzz*); it is also just as common for the intervocalic [p t k] closures to be voiceless. Furthermore, if the buzz from the closure of a naturally produced token of *rabid* is deleted, listeners perceive *rapid*; similarly, inserting buzz into a naturally produced token of *rapid* elicits reponses of *rabid*. Lisker (1978) also shows that listeners always perceive buzz-filled closures as /b/, irrespective of the closure duration.

The *onset frequency of F0* has been shown to be higher following voiceless than voiced stops (House and Fairbanks 1953, Lehiste and Peterson 1961, Mohr 1971, Løfqvist 1975, Umeda 1981). F0 can be higher following voiceless consonants as far as 100 ms from voicing onset (House and Fairbanks 1953, Umeda 1981). When a stop-vowel stimulus is synthesised whose VOT is intermediate between values for clearly voiced and clearly voiceless, the superimposition of a low rising F0 contour favours the perception of a voiced stop; a high falling contour will generally result in the identification of a voiceless stop (Haggard, Ambler and Callow 1970). In an acoustic analysis of 60 CV syllables produced by eight speakers, Haggard, Summerfield and Roberts (1981) found that the onset frequency of F0 is lower for [b d g] than [p t k]. Abramson and Lisker (1984) argue, however, for the primacy of the VOT cue and demonstrate that at extreme VOT values (i.e. clearly voiced and clearly voiceless), there are no F0 onset values, or F0 slopes, which will change listeners' identifications; on the other hand, at extreme F0 values (i.e. voiced: low F0 onset and rising F0; voiceless: high F0 onset and falling F0), extreme VOT values always changes listeners' identifications from voiced to voiceless and vice-versa. In an acoustic study by Edwards (1981) of /$hV_1C_1V_2C_2$/ (where $C_1$ is the target stop), three F0 measures were used to identify $C_1$ as a voiced or voiceless stop: (1) the difference between peak F0 in $V_2$ and the average F0 of the entire $V_1C_1V_2$ sequence; (2) the difference between the final measurable F0 value of $V_1$ and the first measurable F0 value of $V_2$; (3) the duration from F0 onset of $V_2$ to the F0 peak of $V_2$ (on the assumption that the peak was not at the onset). Measure (3) most reliably differentiated between voiced and voiceless stops with durations of 79.3 ms for voiced, and 24.9 ms for voiceless, stops.

Slis and Cohen (1969) report that the intensity of the burst of voiceless stops is greater than that of voiced stops by up to 50%. Similarly, Zue (1976) finds that the average burst intensity of voiceless stops is 2 dB greater than that of voiced stops. Repp (1979) finds that VOT for the distinction between /da/ and /ta/ is affected by the A/V ratio, the ratio of the overall amplitude of aspiration noise to that of the periodic portion of the vowel. Specifically, a 1 dB increase in the ratio shortens the VOT boundary by 0.43 ms (i.e. stops are more likely to be identified as voiceless as the A/V ratio increases).

*Summary*

Figure 2.9 shows a preliminary discrimination tree for the identification of place and voicing from the category oral stop. At the first level of the tree, (1), stops with voicing during the closure are identified as voiced stops (Lisker 1978). These voiced stops will include primarily intervocalic [b d g], and possibly [b d g] which occur between two voiced sonorants. [b d g] which do not have voicing during the closure will be classified with the aspirated voiceless stops [$p^h$ $t^h$ $k^h$] and the unaspirated voiceless stops [p t k]. In their automatic, acoustic-phonetic analyser of continuous speech, Weinstein, McCandless, Mondshein and Zue (1975) show that voicing during closure enables only a small percentage of voiced stops to be identified; but that this is nevertheless a robust feature (i.e. voiceless stops are rarely misclassified as voiced using this cue). At the next level in the tree, (2, 3, 4), a combination of a low VOT, F1 slope and F1 onset frequency (figure 2.8) are used to separate [b d g] and the unaspirated stops [p t k] from the aspirated voiceless stops [$p^h$ $t^h$ $k^h$]. Acoustic cues for the separation of [b d g] from the unaspirated [p t k] are not known to this author and this is why a subsequent classification has not been made.

At the next level of the tree (5, 6) a separation is made into

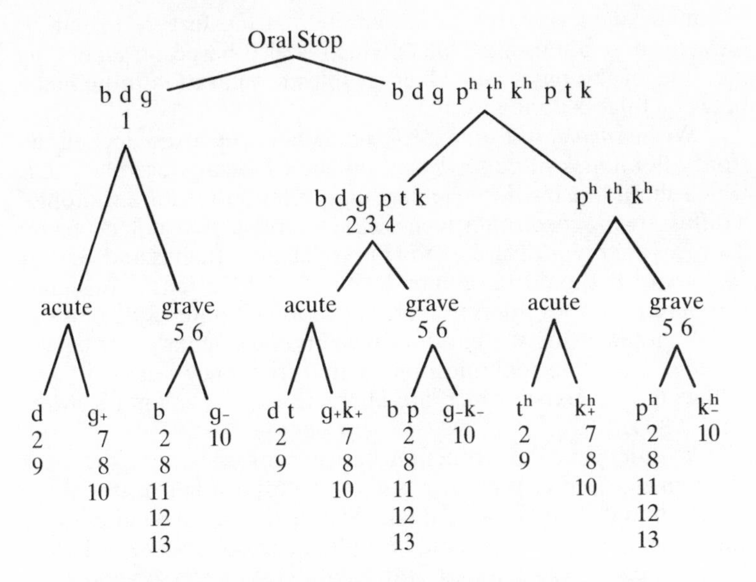


*Figure* 2.9. Place and voice identification within the class oral stop. [b d g] voiced stops; [p t k] unaspirated voiceless stops; [k+ g+] front allophones of /k/ and /g/; [k- g-] back allophones of /k/ and /g/). The relationship between numbers and features is as follows:

1 Voicing during closure
2 Voice onset time low
3 F1 transition rising
4 Voiced F1 onset frequency low
5 Even distribution of energy throughout the burst spectrum (Halle, Hughes and Radley 1956)
6 Ratio $(d-b)/(c-a)$ (see figure 2.2)
7 High mid-frequency energy in burst (Halle, Hughes and Radley 1956)
8 F2/F3 at plosion (see figures 2.5, 2.6)
9 Diffuse-rising burst spectrum (see figure 2.1)
10 Mid-frequency peaks in burst spectrum (Kewley-Port 1983)
11 Low burst energy
12 Change in F2 against change in F3 (see figure 2.7)
13 Burst spectrum is flat or diffuse-falling

*acute* (alveolars and the front allophones of velars) and *grave* (bilabials and the back allophones of velars). Two features are implemented for this purpose: a feature in Halle, Hughes and Radley (1956) which, broadly speaking, measures the ratio of high to low-frequency energy in the burst; and the feature in Lahiri, Gewirth and Blumstein (1984) which is also based on an energy measure of the burst and which enables a separation to be made between bilabials and alveolars.

Within *acute,* a distinction is made between alveolars and the front allophones of velars based on the following features: VOT, which should be less for alveolars than velars; the identification of a diffuse-rising spectrum which is characteristic of alveolars (figure 2.1); an energy measure of the burst (Halle, Hughes and Radley 1956) which should be greater in the 2–4 kHz range for velars compared with alveolars; a calculation of F2 and F3 at plosion, which should enable a separation of velars and alveolars (see figures 2.5 and 2.6); the identification of mid-frequency peaks at, and following, the burst, which are characteristic of velars (Kewley-Port 1983).

Within *grave,* a distinction between bilabials and the back allophones of velars is made based on the following features: VOT, which should be less for bilabials than velars; a calculation of F2 and F3 at plosion, which should enable the allophones of /b/ before back vowels to be separated from velars (see figures 2.5 and 2.6); the intensity of the burst, which should be less for bilabials than velars; a calculation of *change in F2* to the *change in F3* (figure 2.8), which should enable the allophones of /b/ before front vowels to be separated from velars; the identification of a flat or diffuse-falling spectrum at the burst, which is a positive cue for bilabials (figure 2.1); the identification of mid-frequency peaks at, and following, the burst, which are characteristic of velars (Kewley-Port 1983).

FRICATIVES

Fricatives are characterised by a turbulent airstream which occurs when air is channelled through a narrow constriction, possibly striking an object at high velocity (the upper front teeth in the case of [s]). The acoustic consequence of turbulence is a noise source; the spectra of fricatives can be considered as the product of the noise source located supraglottally (except for [h]), which is modified by a resonator. In the case of voiced fricatives, a second source due to the vibrating vocal folds may also be present in the spectrum.

The supraglottal source in fricatives excites primarily the cavity in front of the supraglottal constriction (known as the *front cavity resonator*). Since, in many fricatives, the cross-sectional area of the constriction is very small, this front cavity can be modelled by a tube closed at one end and open at the other and therefore has resonance peaks at quarter-wavelengths of the tube. Since the tube is very short (by comparison with the length of the supraglottal tract), the resonances of such a tube are high in frequency.

Resonances due to the constriction channel itself can also be introduced into the spectrum: in this case, the constriction channel can be modelled by a short tube open at both ends (one into the front cavity, the other into the back cavity) that has resonances at half-wavelengths of the tube (Heinz and Stevens 1961). These resonances are again high in frequency as a result of the shortness of the tube. It is also possible for anti-resonances to be introduced at a quarter-wavelength of the constriction channel (Fant 1970).

Back-cavity resonances are not considered to be dominant in fricative spectra because the narrowness of the constriction channel in relation to its length inhibits excitation of the back cavity by the supraglottal noise source. If, however, the cross-sectional area of the channel increases, paired back-cavity resonances and anti-resonances can be introduced into the spectrum.

*Sibilant / Non-Sibilant Distinction*

It has been recognised that the fricatives [s], [ʃ], the fricated part of [tʃ] and their voiced counterparts share the acoustic property *sibilance* (Ladefoged 1971) which is not characteristic of other fricatives. One of the main acoustic characteristics of sibilants is their greater relative intensity compared with non-sibilants (Ladefoged 1971, Strevens 1960).

Another acoustic characteristic of sibilants is a concentration of energy in a 3–4 kHz band (Strevens 1960) in the 1.5–7 kHz region: the spectra of [f] and [v], by contrast, exhibit an approximately equal concentration of energy in the 1–8 kHz range (Strevens 1960); or else, unlike sibilants, an intensification in the 8–16 kHz range due to the very small front cavity resonator is considered to be characteristic of [θ] and [f] (Heinz and Stevens 1961, Lauttamus 1984). In the experiments by Heinz and Stevens, for example, in which the output of a two-pole, single-zero electrical circuit was matched against fricative spectra, best matches were obtained for [f] when the first two resonances occurred in the 6.8–8.4 kHz and 8.2–12.2 kHz ranges respectively. The ranges of the first two resonances of [ʃ], on the other hand, were 2.2–2.7 kHz and 4.3–5.4 kHz respectively and those of [s] were 3.5–6.4 kHz and 8–8.4 kHz. In a subsequent perceptual experiment, Heinz and Stevens varied the resonant frequency of a single-pole circuit through several values from 2–8 kHz. Identification of /f/ and /θ/ usually occurred at 6.5 kHz or 8 kHz; most /ʃ/ responses were obtained when the resonance was around 2.5 kHz while most /s/ responses were elicited for resonances around 5 kHz.

For some non-sibilants, in particular [h], energy is sometimes concentrated into formant-like bands in the lower frequency region which is not usually considered to be characteristic of sibilants. In [h], the supralaryngeal tract shape is similar to that of the following vowel and so the formant-like concentration of noise in [h] bears a strong resemblance to the vowel's F-pattern (Lehiste 1964); for [f] and [θ], the presence of a formant-like pattern in the lower frequency range would be caused by an increase in coupling to the back cavity resonator; Lauttamus (1984) reports that formant peaks below 4 kHz characteristic of the F-pattern of the vowel are typical of [fV] syllables.

There seem, therefore, to be three main cues which may enable a distinction to be made between sibilants and non-sibilants. Sibilants are greater in intensity than non-sibilants; they are also characterised by spectral dominance in the 1.5–8 kHz range, while [f] and [θ] show an even distribution of energy in this range and a possible intensification above 8 kHz. The spectra of non-sibilants may be characterised by a greater tendency towards concentration into formant-like bands in the lower part of the spectrum, although this is more common for [h] than [f] or [θ], and can also occur in sibilants.

These three cues form the basis for isolating [s], [ʃ], the fricated section of the affricate [tʃ] and probably also the fricated part of a voiceless, stressed alveolar stop from other fricatives. The second cue may also enable the identification of the voiced counterparts of these voiceless sibilants. However, the first cue is more unreliable: since the intra-oral air-pressure in the production of voiced sibilants is likely to be reduced compared with voiceless sibilants in order to maintain a suitable transglottal pressure difference (Ohala 1983), the velocity of air-flow through the constriction channel is also likely to be reduced resulting in a decrease in acoustic intensity (Catford 1977); from this perspective, the overall intensity of voiced sibilants (first cue) may not be substantially different from that of voiceless non-sibilants.

Hughes and Halle (1956) distinguish between the voiceless sibilants and [f] using three features. For the first, energy in the 4.2–10 kHz range ($a$) and energy in the 0.72–10 kHz range ($b$) are measured. For [s], the difference ($a-b$) is less than 2 dB, whereas for [ʃ] and [f], ($a-b$) is greater than 4 dB. (As discussed below, the basis for this measurement is that [s], unlike [ʃ], has a strong concentration of energy above 4 kHz.) In order to distinguish [s] from [f], energy in the 0.72–2.1 kHz ($c$) and 0.72–6.5 kHz ($d$) ranges are measured. For [f], the difference ($c - d$) is less than 5 dB; for [s], on the other hand, ($d$) is greater than ($c$) by at least 10 dB. (This feature works on the basis that [s], unlike [f], has a concentration of energy in the middle of the spectrum.) In order to distinguish [ʃ] and [f], the most intense peak with a 500 Hz bandwidth was located in the 1.5–4 kHz region ($e$) and a measurement was made of energy in the 0.72–1.4 kHz band ($f$). For [ʃ] the difference ($e - f$) was greater than 5 dB, whereas for [f] it was less than 2 dB. (Again, the separation using this feature is successful because [ʃ], unlike [f], has a concentration of energy in the low-mid region of the spectrum.) A summary of these features in terms of a discrimination tree is shown in figure 2.10.

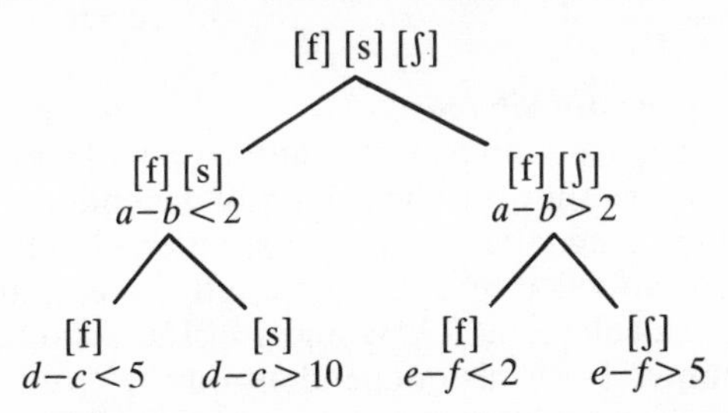


*Figure* 2.10. Discrimination tree based on Hughes and Halle (1956). *a* energy in 720 Hz-10 kHz band; *b* energy in 4.2-10 kHz band; *c* energy in 720 Hz-2.1 kHz band; *d* energy in 720 Hz-6.5 kHz band; *e* energy peak in 1.5-4 kHz band; *f* energy in 720 Hz-1.4 kHz band. All numerical values in dB.

Jassem (1965) locates the first four formants in the noise section of a variety of fricatives and calculates the difference between F4 and F2; for [f], [v], [θ], [ð], [s] and [z], (F4 − F2) was greater than 1.8 kHz, whereas for [ʃ] and [ʒ], (F4 − F2) was less than 1.8 kHz. Furthermore, where A*n* denotes the amplitude of the *n*th formant, A3 was greater than A2 for [ʃ] and [ʒ]. The difference between A4 and A2 enabled a distinction to be made between the alveolar fricatives and non-sibilants: for [s] and [z], A4 was greater than A2, whereas for [f], [v], [θ] and [ð], A4 was less than A2. A summary of these features in terms of a discrimination tree is shown in figure 2.11. The experiment was based on natural speech produced by native speakers of Stockholm Swedish, American English and non-regional Polish (three speakers, one of each language) producing primarily CVC and CV syllables.

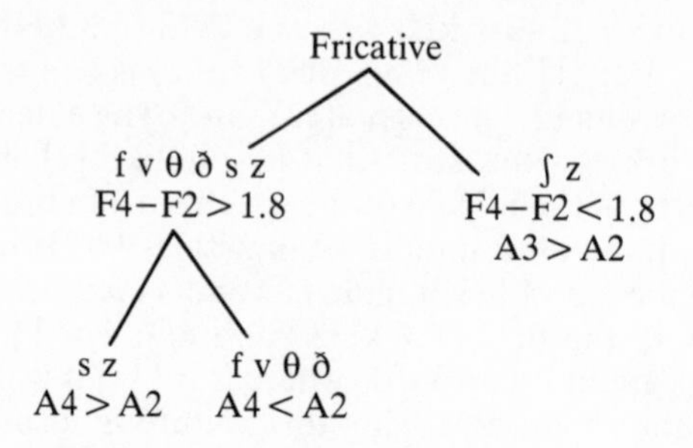


*Figure* 2.11. Discrimination tree based on Jassem (1965). F2, F4 second and fourth energy peaks in noise spectra; A*n* intensity of noise of the *n*th peak.

*Place Distinction within Sibilants*

On the basis of a noise source located at the upper front teeth, the resonances of [s] are shown to be primarily dependent on the half-wavelength of the tongue-teethridge constriction (Fant 1970); when this channel is modelled by a tube of length 2.5 cm, the first two resonances are calculated at 7 kHz and 14 kHz. As discussed above, the spectrum of [s] is also dependent on the front cavity resonance extending from the anterior opening of the constriction channel to the lips which can be modelled by a tube closed at one end of length 1 cm (Heinz and Stevens 1961): the first resonance in this case occurs at 8.6 kHz. Anti-resonances due to the cavities behind the source are introduced into the spectrum of [s]: the main anti-resonance occurs at 3 kHz (Fant 1970).

These calculations lead to the prediction that there is greater energy in the spectrum of [s] in the 3–8 kHz range compared with energy below 3 kHz; furthermore, a resonance due either to the constriction channel or the front cavity should be detectable in the 6–8 kHz region.

There are several experiments which are consistent with these theoretical calculations. In a spectrographic study of isolated fricatives, Strevens (1960) reports a sharp rise in intensity for [s] above 3.5 kHz. In an LPC analysis of voiced and voiceless alveolar and palato-alveolar fricatives produced by two speakers in an [FV] context in which [V] was one of [ɑ i u], Yeni-Komshian and Soli (1983) report that the spectra of [s] and [z] rise gradually and reach a peak at 5.5 kHz for male speakers and a similar result is reported in Soli (1981); in the female speaker, there was a gradual rise to 6.7 kHz in [s] and to 7.4 kHz in [z] – that the spectral centre of gravity of sibilants is higher for female, than male, speakers is also reported by Schwartz (1968).

The main resonance of [ʃ] occurs at a lower frequency primarily because of the longer front cavity accentuated by slight lip protrusion in the production of English RP [ʃ] (Jones 1972). From Fant's (1970) data, the first main resonance of [ʃ] is predicted to occur at 2.7 kHz; the spectrum lacks energy below this region in part due to the presence of an anti-resonance at 1 kHz; Heinz and Stevens (1961) calculate the first anti-resonance at 900 Hz; from a two-pole, single-zero circuit model, Fujisaki and Kunisaki (1977) calculate the first anti-resonance of [ʃ] at 1280 Hz. The results of some experimental studies are consistent with most of these calculations. In the spectrographic study referred to above, Strevens (1960) finds that [ʃ] is characterised by an absence of energy below about 1.6 kHz and Lauttamus (1984) reports a major trough in the spectrum at around 1 kHz. Soli (1981) finds that [ʃ] is characterised by a major resonance at 2.4 kHz with a relatively flat, or downward sloping, spectrum at higher frequencies. Like Lacerda (1982), who reports the main resonance of [ʃ] at 3 kHz, Manrique and Massone find the main resonance of [ʃ] to be located in the 2.5–5 kHz region; Yeni-Komshian and Soli (1983) find the peak resonance of palato-alveolars at 2.3 kHz for male speakers and 3.6 kHz for female speakers.

There is experimental evidence to suggest that the spectral characteristics of [s] and [ʃ] are modified due to anticipatory and progressive coarticulatory effects of adjacent segments. In an LPC study of naturally produced [FV] syllables (where F is any fricative and V a vowel), Soli (1981) reports the occurrence of F2 peaks in the fricative noise (that connect with F2 of the following vowel) at 1.5, 1.8 and 1.5 kHz for [sɑ], [si] and [su] respectively. For naturally produced [ʃɑ], [ʃi] and [ʃu] syllables, emergent F2 peaks in the noise portions are reported at 1.7, 1.9 and 1.7 kHz respectively. Mann and Repp (1980) find that the F2 vowel onset frequencies are lower in natural [sV] compared with [ʃV] syllables while F3 vowel onset frequencies are higher for [sV] than [ʃV] (average F2 vowel onsets of [sa], [ʃa], [su], [ʃu] are 1506, 1573, 1501 and 1724 Hz; average F3 vowel onsets of these four syllables are 2648, 2434, 2520 and 2232 Hz respectively).

It has long been known that the spectral centre of gravity of the fricative noise of [s], and to a lesser extent [ʃ] (since [ʃ] is often inherently rounded), is shifted down in frequency due to lip-rounding before a rounded vowel (Fant 1970, Lindblom and Sundberg 1971). It is perhaps not surprising to find, therefore, that when the fricative noise from [su] is excised and presented to listeners for classification as /s/ or /ʃ/, listeners often misclassify it as /ʃ/ (Yeni-Komshian and Soli 1983). Relatedly, when a continuum of fricative noise from /su/ to /ʃu/ is synthesised, a given token is more likely to be identifed as /s/ than a fricative-noise token with identical spectral characteristics from a /sa-ʃa/ continuum (Mann and Repp 1980). Informally, this is because, in the /su-ʃu/ continuum, listeners attribute a certain degree of spectral centre of gravity lowering to lip-rounding. Therefore, prior to phonemic classification as /s/ or /ʃ/, a perceptual correction factor is applied which raises the frequency of the fricative noise, thus increasing the probablility of classification as /s/. On the other hand, the perceptual correction factor is not applied in the /sa-ʃa/ continuum. Since /a/ is unrounded, a token from the /sa-ʃa/ continuum is more likely to be classified as /s/ than the same fricative noise from the /su-ʃu/ continuum. Interestingly, Mann and Repp find that this perceptual correction factor disappears when the same continua are synthesised but with a silent gap intervening between the fricative-noise and following synthetic vowel. This is presumably because if a silent gap intervenes in this way, the spectral centre of gravity lowering in the /su-ʃu/ continuum can no longer be attributed to the anticipatory coarticulatory effects of /u/.

The effects of coarticulation, in particular the shift in centre of gravity of the fricative noise, are likely to result in confusion in the automatic identification of the sibilant as either /s/ or /ʃ/. Fujisaki and Kunisaki (1977) suggest that identifying the first anti-resonance may enable some of the confusion due to coarticulatory influences to be resolved. These authors propose that while front-cavity length may vary as a result of coarticulatory influences, the length of the constriction channel does not vary substantially for different vowels in /FV/ syllables. Since the frequency of the first anti-resonance depends on the quarter-wavelength of this channel (Fant 1970), the frequency of this anti-resonance should not vary substantially when [s] or [ʃ] occur in the context of different following vowels. Using a two-pole, single-zero circuit, these authors find the best matches to fricative spectra are obtained when the anti-resonance (FZ), first resonance (FP1) and second resonance (FP2) occur at 3370, 4170 and 5000 Hz respectively; for [ʃ] the corresponding values are 1280, 2480 and 3580 Hz. For these sibilants produced in various contexts, separation between [s] and [ʃ] was obtained entirely in the FZ-FP1 plane: all tokens of /s/ were separated from those of /ʃ/ by the equation

$$0.78\,\mathrm{FZ} + 0.63\,\mathrm{FP1} - 3.59 = 0.0 \qquad (2.1)$$

*Fricative / Affricate Distinction*

The cues for the within-class distinction of sibilants discussed above are likely to result in a distinction between [s] on the one hand and both [ʃ] and the affricate [tʃ] on the other, since the fricated part of the palato-alveolar affricate [tʃ] is considered to be produced at the same place of articulation as [ʃ]; the cues in this section apply primarily to the differentiation between [ʃ] and [tʃ], and the distinction between [ʒ] and [dʒ]. Some of the cues discussed in this section may also apply to the separation of [tˢ] from [s], if, after application of the cues discussed above, the fricated part of a strongly released alveolar stop was misclassified as [s]. Two of the main cues for distinguishing [tʃ] from [ʃ] are the *presence of a closure,* which is characteristic of [tʃ] (especially intervocalically) but not of [ʃ], and the *duration of the fricative noise,* which is greater for [ʃ] than [tʃ]. Repp Liberman, Eccardt and Pesetsky (1978) have demonstrated a trading relationship between these two cues: the shorter the duration of the noise, the shorter the silence necessary to elicit [tʃ] responses. Repp *et al.* and Dorman, Raphael and Isenberg (1980) have also shown that as tempo increases, a greater duration of silence is required to elicit [tʃ] responses for equal durations of fricative-noise.

Gerstman (1957) first showed that the *rise time* was less for affricates than fricatives, where rise time is defined as the duration from the onset of frication to the time of the peak amplitude of friction noise. A similar result has been obtained by Howell and Rosen (1983) who used natural speech stimuli from a passage and various isolated words produced by two male and two female speakers. For most stimuli in Howell and Rosen's study, oscillographic traces showed that the onset of affricates was characterised by a burst, followed by a short interval of silence followed by frication. Rise time was defined as the interval from the onset of frication to the point at which frication reached its maximum level. Mean rise times for [ʃ] and [tʃ] in the connected passage were 76 ms and 33 ms respectively. The authors also found that the rise times were greatest for affricates word-finally (compared with affricates in other positions) and shortest for word-medial fricatives (compared with fricatives in other positions). There was no significant difference in duration between word-initial and word-medial affricates, nor between word-initial and word-final fricatives. In contrast to rise time, there was no significant difference between affricates and fricatives for the duration from the peak rise time to the end of frication. For voiced affricates, rise-time measurements of frication were only made for syllable initial affricates in nonsense syllables: average values were 49 ms and 90 ms for voiced affricates and voiced fricatives respectively.

Dorman, Raphael and Isenberg (1980) have identified five possible cues for the distinction by listeners between /tʃ/ and /ʃ/ syllable finally: *rise time of frication, duration of silent interval, duration of frication, presence / absence of release burst* and *formant transitions* in the offset of the preceding vowel. The authors found a trading relationship between the presence/absence of the burst and duration of the silent interval: when the burst was removed from a naturally produced token of *ditch*, the duration of the silent interval had to be increased to elicit *ditch* (as opposed to *dish*) responses. There was also a trading relationship between duration of the silent interval and rise-time duration. When the rise-time was reduced to 0 ms (by splicing out the initial portion of the frication), shorter closure durations were required to elicit responses of *ditch* compared with the closure durations required for a rise-time of 35 ms. The authors also note that that there are closing formant transitions indicative of complete vocal tract closure in *ditch* which are absent in the case of *dish*. Isenberg (1978) reports that the duration of the preceding vowel may be an important cue since it is shorter preceding [tʃ] than [ʃ].

*Place-Distinction within Non-Sibilants*

Lauttamus (1984) reports that the distribution of energy in the 7–16 kHz band may provide a basis for distinguishing [f] from [θ]: the resonance of greatest intensity generally occurs above 9 kHz, often at 10 kHz and sometimes as high as 12 kHz for [f]; for [θ], on the other hand, the main resonance is slightly lower at around 8 kHz; the higher frequency resonance for [f] is presumably attributable to the shorter front-cavity resonator compared with [θ].

An early study by Harris (1958) indicates, however, that formant transition cues may form a more reliable basis for differentiating between labiodental and dental fricatives in [FV] syllables. In this experiment, the noise portion in natural /FV/ syllables was separated and recombined with various vowels (thus, in one such token, the noise from [f] might be combined with a vowel that had been spliced from a [θV] syllable). In the alveolar and palato-alveolar fricatives, listeners' identifications were determined by the noise (i.e. when [s] noise was combined with a vowel from an [ʃV] syllable, listeners identify the noise as /s/); for labiodental and dental fricatives, on the other hand, listeners' identifications were directed by the vowel ([f] noise combined with a vowel from a [θV] syllable was labelled as /f/). Harris (1958) concluded that formant transitions into the vowel were the prime cue that enabled /f/ to be differentiated from /θ/. However, LaRiviere, Winitz and Herriman (1975) have since show that when all fricative-vowel formant transitions are removed in [FV] syllables, listeners can still correctly identify the transitionless noise as labiodental or dental in the syllables /fa/, /fu/ and /θu/. In a synthesis and labelling experiment, Delattre, Liberman and Cooper (1961) report that the distinction between /v/ and /ð/ before /u/ is cued primarily by F2 (level for /v/ responses, strongly falling for /ð/ responses); by the F3 transition before /i/ (rising for /v/ responses); and by both F2 and F3 before /e/ (for /v/ responses, the difference between the vowel onset frequencies of F2 and F3 is greater than for /ð/). Lauttamus (1984) reports that the F2 transition of the acoustic vowel onglide in [fV] syllables was rising (52%), had no identifiable transition (44%) and was falling (4%). The F2 transition in the onglide of [θV] syllables was rising (30%), had no identifiable transition (9%) and was falling (61%). In [fV] syllables, F3 in the onglide was rising (71%), had no reliable transition (29%) and was falling (0%); for [θV] syllables, F3 in the onglide was rising (25%) had no reliable transition (33%) and was falling (42%). These results suggest that [f] may be characterised a rising F2 and rising F3 and [θ] by a a falling F2 and falling F3 in the onglide.

There are very few detailed studies on the acoustic characteristics of [h]; but, as discussed above, there seems to be general agreement in the literature that the formant-like banding characteristic of the noise of [h] resembles the F-pattern of the following vowel. In a detailed spectrographic study by Lehiste (1964), F1 of [h] is shown to be weak in comparison with other fricative-noise formants. The reason for this may be because of coupling of the vocal tract to the subglottal system which is known to attenuate F1. F2 was greatest in intensity in the noise especially in [hə], [ha] and [hou] syllables. F2 of [h] was reported to be slightly higher than F2 of the vowel targets in [hi], [he], [ho], [hou] and [hu] syllables. F3 did not change appreciably from [h] to the vowel target in any syllable.

*Voiced/Voiceless Distinction*

Unlike voiceless fricatives, voiced fricatives are sometimes characterised by the simultaneous occurrence of noise and periodicity (Raphael 1972). However, the periodic component is often absent in the realisation of phonologically voiced fricatives; in this case, one of the cues which distinguishes voiceless fricatives from their voiced counterparts is the *duration of the noise,* which is less for voiced fricatives (Abbs and Minifie 1969; Cole and Cooper 1975). Related to this cue is Massaro and Cohen's (1976) measure of *voice onset time,* which is less for [s] than [z] in [FV] syllables (voice onset time was defined as the interval between the onset of frication and onset of periodicity). Massaro and Cohen (1976) also show that the *onset frequency* of F0 was lower in [zV] than [sV] syllables. A trading relationship between F0 vowel onset and VOT was also found: the lower the F0 vowel onset, the greater VOT must be to elicit /s/ responses. Massaro and Cohen also show that F0 vowel onset, and not shape of the F0 contour, was the important perceptual cue for /s/-/z/ differentiation. Finally, some experiments suggest that the *amplitude of the noise component* is less for voiced, than voiceless, fricatives (Soli 1981).

Haggard's (1978) experiments on the devoicing of voiced fricatives is relevant to the identification of cues that underlie the voiced/voiceless distinction in fricatives. Haggard finds that there was a higher incidence of devoicing when the voiced fricatives [v ð z ʒ dʒ] (produced by twenty-five male and seven female British English speakers) followed a voiced stop than when they occurred intervocalically; stress seemed to have little effect on the degree to which fricatives devoice and devoicing was found to be virtually universal when these fricatives preceded voiceless stops and when they occurred utterance finally. Clearly, if phonologically voiced fricatives devoice substantially – and Haggard's analysis seems to indicate that this is the case – the cue of simultaneous occurrence of noise and periodicity is likely to be of limited value for their identification. Haggard also suggests that some realisations of /v/ and /ð/ are fully voiced but not fricated: this would seem to suggest that these phonologically voiced non-sibilants are, in fact, sometimes realised phonetically as voiced approximants – in this case, not only are many of the cues discussed in this section inapplicable to their identification, but it is also quite likely that at a broad class level of analysis in speech recognition they will have been classified as *sonorants* rather than *fricatives* (Weinstein, McCandless, Mondshein and Zue 1975).

In a study in which [s] and [z] were interchanged in naturally produced tokens of [jus] (*use*, noun) and [juz] (*use*, verb), Denes (1955) showed that as the ratio of the duration of the noise to the duration of the preceding vowel decreases, listeners tend to perceive /z/. More specifically, the vowel duration ranged from 40–80 ms in /jus/ responses and from 120–200 ms in /juz/ responses; the fricative duration ranged from 200–380 ms for /jus/ responses and from 100–180 ms for /juz/ responses; similar results have been found in many other studies, e.g. Derr and Massaro (1980). When a /jus/-/juz/ continuum was presented to American, French, Swedish and Finnish subjects, Flege and Hillenbrand (1986) found that changes in vowel duration had a substantially greater effect on voicing judgements than comparable changes in fricative duration: averaging across all 104 subjects, /z/ responses increased by 77% as the duration of the vowel was increased by 200 ms; but /z/ responses increased by only 15% as fricative duration was shortened by the same amount. Soli (1982) suggests that the ratio defined by Denes (1955) cannot be a primary perceptual cue since stimuli with a particular duration ratio do not receive the same number of /jus/ responses regardless of the absolute duration of their vowels or noise. Neither is the duration of the noise a primary cue, since when the duration of the vowel is held constant, varying the duration of the noise has little effect on perception of /s/ or /z/. Instead, Soli finds the most reliable cue for the perceptual distinction between /s/ and /z/ in /ju-/ syllables is the ratio of the duration of the transition of [j] to the sum of the duration of transition of [j] and the duration of steady-state [u]. When subjects produced [jus] and [juz] embedded in passages at two different tempos (conversational and rapid), the results showed that the duration of [u] in [juz] and [jus] varied a great deal across the two tempos. However, this ratio metric showed considerably less variability from the conversational to the rapid tempos.

*Summary*

A summary of the cues which could be used for the automatic identification of place and voicing within the class *fricative* is shown in the form of a discrimination tree in figure 2.12.

At the top of the tree, all fricatives with simultaneous noise and pulse-excitation are identified as voiced (1): this may include some intervocalic fricatives and fricatives that occur between sonorants. Since, however, many phonologically voiced fricatives are realised without voicing in the noise part of the fricative, this feature will by no means separate all phonologically voiced from phonologically voiceless fricatives.

At the next level of the tree, a division is made into the alveolar fricatives and non-sibilants on the one hand and palato-alveolar fricatives, palato-alveolar affricates and non-sibilants on the other (i.e features 2, 3 and 4 are designed to separate alveolars from palato-alveolars irrespective of the way in which non-sibilants might be classified). Three features are implemented for this purpose:

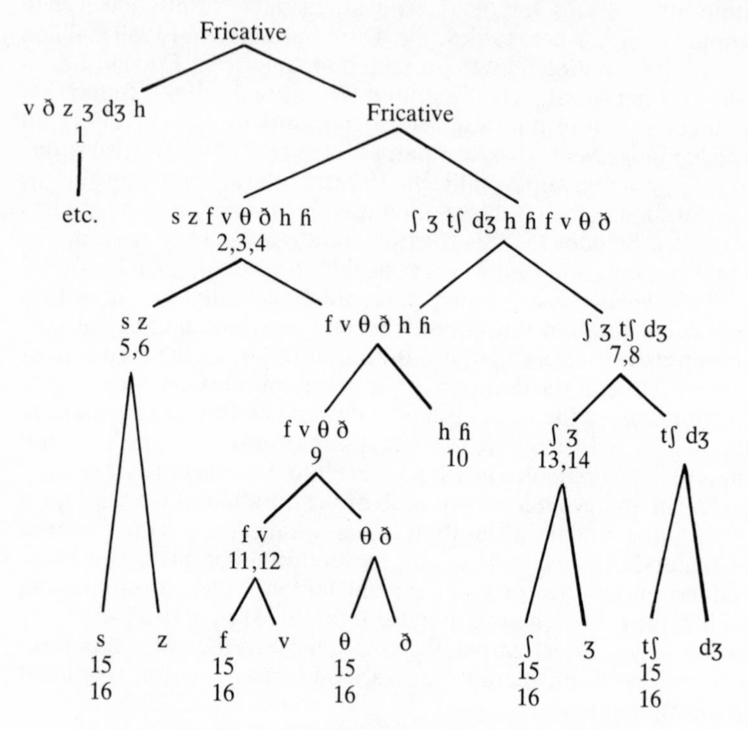


*Figure* 2.12. Discrimination tree for fricatives. The relationship between numbers and features is as follows:

1 Simultaneous pulse excitation and noise
2 Difference in energy in the 720 Hz-10 kHz and 4.2-10 kHz bands (Hughes and Halle 1956)
3 Difference in frequency between fourth and second resonance (Jassem 1965)
4 Frequency of first anti-resonance (Fujisaki and Kunisaki 1977)
5 Energy difference in the 720 Hz-6.5 kHz and 720 Hz-2.1 kHz bands (Hughes and Halle 1956)
6 Difference between A4 and A2 (Jassem 1965)
7 Energy difference between the most intense peak in the 1.5-4 kHz band and total energy in the 720 Hz-1.4 kHz band (Hughes and Halle 1956)
8 Difference between A3 and A2 (Hughes and Halle 1956)
9 Energy in the 6-16 kHz band is high
10 Resonances in noise continuous with F-pattern of a following sonorant
11 Energy in the 8.5-16 kHz band is high

the difference between energy in a 0.72–10 kHz band and a 4.2–10 kHz band, which should be greater for alveolars than palato-alveolars (Halle and Hughes 1956); the difference in frequency between the fourth and second resonance, which should be greater for alveolars than palato-alveolars (Jassem 1965); and the first anti-resonance, which should be higher in frequency for alveolars than palato-alveolars (Fujisaki and Kunisaki 1977).

At the next level alveolar fricatives and non-sibilants are separated, based on two features (5, 6): the difference between energy in the 0.72–6.5 kHz band and the 0.72–2.1 kHz band, which should be greater for alveolars than non-sibilants (Hughes and Halle 1956); and the energy in the fourth resonance, which should be greater than that of the second resonance for alveolars (Jassem 1965). At the same level, a separation between palato-alveolars and non-sibilants is made on the basis of two features (7,8): a peak (bandwidth 500 Hz) in the 1.5–4 kHz range, which should be greater than the energy in the 0.72–1.4 kHz range for palato-alveolars (Hughes and Halle 1956); and the energy in the third resonance, which should be greater than the energy in the second resonance for palato-alveolars (Jassem 1965).

Within non-sibilants, [h] and [ɦ] are separated from [f v θ ð] based on two features (9, 10): the energy in the 6–16 kHz region, which should be less for [h] and [ɦ] than the other non-sibilants; and lack of change in frequency from the resonances in the fricative-noise to the F-pattern of the following vowel target, which is characteristic of [h]. At the next level, a distinction between labiodentals and dentals is made based on two features (11, 12): energy in the 8.5–16 kHz band, which may be greater for labiodentals than dentals (Lauttamus 1984); and the slopes of F2 and F3 in the acoustic vowel onglide (if present), which should both be rising following labiodentals.

---

12 F2 and F3 in a following acoustic vowel onglide are both rising, and/or F2 and F3 in a preceding acoustic vowel onglide are both falling
13 Rise time of friction noise is high
14 Ratio of duration of friction noise to duration of preceding closure is high
15 Duration of friction noise is high
16 High energy in friction noise

With the exception of the voiced/voiceless distinction (15,16) the same sub-classifications are performed on the voiced fricatives (1) on the left branch at the top of the tree as on the right sister branch shown, labelled Fricative.

Within the palato-alveolar class (far right of the tree), a division is made between fricatives and affricates based on two features (13, 14): rise time, which should be greater for fricatives than affricates (Gerstman 1957, Howell and Rosen 1983); and second, if a closure precedes the fricative noise, the ratio of the duration of the fricative noise to the duration of the preceding closure will be greater for fricatives than affricates.

Finally, a division is made into voiced and voiceless at all levels (except for the [h]/[ɦ] distinction, since this is not phonemically relevant) based on two features (15, 16): the duration of the fricative noise, which should be greater for voiceless, than voiced, fricatives; the energy in the fricative noise, which should be greater (over a given time interval) for voiceless, compared with voiced, fricatives.

### Nasal Consonants

The acoustic structure of nasal consonants and nasal vowels is complicated by the presence of a side-branching resonator which introduces anti-resonances and its own set of resonances into the spectrum. The anti-resonances are introduced because the side-branching chamber acts a filter and selectively absorbs energy from the main tube at frequencies which are dependent, in part, on the side-chamber's own resonant frequencies. Thus, if the side-chamber has a natural resonance of *R* Hz, an anti-resonance close to *R* Hz is to be expected in the combined spectrum of both tubes; if one of the main tube's resonances coincides with this anti-resonance, it will be severely attenuated and possibly even cancelled. At the same time, a side-chamber *resonance* can occur close to the frequency of *R* Hz. This implies that the anti-resonances and resonances are always *paired* (Fant 1970), i.e. for each anti-resonance introduced by the side-chamber into the spectrum, a corresponding resonance will also be introduced. When the nasal tract is coupled to the oral tract in the production of nasal vowels, there is evidence to suggest that these paired anti-resonances and resonances originate at the same frequency and then split apart as the size of the velopharyngeal aperture increases (Fant 1970; Abramson, Nye, Henderson and Marshall 1981; Hawkins and Stevens 1985).

In speech production, either the nasal cavity or the mouth cavity can act as a side-chamber relative to the other (Laver 1980). Provided that the exit to the cavity is smaller than the entrance, whichever cavity has the smaller exit becomes the side-chamber. In nasal consonants, therefore, the closed oral cavity acts a side-chamber to the main nasal-pharyngeal tube. In most nasalised vowels, on the other hand, the nasal cavity acts a side-branching chamber to the oral cavity.

Acoustically, nasal consonants are dominated primarily by nasal resonances, i.e. the resonances of the nasal-pharyngeal tube, and mouth-cavity anti-resonances. The paired mouth-cavity resonances are not considered to be dominant in the spectrum, partly because they are damped by the complete closure of the oral tract. Spectra of nasal vowels, or indeed any nasalised sounds (that is with the nasal cavity as a side-branch to the oral cavity), are dominated by mouth-cavity resonances and nasal resonance/anti-resonance pairs. Unlike nasal consonants, in which the mouth-cavity is closed at one end, the mouth-cavity resonances are clearly present in spectra of nasal vowels, since the anterior aperture of the mouth cavity facilitates their radiation to the air.

The first five resonances of the nasal-pharyngeal tube (with the oral cavity, i.e. mouth shunt, removed) have been calculated by Fant (1970) at 300 Hz, 1 kHz, 2.2 kHz, 2.9 kHz and 4 kHz. These resonances are known as *nasal formants* (labelled N1, N2, . . ., N*n*) and are approximated in the production of a uvular nasal in which the complete lingual-uvula closure seals the oral cavity from the nasal-pharyngeal tube. Since the side-chamber is sealed in this way, no anti-resonance/resonance pairs due to the mouth cavity should be introduced into the spectrum. However, even in the production of a uvular nasal, a *cul-de-sac* exists behind the tongue and between the faucal pillars which is sufficient to introduce an oral anti-resonance into the system above 3 kHz (as well as a paired resonance if it is not damped by the anti-resonance). Laver (1980) raises the interesting possibility that the air in the oral cavity in front of the tongue closure may be excited since the surface of the front of the tongue vibrates in the production of voiced uvular and velar nasals; if this is the case, another resonance due to this cavity would be introduced.

Studies by Curtis (1942), Delattre (1954), Fujimura (1962), Fujimura and Lindqvist (1971), Hattori, Yamamoto and Fujimura (1958), House and Stevens (1956) and Potter, Kopp and Green (1966) also locate N1 in the 200–300 Hz range. Smith (1951), House and Stevens (1956) and Joos (1948) report N2 around 1 kHz and Smith (1951), Delattre (1954) and Tarnóczy (1948) report N3 in the 2 kHz range, in agreement with Fant's calculations. Hattori, Yamamoto and Fujimura (1958) note that the spectrum of the nasal-pharyngeal tube may show considerable variation from speaker to speaker due to large variation in the shape of the nasal cavity.

As the lingual constriction moves forward in the mouth, the length of the side-branch is increased and the anti-resonance shifts down in frequency (Hattori *et al.* 1958, Fant 1970, Fujimura 1962). This implies that the frequency location of the anti-resonance is highest for velars and lowest for bilabials (with a neutral tongue configuration): since, of course, the tongue can adopt a variety of different configurations dependent on the following vowel in the production of [mV] syllables, the anti-resonance for [m] will vary accordingly.

Several authors report that the murmur of nasal consonants (the murmur is defined as the interval in the acoustic signal that corresponds to complete oral tract closure with the oral tract coupled to the nasal-pharyngeal tract) is characterised by *N1 at 250–300 Hz of high intensity* compared with the rest of the spectrum (Nakata 1959, Fant 1970, Fujumura 1962, Fujimura and Lindqvist 1971). Nakata (1959) also considers that the increased damping of formants compared with non-nasal sonorants can lead to a *broadening of the resonance bandwidths.* House (1957) reports that, compared with oral vowels, nasal consonants are *low in energy.* Fujimura (1962) reports that the murmur is characterised by an *even distribution of energy in the range 0.8–2.3 kHz* and comments that this feature, when combined with very low and intense N1, cannot be found in other sonorants. Related to the cues of a low N1 of high intensity and broad bandwidth is Mermelstein's (1977) measure of the centre of gravity of spectrum in the 0–500 Hz band, which is found to be lower for nasal consonants compared with other sonorants.

For [m], Fant's theoretical calculations predict the first anti-resonance at 800 Hz while in the spectra of natural speech produced by a Russian speaker the anti-resonance occurred at 550 Hz. Hattori, Yamamoto and Fujimura (1958) also report that for [m] produced with a neutral tongue configuration, the first anti-resonance occurs with a centre frequency of 750 Hz and bandwidth of 500 Hz. Fujimura (1962) reports the centre frequency of the first anti-resonance at around 1 kHz for [m] produced with a neutral tongue configuration. For [n], Fant's calculations show the frequency of the first anti-resonance to be considerably higher than that of [m] at 1.8 kHz. Hattori *et al.* (1958) report a centre frequency for the first anti-resonance of [n] at 900 Hz with a bandwidth of 800 Hz while in Fujimura (1962), the centre frequency for the anti-resonance of [n] is reported at 1.7 kHz. With regard to [ŋ], Hattori *et al.* (1958) find no absorption in the spectrum while Fujimura (1962) predicts the first anti-resonance of [ŋ] to occur above 3 kHz. There are no studies to the knowledge of this author that have investigated the frequency of the second anti-resonance, although Fujimura (1962) predicts that it cannot occur less than 3 kHz for [m] and would be considerably higher in frequency for [n].

A review of studies of the acoustic structure of nasal consonants in several languages (Czech and German: Romportl 1973; English: Fujimura 1962; Hungarian: Magdics, 1969; Polish: Dukiewicz 1967; Russian: Fant 1970; Swedish: Fant 1973) by Recasens (1983) concurs with the studies in the previous paragraph that the centre frequency of the first anti-resonance is highest for [ŋ], intermediate for [n], and lowest for [m].

The bandwidth of the anti-resonance deserves some comment since it may provide a potential cue for differentiating [m] from [n]: Hattori *et al.* (1958) report that the bandwidth of the anti-resonance is around 400 Hz greater for [n] than [m]. Fujimura (1962) reports half-power bandwidths of the first anti-resonance of intervocalic [m] and [n] at 80 Hz and 600 Hz respectively: Fujimura agrees, therefore, with Hattori *et al.* that the bandwidth is greater for [n] than [m] although there is a discrepancy between the two studies on the values quoted for the bandwidths.

The *oral resonance* for [m] is calculated at 900 Hz by Fant (1970), in agreement with the value reported by Ochiai, Fukumura and Nakatami (1957); this oral resonance may coincide with the anti-resonance for [m] between 500 Hz and 1 kHz and should therefore be weakened. The anti-resonance may presumably also weaken N2 at 1 kHz in which case the second resonance detectable on spectrograms may be N3 at around 2 kHz. When /m/ is realised as a palatal [$m_j$], the oral resonance at 900 Hz shifts upwards in frequency to around 1.7 kHz (Fant 1970) which is close to N3. However, for [$m_j$], the anti-resonance also shifts upwards in frequency to around 2 kHz (Fant 1970, Hattori *et al.* 1958) and so intensification of the spectrum in this region is unlikely.

With regard to [n], Fant (1970) predicts an oral resonance at 1.4 kHz; this is in close agreement with his measured data which shows the oral resonance at 1.2 kHz. Since the anti-resonance of [n] occurs at 1.8 Hz, spectrograms may show resonances at 300 Hz (N1), 800 Hz (N2), 1.4 kHz (oral resonance) and 2 kHz (N3). However, recall that some of the studies discussed above found the bandwidth of the anti-resonance of [n] to be as wide as 600 Hz which would mean that some of these resonances may well be severely damped. As in the case of [m], a lack of energy on spectrograms for [n] above 250 Hz up to at least 2 kHz does not seem impossible in the light of these findings. Fant (1970) reports that when /n/ is realised as a palatal [ $n_j$ ], the oral resonance shifts upwards in frequency to around 2 kHz. This oral resonance is attenuated by the anti-resonance which, as in the case of [ $m_j$ ], also shifts upwards in frequency to 2 kHz.

It may be said, therefore, that both [ m ] and [ n ] are characterised by nasal formants N1–N5 at approximately the frequency values quoted for the uncoupled nasal-pharyngeal tract and an oral resonance/anti-resonance pair which is lower for [ m ] than [ n ]; furthermore, this resonance/anti-resonance pair varies in frequency with F2 of a following oral vowel. The oral resonance may also show some continuity with the vowel's second formant: thus, the oral resonance is weak in the nasal murmur itself but, as the closure is released, it will increase considerably in intensity as it merges with F2 of the vowel onglide. The slope of the oral resonance offset/F2 onglide may have some parallels with the F2 transitions and loci discussed in connection with oral stops (Liberman, Delattre, Cooper and Gerstman 1954; Delattre, Liberman and Cooper 1955). Thus, in [ mV ] syllables, the oral resonance offset/F2 onglide should rise while in [ nV ] syllables it should be falling or flat; in the case of [ Vŋ ] syllables, the F2 offglide/oral resonance onset should be rising or flat. The identification of this transition is hampered by the problem (discussed above) that an apparent second resonance may in fact be N3 if N2 and the oral resonance have been damped by the anti-resonance; similarly, if the oral resonance and N3 are damped, N2, not F2, may be the second visible resonance. Until a robust method is found both for identifying the frequency location of the anti-resonance and differentiating oral from nasal resonances, it seems unwise for automatic speech recognition systems to depend on F2 transitions as a cue for distinguishing [ m ] from [ n ].

There have been several experiments which have attempted to assess how listeners respond to murmur and transition cues for the identification of place of articulation within nasal consonants. On the basis of a review of studies (Dukiewicz 1967, Henderson 1978, House 1957, Malécot 1956, Nakata 1959) of the perceptual importance of nasal murmurs, Recasens (1983) concludes that only [ m ] murmurs are reliably identified when they are isolated from the surrounding context. [ n ] murmurs appear to give around 50–60 % correct responses; however, listeners often classify the stimulus incorrectly when [ ŋ ] murmurs are presented in isolation. Recasens goes on to suggest that [ m ] may be reliably identified using the murmur alone because of its low N1 and low oral anti-resonance which are not characteristic of other nasal consonants.

In a synthesis and labelling experiment, Malécot (1956) showed that when the murmur and transitions from naturally spoken syllables were cross-spliced providing conflicting cues to place of articulation, listeners' judgements were generally guided by the transitions, not the murmur. In an experiment by Recasens (1983), variable F2/F3 transitions were combined with three fixed murmur patterns that were believed to be optimal for the perception of /m/, /n/ and /ŋ/ in /aN/ syllables (/N/ stands for nasal consonant). When optimal murmurs are attached to transitions in this way, the results showed, in agreement with those of Malécot, that category judgements could not be reliably predicted on the basis of murmurs alone (i.e. when transitions conflicted with the murmurs, listeners' judgements were often directed by the transitions). Similarly, when variable murmurs where attached to transitions that were believed to be optimal for [m], [n] and [ŋ], category judgements were determined largely by transitions. The results of an experiment by Kurowski and Blumstein (1984), on the other hand, suggest that murmurs are as important as transitions for identifying place of articulation. In this experiment, a male speaker produced CV syllables consisting of [m n] followed by one of the vowels [i e a o u]. When parts of these syllables were spliced such that they contained either just six pulses of the murmur or six pulses of the transition, the proportion of correct identifications of the nasal consonants was about the same in both cases, suggesting that murmurs provide as important place cues as transitions. The highest identification score was obtained when the last three pulses of the murmur and the first three pulses of the transition were presented together, which suggests that both murmurs and transitions together are important cues for place of articulation.

*Nasal Vowels*

Several authors have reported that one of the prime cues that differentiates oral vowels from their nasal counterparts is a difference in the spectrum around the F1 region (Smith 1951; Delattre 1954; Hattori, Yamamotto and Fujimura 1958; House and Stevens 1956; Fant 1970; Hawkins and Stevens 1985). On the basis of experiments with articulatory synthesis, Abramson, Nye, Henderson and Marshall (1981) report that a nasal resonance/anti-resonance pair is introduced around 400 Hz which splits apart as the size of the velopharyngeal aperture increases. Since F1 of [i] is low at around 250 Hz, the nasal resonance (henceforth N1) and anti-resonance (henceforth NZ1) pair in [ĩ] are above F1 at 800 Hz and 1.4 kHz respectively when the velopharyngeal aperture is at a maximum. For [ã], whose F1 is around 900 Hz, N1 and NZ1 are lower than F1 at 600 Hz and 800 Hz respectively for maximum coupling. For [ʌ̃], F1, NZ1 and N1 occur at 500 Hz, 800 Hz and 900 Hz respectively. On the basis of synthesis and labelling experiments of the vowels [i e ɑ o u] on a nasal/non-nasal continuum following the dental consonant [t̪], Hawkins and Stevens (1985) found that the preferred F1, N1, NZ1 combination was found to be where NZ1 occurs approximately at the midpoint between F1 and N1. Listeners found the most natural nasal vowels occurred when the spacing between NZ1 and N1 was in the range of between 75–110 Hz. As in the experiment by Abramson *et al.* (1981), N1 and NZ1 emerged together at 400 Hz and then split apart as the vowel became progressively more nasal.

If F1, NZ1 and N1 occur close together, which is particularly the case for the nasal mid-vowels [ẽ] and [õ], Hawkins and Stevens (1985) note that F1 and N1 may coalesce: this can give the impression that the bandwidth of first resonance has broadened and that its centre frequency has shifted. In addition, the intensity of the merged F1/N1 is less than F1 of oral vowels for at least two reasons: first, energy is absorbed due to the presence of NZ1 between F1 and N1; second, the intensity of all nasal formants is low compared with oral formants due to the comparatively low acoustic impedance of the nasal tract which damps resonance peaks (Nakata 1959). That the bandwidth of the first (combined oral and nasal) resonance is greater that the bandwidth of F1 of the corresponding oral vowel has been reported by Dickson (1962) and House and Stevens (1956); Delattre (1954), Dickson (1962) and House and Stevens (1956) have shown that the first resonance of nasal vowels is lower in intensity than F1 of the corresponding oral vowel, and House and Stevens (1956) report that the first resonance of vowels shifts as vowels are nasalised.

If N1 is clearly separated from F1, an additional nasal resonance should appear in the spectra of nasal vowels compared with oral vowels and this has been found by Dickson (1962) and Joos (1948); Smith (1951) finds additional nasal resonances at 400–500 Hz and 2169–3069 Hz.

Some authors (e.g. Delattre 1954, Fujimura 1962) have reported that a nasal vowel is characterised by the intensification of the spectrum around 250 Hz, and this appears to be contradictory to some of the experiments reported above which report the introduction of a nasal resonance/anti-resonance pair from 400 Hz. According to Hattori *et al.* (1958), the increase in spectral energy in this region is the fundamental resonance of the nasal-pharyngeal cavity (i.e. N1). Hattori *et al.* (1958) note that its paired anti-resonance occurs at 500 Hz for nasal vowels.

Finally, several authors (Bloomer 1953; Harrington 1944; Nusbaum, Foley and Wells 1935 – all cited in House and Stevens 1956) have reported that there is a relationship between the size of the velopharyngeal aperture and the height of the vowel in nasal vowels: in general, the closer (phonetically) the vowel, the smaller the velopharyngeal aperture in the production of its nasal counterpart. McDonald and Baker (1951) (cited in Abramson, Nye, Henderson and Marshall 1981) explain this phenomenon in terms of the need to maintain a balance between oral and nasal resonances. The ratio of oral to nasal resonance depends on the relative sizes of the velopharyngeal aperture and the entrance into the oral cavity from the pharynx. Since the posterior oral tract opening is greater for open vowels, the size of the velopharyngeal aperture has to be greater than for close vowels, if the ratio of oral to nasal resonance is to be the same in both cases. This also implies that a certain degree of nasal coupling will be tolerated when the speaker produces an open oral vowel; but the same degree of nasal coupling would result in the perception of nasalisation in producing a close vowel. House and Stevens (1956) offer a similar explanation in terms of the ratio of acoustic impedances of the nasal tract and oral tract. Alternative explanations for the relationship between vowel height and size of the velopharyngeal port have been proposed based on a muscular connection between the tongue and velum which causes the velum to be lowered slightly in the production of an open vowel (Moll 1962). However, this theory does not seem to be consistent with electromyographic and cinefluorographic experiments reported in Lubker (1968) and Bell-Berti (1976, 1980). (See Abramson *et al.* 1981 for a more detailed discussion.)

Some data by Summerfield (1987), based on the implementation of a pole-zero extraction algorithm by Yegnanarayana (1981), is relevant in the light of the theory developed in McDonald and Baker (1951) and House and Stevens (1956). When an English speaker was instructed to produce the syllables [pal], [pɑl], [pil], [pul], an anti-resonance is clearly detectable in open vowels but is less prominent in the two close vowels; this preliminary data would seem to support the view that, for the same context, the degree of nasal coupling to the oral tract is greater for open compared with close vowels.

Figure 2.13 shows a preliminary discrimination tree for the identification of place of articulation of nasal consonants. As in Mermelstein (1977), the analysis is performed on a section of the

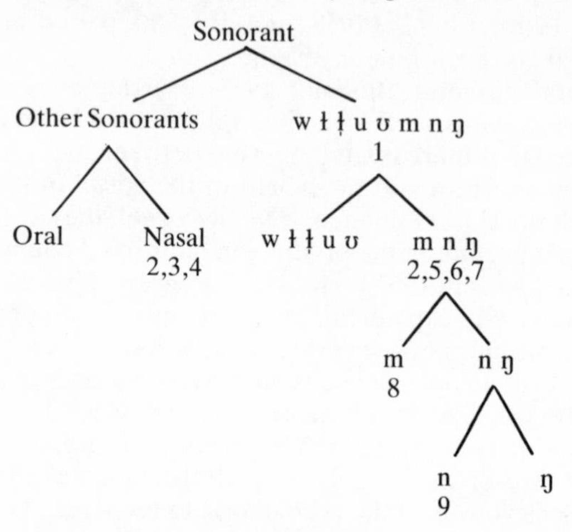


*Figure* 2.13. Discrimination tree for nasals. The relationship between numbers and features is as follows:

1 Ratio of energy in 0-500 Hz band to 500 Hz-8 kHz band is high
2 Bandwidths are high
3 Anti-resonance in the 400 Hz-2 kHz range
4 'Resonance density' high
5 Anti-resonance in the 700 Hz-4 kHz range
6 Low centre of gravity in the 0-500 Hz band
7 Energy in the 500 Hz-8 kHz band is low
8 Frequency of first resonance low and frequency of anti-resonance low
9 Frequency of first anti-resonance low and bandwidth of first anti-resonance high

speech waveform which is both voiced and non-fricative, labelled *sonorant* at the top of the tree. The first feature (1) is the ratio of energy in the 0–500 Hz region to energy in the entire spectrum; this ratio should be greater for all nasal consonants since N1 is low in frequency, high in intensity and has a broad bandwidth. This feature will probably also result in the detection of [w], the dark allophones of /l/, [u] and [ʊ], all of which have a low F1 close to F2. The observation in Tarnóczy (1948), that nasals have a formant structure that closely resembles [u], seems particularly relevant in this respect. It is quite possible that some nasalised vowels will have been classified together with 'other sonorants' at the first level of the tree since the low energy in the 0–500 Hz band for [ã], for example, and the high density of formants above 500 Hz characteristic of nasal vowels could result in a low value for this ratio metric. Therefore, another test is performed within 'other sonorants' in which nasalised vowels are separated from oral sonorants based on three features (2, 3, 4): the average bandwidths of the resonances, which should be greater for nasalised vowels; the presence of an anti-resonance in the 0.4–2 kHz range, which is characteristic of nasalised vowels; and the *number* of detectable resonances in the 0–3 kHz band ('resonance density'), which should be greater for nasalised vowels compared with oral sonorants.

At the next level, four features separate the nasal consonants from [w], [ʊ], [u], [ɫ] and [ɫ̩] (2, 5, 6, 7): the bandwidths of the resonances, which should be greater for nasal consonants; the presence of an anti-resonance in the 0.7–4 kHz range, which is characteristic of nasal consonants; the spectral centre of gravity of energy in the 0–500 Hz band, which should be lower for nasal consonants (Mermelstein 1977); and the energy in the 0.5–8 kHz band, which is lower for the nasal consonants.

At the next level of the tree, [m] is separated from [n] and [ŋ] if the frequency of the first resonance and the frequency of the first anti-resonance are low. At the bottom of the tree, [n] is separated from [ŋ] if the frequency of the first anti-resonance is low and if the anti-resonance bandwidth is high.

### Approximants

Acoustically, most of the allophones of approximant consonants bear the closest similarity to allophones of vowels and diphthongs and are characterised primarily by a pulse-excited formant-pattern. There is no single acoustic cue which enables approximant consonants to be differentiated from other oral sonorants, nor are there reliable acoustic cues for isolating either the liquids /l/ and /r/ as a class, or the glides /w/ and /j/. Consequently, the acoustic characteristics of each of the approximant consonants are treated separately below; in the final section, a discrimination tree of approximant consonants is developed which also takes into account the acoustic characteristics of vowels and diphthongs.

#### *Laterals*

In RP and many dialects of English, /l/ has two main allophones: clear [l] which occurs before vowels and /j/ (*leaf, law, fly, million, along, follow, value, feeling, silly*); and dark, or velarised, [ɫ] which occurs before consonants and prepausally (*feel, help, shelf, already, alter, algebra*). Clear [l] is often partially voiceless (*fly, play, clue*) following voiceless consonants and dark [ɫ] is also syllabic word-internally following postvocalic consonants (*table, trouble, little*). In all cases, the primary articulation is alveolar; but this can vary between dental (*health*) and retroflex (*already*). Since clear [l] and dark [ɫ] have different acoustic properties, they are analysed separately below. The allophonic distribution of /l/ is affected in a further complicating way by anticipatory and progressive coarticulation across word-boundaries; wherever necessary, examples of this type of coarticulation will be given together with the results of any relevant acoustic studies.

Table 2.5 gives average formant frequency values of [l] in various studies; the results in Lehiste (1964) are derived from a spectrographic study of five speakers producing various CVC syllables (V includes American English vowels and diphthongs); the calculations in Nolan (1983) are based on a spectrographic study of fifteen subjects (RP) producing a variety of /l/ initial monosyllabic words in isolation; the studies in Bladon and Al-Bamerni (1976) and Al-Bamerni (1975) are based on a spectrographic analysis of four speakers (RP) producing /l/ initial words (the results in Al-Bamerni (1975) are taken from Nolan (1978)); finally, Dalston (1975) made spectrographic measurements of /l/ initial words produced by six male, and four female, adult speakers (American English). The data from Fant (1970) are based on calculations from a vocal tract model and acoustic measurements from a Russian speaker.

There seems to be some agreement across the studies in locating F1 of [l] in the 250–400 Hz range. F1 is therefore slightly higher than the first resonance of nasals, especially in RP; this is consistent with the synthesis experiments of O'Connor, Gerstman, Liberman, Delattre and Cooper (1957), who show that if F1 is located at 240 Hz, but the higher formants are appropriate for synthesis of [l], listeners nevertheless perceive /n/; for the perception of /l/, F1 has to occur at a frequency of at least 360 Hz. Both Lehiste's and Nolan's studies (table 2.5) show that F1 of [l] exhibits a range of less than 100 Hz. A further inspection of Lehiste's and Nolan's data shows that F1 rises from [l] to all vowel targets except [i]. Fant (1970) notes that the low F1 value in his calculated data (table 2.5) was due to underestimating the width of the lateral passage. In a second calculation, in which the minimum cross-sectional area of the lateral passage was increased and in which the pharyngeal passage was narrowed, calculated F1 rose to 290 Hz. Finally, as Fant's calculated and measured data show, the effect of palatalisation is to shift F1 down in frequency to less than 250 Hz.

It is more difficult to generalise about average F2 values of [l]

Average values

| | Leh | Al-B | B&Al | Dal(M) | Dal(F) | Nol | Fant[l] | Fant[$l_j$] |
|---|---|---|---|---|---|---|---|---|
| F1 | 295 | 365 | | 344 | 365 | 360 | 350 (220) | 230 (210) |
| F2 | 950 | 1305 | 1326 | 1179 | 1340 | 1350 | 800 (850) | 1600 (1700) |
| F3 | 2610 | 2780 | | 2523 | 2935 | 3050 | 2000 (2300) | 2300 (2500) |
| FZ | | | | | | | 1800 (2100) | 2600 (3600) |
| F4 | | | | | | | 2700 (2900) | 3100 (3050) |

Ranges of values

| | Lehiste | Nolan |
|---|---|---|
| F1 | 250 (/u,ou/) – 330 (/ɑ/) | 340 (/i,o,u/) – 410 (/æ/) |
| F2 | 825 (/ɜ/) – 1185 (/i/) | 1230 (/ɒ/) – 1510 (/i/) |
| F3 | 2530 (/u/) – 2690 (/oi/) | 2980 (/u/) – 3130 (/ʌ/) |

*Table* 2.5. The upper table shows average formant frequency values (Hz) for steady-state [l] in the following studies: Leh, Lehiste (1964); Al-B, Al-Bamerni (1975); B&Al, Bladon and Al-Bamerni (1976); Dal(M), male subjects in Dalston (1975); Dal(F), female subjects in Dalston (1975); Nol, Nolan (1983); Fant[l], values for [l] (upper values calculated, lower values in brackets measured) in Fant (1970); Fant[$l_j$] values for palatalised [$l_j$] (upper values calculated, lower values in brackets measured) in Fant (1970). FZ, frequency of first anti-resonance; F4 in Fant[l] and Fant[$l_j$] corresponds to F3 in the other studies.

The lower table shows the minimum and maximum formant frequency values for steady-state [l] and the vowel contexts in which they occurred.

since, unlike F1 and F3, F2 of [l] varies considerably with F2 of the following vowel. Thus, since F2 of [i] is high, F2 of [l] is also higher than F2 of other allophones of [l] (Bladon and Al-Bamerni 1976, Lehiste 1964, Nolan 1983); in Nolan's data, F2 of [l] before [i] is at 1510 Hz, which is in close agreement with the corresponding F2 value in Bladon and Al-Bamerni's data[2] (1600 Hz, as estimated from their figure 2.2). Similarly, since back rounded vowels have a low F2, F2 of [l] before these vowels is also low. Like Nolan (1983), Bladon and Al-Bamerni find F2 of [l] lowest before [ɒ] at around 1100 Hz (1230 Hz in Nolan's data). Across all vowels, both Lehiste and Nolan find F2 of [l] has a range of just under 300 Hz (table 2.5); in Bladon and Al-Bamerni, the range is slightly higher at around 500 Hz. Table 2.5 shows that the mean F2 of [l] in Lehiste's results is lower than mean F2 of [l] in Nolan's data: the difference in these mean frequencies may occur because American /l/ is realised as a darker allophone than RP /l/ (Jones 1972, cited in Nolan 1983). A further inspection of Lehiste's data shows that, because of the comparatively low F2 of [l], F2 rises from [l] to all vowel targets, but no such generalisation is possible in Nolan's study.

Fant calculates an anti-resonance due to the shunting effect of the mouth cavity behind the tongue blade that be modelled by a tube closed at its far end; its frequency is calculated at 2 kHz and depends on $c/4l$, where $l$ is the length of this tube and $c$ the velocity of sound. The effect of the anti-resonance is to attenuate considerably, or even cancel, the third resonance: therefore, F3 in all studies other than Fant's in table 2.5 is in reality the fourth resonance and corresponds to F4 in Fant's data.

The studies in table 2.5 locate 'pseudo-F3' in the 2.5–3 kHz range. In Lehiste's data, 'F3' falls to every vowel except [i] (no F3 values for the vowel target are given in Nolan's data). Since the third resonance of [l] visible on spectrograms is usually in reality F4, it may show a greater continuity with F4, rather than F3, of the following vowel: in this case, if F3 of the following vowel is continuous with a damped, or cancelled, F3 of [l], F3 of the vowel may appear to be discontinuous with any of the resonances of the preceding [l].

As discussed above, a large proportion of clear [l] may be voiceless due to the progressive coarticulatory effects of preceding voiceless consonants. Table 2.6 shows the proportion of voicelessness of [l̥] in Bladon and Al-Bamerni (1976) expressed as a percentage of its steady-state duration. The study shows that /l/ is realised as an entirely voiced allophone following the unaspirated [p] of words such as *splint* and that /l/ can also be realised as partly voiceless due to the coarticulatory effects across word boundaries as in *damp linen*. The results of Bladon and Al-Bamerni's study also show that F2 of [l̥] in /#ClV/ sequences exhibits the same range of variation as [l] in /#lV/ sequences (around 500 Hz): in both cases, F2 is highest preceding [i] and lowest preceding [ɒ]. However, the results also show that F2 of [l̥] is consistently 100–200 Hz lower

| Context | Example | % voiceless |
|---|---|---|
| #S[l]V | play, clay | 58 |
| S.[l]V | atlas, duckling | 48 |
| S#[l]V | damp linen | 39 |
| #F[l]V | flow | 22 |
| F.[l]V | athlete, roughly | 22 |
| F#[l]V | fresh lemon | 21 |
| #sS[l]V | splint | 0 |

*Table* 2.6. The extent of voicelessness in the realisation of /l/ in various contexts expressed as a percentage of its steady-state duration in Bladon and Al-Bamerni (1976). '#' word boundary; '.' syllable boundary; 'S' stop; 'F' fricative.

than F2 of [l] before the same vowel. No indication is given in Bladon and Al-Bamerni's study of the point in [l̥] at which F2 was measured.

Bladon and Al-Bamerni report that prevocalic /l/ can be realised as an allophone which is intermediate between clear and dark in /Cl#lV/ (*terrible leak*) and /Vl#lV/ (*I'll lead*) due to the left-to-right (progressive) coarticulatory effects of the first /l/ which is realised as a dark [ɫ]. F2 of dark [ɫ] is usually considerably lower than F2 of clear [l]. Thus, the allophone of the second /l/ in *terrible leak* has an average steady-state F2 value of 978 Hz, which is around 350 Hz lower than the average steady-state of [l] in /#lV/ syllables at 1326 Hz. In /Vl#lV/ (*I'll lead*) contexts, F2 of the allophone of the second /l/ is around 300 Hz lower at 1098 Hz, and in /VllV/ contexts (e.g. *vilely*) average steady-state F2 of the allophone of the second /l/ is around 400 Hz lower than [l] in /#lV/ contexts, at 989 Hz. Bladon and Al-Bamerni stress that the two adjacent /l/s do not merge into a single /l/, i.e. 'there was a noticable spectral shift separating two distinct segments corresponding to two successive /l/ tokens, rather than either a smoothly changing spectrum, or a single unchanging /l/'. The acoustic characteristics of the allophone of the first /l/ has many of the acoustic attributes of the dark [ɫ] discussed below.

Table 2.7 shows ranges of F2 for intervocalic /l/ in various contexts from Bladon and Al-Bamerni (1976). In monomorphemic words (*bullet, Ely, silly*), the F2 range is the same as [l] in /#lV/

| Context | Example | F2 range |
|---|---|---|
| 1. /VlV/ | bullet | 1600 (/i/) – 1110 (/o/) |
| 2. /Vl+V/ | feeling | 1500 (/i/) – 990 (/o/) |
| 3. /Vl#V/ | feel it | 1480 (/i/) – 1000 (/ɑ/) |
| 4. /Vl\|#V/ | feel ill | 1428 (/ɪ/) – 970 (/o/) |
| 5. /Cl+V/ | shovelling | 1160 |
| 6. /Cl#V/ | shovel it | 1115 |
| 7. /Cl\|#V/ | shovel in | 1000 |
| 8. /#lV/ | leaf, lot | 1600 (/i/) – 1110 (/ɒ/) |
| 9. /Vl#/ | feel | 1000 (/i/) – 700 (/u/) |
| 10. /Cl#/ | shovel | 790 |

*Table* 2.7. Minimum and maximum values of F2 (Hz) for various allophones of [l] and the following context in which they occurred, from Bladon and Al-Bamerni (1976). The following boundary contexts were investigated: 1 monomorphemic; 2, 5 internal morpheme boundary +; 3, 6 internal word boundary #; 4, 7 internal word and foot boundary |#. 8, 9 and 10 are shown for comparison. Many of the values are estimated from figures in their study.

sequences (*leaf, law*); furthermore, since the following vowel exerts a similar influence on F2 of [l] in these contexts, it can be concluded that /l/ in such monomorphemic words is realised as a clear allophone. In examples 2, 3 and 4 in table 2.7, intervocalic /l/ occurs in a word in which /l/ would be realised as a dark [ɫ] if the word occurred before a pause (*How do you feel?*) or a stop (*it feels cold*). In example 2, intervocalic /l/ occurs before a morpheme boundary; in example 3, before a word boundary and in example 4, before a word and foot boundary. It was stated previously that /l/ should be realised as a clear [l] intervocalically and this impression is consistent with the ranges of F2 in examples 2, 3 and 4 in table 2.7 which are similar to the F2 range of [l] in /#lV/ (example 1: *leaf, law*). However, the range is progressively compressed in examples 2, 3 and 4 which implies that, although /l/ in these contexts is influenced by the following vowel in the same way as [l] in an /#lV/ context, the addition of boundaries between the lateral and the second vowel can result in a slightly darker realisation. In examples 5, 6 and 7, intervocalic /l/ occurs in a word in which /l/ is realised as a syllabic, dark [ɫ̩] when the word occurs before a pause (*pick up the shovel*) or before a stop (*shovel coal*). Again, F2 of examples 5, 6 and 7 is higher than F2 of syllabic [ɫ̩] (example 10); but since ranges are not reported in Bladon and Al-Bamerni, it is difficult to assess whether the allophones in these cases should be classified as clear or dark.

Finally, there is some experimental evidence to suggest that the duration of the steady-state of [l] and the duration of the transition from [l] to a following vowel target may be important cues for isolating /l/ from the other approximant consonants. O'Connor, Gerstman, Liberman, Delattre and Cooper (1957) were among the first to note that an abrupt change in F1 frequency was characteristic of [lV] sequences. A detailed study of such steady-state and transition durations is reported in Dalston (1975) who shows that the steady-state duration of F1, F2 and F3 is greater for [l] than both [r] and [w]; the transition durations of F1, F2 and F3 were found to be shorter for [l] than both [r] and [w] and the F1 transition rate (i.e. slope of transition in Hz/ms) was found to be greater for [l] than both [r] and [w]. In a related study, Polka and Strange (1985) synthesised a continuum from *lock* to *rock* by varying F2 and F3 onset frequencies from 1067–1207 Hz and 1477–2594 Hz respectively; the duration of F1 (at 350 Hz) was varied in a 10-step continuum from 10 ms steady-state and 55 ms transition duration (*rock* endpoint of the continuum) to 55 ms steady-state and 10 ms transition duration (*lock* endpoint of the continuum). The results of a labelling experiment showed a trading relationship between the spectral and temporal cues: for stimuli that had spectral characteristics of a most extreme /r/ token (i.e. F2 at 1067 Hz, F3 at 1477 Hz) and temporal characteristics of a most extreme /l/ token (55 ms steady-state F1, 10 ms F1 transition duration), listeners tended to perceive /l/ rather than /r/.

In the study by Lehiste (1964) discussed in the preceding section, the same five speakers produced a variety of /CVl/ syllables. Average formant frequency values of steady-state F1, F2 and F3 of [ɫ] and their ranges are given in table 2.8. A further examination of Lehiste's data also shows that F2 falls from all vowel targets to [ɫ]. Based on an estimate from Bladon and Al-Bamerni (1976), F2 of [ɫ] in their /-Vl#/ contexts ranges from 700 Hz (/Cul#/) to 1000 Hz (/Cil#/). Both Lehiste and Bladon and Al-Bamerni are in agreement in showing that the coarticulatory influence of preceding vowels on [ɫ] is less than the anticipatory coarticulatory influence of following vowels on clear [l]: thus F2 of [ɫ] exhibits a range of 215 Hz compared with 360 Hz for F2 of [l] in Lehiste's study. Finally, Bladon and Al-Bamerni showed that the anticipatory effect of [l] on [ɫ] in /Vl#lV/ sequences such as *I'll lead* were minimal: the mean change in F2 of [ɫ] due to [l] was only 25 Hz.

Values of steady-state F1, F2, F3 and their ranges for syllabic,

| | 'Dark' [ɫ] | | 'Dark' syllabic [ɫ̩] | |
|---|---|---|---|---|
| | Mean | Range | Mean | Range |
| F1 | 455 | 400 (/ʊ/) – 535 (/æ/) | 430 | 395 (/b/) – 465 (/ʃ/) |
| F2 | 795 | 655 (/u/) – 870 (/æ/,/ɑ/) | 785 | 715 (/θ/) – 835 ([g+],/ʃ/) |
| F3 | 2585 | 2485 (/ɜ/) – 2720 (/ɪ/) | 2380 | 2265 (/f/) – 2470 (/d/) |

*Table* 2.8. Average, minimum and maximum values of F1, F2 and F3 (Hz) of dark/syllabic [ɫ] and the preceding context (in brackets) in which they occurred, from Lehiste (1964).

dark [ɫ̩] (e.g. *table*) from Lehiste (1964) are also given in table 2.8, which shows that these values and ranges are very similar to those of [ɫ] in /Vl#/ syllables, although mean F3 of [ɫ̩] is slightly lower than that of [ɫ]. Inspection of Lehiste's data also shows that F2 and F3 transitions from all preceding consonants are falling and rising respectively to the target formant frequency values of [ɫ̩]. In the spectrographic study by Bladon and Al-Bamerni (1976), the average F2 steady-state of [ɫ̩] following /p/, /b/, /f/, /v/ (e.g. *apple, able, awful* and *shovel* respectively) was 790 Hz, which agrees almost entirely with Lehiste (1964). There is again agreement between the study of Lehiste and that of Bladon and Al-Bamerni on the small coarticulatory influence of preceding consonants on syllabic [ɫ̩]: in Lehiste's study the F2 range (table 2.8) is 120 Hz, which is considerably less than the corresponding ranges for /-Vl#/ (215 Hz) and /#lV-/ contexts (360 Hz). Similarly, in a /Cl#lV/ context, such as *terrible leak*, Bladon and Al-Bamerni find that right-to-left (anticipatory) coarticulation of the second /l/ on the preceding syllabic [ɫ̩] is negligible, causing a shift in the average steady-state of F2 of only 13 Hz (in marked contrast to the considerable left-to-right effects of syllabic [ɫ̩] on the allophone of /l/ in *leak*). Finally, Lehiste (1964) notes that the intensity of [ɫ̩] in comparison with other sonorants is high: the intensity measure may be important in differentiating [ɫ̩] from [u], which otherwise has a somewhat similar formant pattern.

Thus the identification of two allophones clear [l] and dark [ɫ̩] lies at the heart of the automatic extraction of /l/ from the acoustic waveform. The primary acoustic features of RP [l] could include:

(1) F1 in the range 200–400 Hz
(2) F2 in the range 950–1500 Hz
(3) FZ around 2 kHz
(4) F4 ('pseudo F3') in the range 2700–3200 Hz
(5) F3 of adjacent vowels may be discontinous with F3 of [l].

These features are designed to detect /l/ both preceding all vowels, diphthongs and /j/ irrespective of intervening syllable/morpheme/word/foot boundaries. However, the allophones of /l/ in /Cl#V/ (*shovel in*), /Vl#V/ (*feel ill*), /Cl#lV/ (*awful leader*) and /Vl#lV) (*I'll lead*) sequences may fall outside these ranges primarily as a result of their low F2.

The primary acoustic features of the dark allophones of /l/ ([ɫ] and [ɫ̩]) could include:

(1) F1 in the range 350–550 Hz
(2) F2 in the range 650–850 Hz
(3) F3 in the range 2200 Hz - 2700 Hz
(4) preceding F2 transition: sharply falling into the target [ɫ] or [ɫ̩]
(5) preceding F3 transition rising to the target of [ɫ]
(6) high intensity (at least of [ɫ̩]) compared with other sonorants

These features are designed to detect /l/ preceding all consonants and prepausally.

Finally, Fant (1970) notes that the allophones of /l/ may be differentiated from other sonorants by a clustering of no less than five formants in the 2.25–5 kHz range at an average spacing of around 800 Hz.

*Post-alveolar approximants*

In RP, /r/ is realised as a voiced post-alveolar frictionless continuant prevocalically and intervocalically, but does not surface postvocalically before pauses and consonants. Following voiceless consonants, the onset of the realisation of /r/ can be voiceless (*pray, crew, shrew, dark red*); but /r/ is probably fully voiced following unaspirated stops (*spray*). /r/ can also be realised as an alveolar, or dental, tap following dental fricatives (*three, brethren*). For some speakers of RP, /r/ surfaces with lip-rounding (even before unrounded vowels) and with a minimal participation of the tongue-tip: in this case, Gimson (1980) suggests a realisation as [w] although Nolan (1983) considers a realisation as a rounded labiodental approximant [ʋ] more probable.

Table 2.9 shows average formant frequency values for the allophones of prevocalic /r/ in three of the studies discussed previously. These suggest that F1 is located in 250–350 Hz range, which is slightly lower compared with the F1 range of [l]; Nolan's figures on F1 are around 50 Hz higher than Lehiste's which may be due to dialect differences between RP and American English. In Lehiste's data, F1 rises from [ɹ] to all vowel targets, while in Nolan

| | Lehiste | | Nolan | | Dalston | |
|---|---|---|---|---|---|---|
| | Mean | Range | Mean | Range | Mean (M) | Mea (F) |
| F1 | 280 | 255 (/ɪ/) – 300 (/o/) | 320 | 300 (/u/) – 340 (/e/,/æ/) | 348 | 35[illegible] |
| F2 | 950 | 870 (/ou/) – 990 (/ə/) | 1090 | 1000 (/o/) – 1160 (/i/) | 1061 | 116[illegible] |
| F3 | 1350 | 1310 (/æ/,/ai/) – 1400 (/ɪ/) | 1670 | 1580 (/o/) – 1750 (/i/) | 1546 | 207[illegible] |

*Table* 2.9. Average, minimum and maximum values of F1, F2 and F3 (Hz) of prevocalic [ɹ] and the following context (in brackets) in which they occurred, from Lehiste (1964), Nolan (1983) and Dalston (1975).

(1983), F1 rises from [ɹ] to all vowel targets except [i]. A comparison with the acoustic characteristics of the allophones of /l/ (tables 2.5 and 2.8) shows that mean F2 of [ɹ] is intermediate between mean F2 of the clear and dark allophones of /l/ and exhibits a small range which is comparable to that of [ɫ]: in both Lehiste's and Nolan's data, the F2 range of [ɹ] is only 150 Hz which suggests that, unlike [l], the anticipatory coarticulatory effects of following vowels on [ɹ] is minimal. All three studies in table 2.9 confirm that F3 of [ɹ] is low. Again, Nolan's and Lehiste's study show a small range for F3 of [ɹ] at around 150 Hz which is lower than the F3 ranges of any of the allophones of /l/. The slightly higher average F3 in Lehiste's study compared with Nolan's is probably again attributable to dialect differences. Not surprisingly, F3 from [ɹ] to all following vowel targets is strongly rising in Lehiste's study (F3 values of the vowel target are not reported in Nolan 1983).

*Palatal Approximants*

In RP, /j/ only surfaces prevocalically as a palatal approximant; like [l] and [ɹ], the onset of the realisation of /j/ is partially voiceless following voiceless consonants (*cute, few, take you*). The sequence /hj/ is often realised as a voiceless palatal fricative [ç] (*huge*). The sequences /t(#)j/, /d(#)j/can be realised as affricates (*tune, dune, at your, did you*).

Table 2.10 shows average values of the first three formants of the allophones of /j/ in /#jV/ sequences produced by the same five speakers in Lehiste (1964). F1, F2 and F3 were all rising, falling and falling respectively from [j] to following vowel targets in this study.

Within the class sonorant, [j] is most likely to be confused with

| | Mean | Range |
|---|---|---|
| F1 | 259 | 210 (/ʊ/) – 290 (/æ/) |
| F2 | 1972 | 1870 (/o/,/ʊ/) – 2100 (/i/) |
| F3 | 2820 | 2620 (/ɜ/) – 3050 (/i/) |

*Table* 2.10. Average, minimum and maximum values of F1, F2 and F3 (Hz) of [j] and the following context (in brackets) in which they occurred, as estimated from Lehiste (1964, fig.4.6).

[i] and [ɪ] that occur in words such as [si] (*see*), [sɪn] (*sin*), [meɪ] (*may*), [maɪ] (*my*), [tʰoɪ] (*toy*). However, in the same study by Lehiste, F1 and F3 of [j] were found to be lower and higher respectively than F1 and F3 of the allophones of /i/, /ɪ/, /ei/, /ai/ and /oi/ in /CVC/ and /CV/ words produced by five speakers. Lehiste also finds that [j] and [i] are well-separated in the F1-F2 plane since F1 and F2 of [j] are both lower than F1 and F2 of [i].

*Labial-velar approximants*

Table 2.11 shows average F1, F2 and F3 values of [w] in Lehiste (1964), Dalston (1975) and Mack and Blumstein (1983). The study by Mack and Blumstein was based on an LPC analysis of [wV] syllables in which the vowel was one of [i e ɑ o u] produced by two male speakers. All studies confirm that [w] is characterised by a low F1 close to F2 which are well separated from F3. Lehiste (1964) finds F1 rises from [w] to all vowel targets except [u] for which the F1 transition is level. In Mack and Blumstein's study, F1

| | Lehiste | | Mack and Blumstein | | Dalston | |
|---|---|---|---|---|---|---|
| | Mean | Range | Mean | Range | Mean (M) | Mean (F) |
| F1 | 305 | 250 (/ɑ/) – 320 (/æ/) | 400 | 328 (/i/) – 478 (/ɑ/) | 336 | 337 |
| F2 | 630 | 580 (/ei/,/o/) – 660 (/ɜ/) | 806 | 754 (/u/) – 849 (/i/) | 732 | 799 |
| F3 | 2180 | 2030 (/ou/) – 2370 (/ə/) | 2217 | 2044 (/i/) – 2404 (/ɑ/) | 2290 | 2768 |

*Table* 2.11. Average, minimum and maximum values of F1, F2 and F3 (Hz) of [w] and the following context (in brackets) in which they occurred, from Lehiste (estimated from fig.4.1 in Lehiste 1964), Mack and Blumstein (1983) and Dalston (1975).

rises to all vowel targets, but is only slightly rising to the targets of [i] and [u] (+21 Hz and +13 Hz respectively). F2 rises from [w] to all vowels in both Lehiste (1964) and Mack and Blumstein (1983). Excluding [ɜ], which has a particularly low F3 due to 'r'-colouring, F3 rises from [w] to all vowel targets in Lehiste's study; in Mack and Blumstein's study, F3 rises to [i], [e] and [ɑ], but falls to [o] and [u]. Lehiste also notes that there is considerably more variation in F3 due to the coarticulatory influence of following vowels: as table 2.11 shows, the F3 range of [w] in Lehiste's study is 340 Hz (360 Hz in Mack and Blumstein's study) while the F1 and F2 ranges are 70 Hz (Mack and Blumstein: 150 Hz) and 80 Hz (Mack and Blumstein: 95 Hz) respectively. Lehiste also finds that F3 is considerably weaker, and often more difficult to identify from spectrograms, than F1 and F2.

Within the class sonorant, [w] is most confusable with [u] (*pool, who*) [ʊ] (*book*) and the second segments in [aʊ] (*now*) and [oʊ] (*so*). In the same study by Lehiste, F1 and F2 of [w] were lower than F1 and F2 of all these vowels in /CVC/ and /CV/ syllables produced by five speakers. There is also the possibility of confusion of [w] with [ɫ] and [ɫ̩], which have a similar formant structure. However, a comparison of the formant values for these segments (tables 2.8 and 2.11) shows that F1 and F3 of [w] are both lower than F1 and F3 of [ɫ] and [ɫ̩]. Furthermore, F3 of [ɫ] and [ɫ̩] is likely to be greater in intensity than F3 of [w].

Much research in the last thirty years has focused on identifying the minimal cues necessary for the perceptual differentiation of /b/ from /w/ in /CV/ syllables (Liberman, Delattre, Gerstman and Cooper 1956; Diehl, 1976; Mack and Blumstein 1983; Schwab, Sawusch and Nusbaum 1981; Shinn and Blumstein 1984; Shinn, Blumstein and Jongman 1985): [b] and [w] both have a labial place of articulation and so the formant transitions from these segments to following vowel targets are likely to be quite similar.

Mack and Blumstein (1983) and Shinn and Blumstein (1984) consider that the intensity contour plays an important role in the perceptual differentiation of /b/ from /w/ in /CV/ syllables. [b], unlike [w], is characterised acoustically by a burst, and a sharp rise in intensity at the burst. Since there is no burst in [w], the intensity rises gradually and at an approximately constant rate from the onset of periodicity to the vowel target. Shinn and Blumstein (1984) found the first consonant is identified as /b/ when listeners are presented with synthetic stimuli that have formant transitions appropriate for [wa], but an amplitude contour characteristic of [ba]. However, listeners perceive /v/, rather than /w/, when stimuli are presented that have an amplitude contour characteristic of [w] superimposed on formant transition appropriate for [ba]. On the basis of these experiments, Shinn and Blumstein argue that the presence of a rapidly changing amplitude contour does not simply underlie the perceptual differentiation of stops from glides, but separates perceptually non-continuants from continuants.

| | i | | e | | ɑ | | o | | u | | Mean | |
|---|---|---|---|---|---|---|---|---|---|---|---|---|
| | b | w | b | w | b | w | b | w | b | w | b | w |
| F1 | 344 | 328 | 471 | 413 | 575 | 478 | 481 | 435 | 361 | 345 | 446 | 399 |
| F2 | 1892 | 849 | 1648 | 799 | 1100 | 840 | 1066 | 785 | 1050 | 754 | 1351 | 805 |
| F3 | 2472 | 2044 | 2336 | 2156 | 2437 | 2404 | 2318 | 2228 | 2463 | 2296 | 2405 | 2226 |

*Table* 2.12. The formant frequency values (Hz) at transition onset in [bV] and the point at which F2 onsets in [wV] syllables, from Mack and Blumstein (1983).

Mack and Blumstein (1983) also suggest that the formant frequencies at the onset of periodicity are important cues for the differentiation of [b] from [w]. In addition to computing data on the first three formant frequencies in [w] initial syllables (table 2.11), Mack and Blumstein also calculated F1, F2 and F3 onset frequencies in [b] initial syllables for the same speakers and same set of vowels. As table 2.12 shows, the average F2 onset of [b] is 564 Hz higher than the average F2 onset of [w]: the F1 (47 Hz) and F3 (179 Hz) differences between [b] and [w] are less marked.

Some studies have suggested that the slope (Liberman, Delattre, Gerstman and Cooper 1956) and duration of the formant transitions (Schwab, Sawusch and Nusbaum 1981) are important cues in the perceptual distinction of /b/ from /w/ (as figure 2.14 shows, for a constant formant onset frequency and constant vowel target, slope and duration necessarily covary since slope is defined as the change in formant frequency per unit of time). A further inspection of Mack and Blumstein's data (table 2.13) does indeed confirm that average F1 and F2 transition durations are greater for [w] than [b]; this is consistent with the result of Schwab *et al.* (1981) who showed that formant transition duration may be an important cue in the perceptual differentiation of /b/ from /w/. However, although the formant transition durations are greater (table 2.13), the onset frequencies of [w] (table 2.11) are lower; as a result, the slopes of the formant transitions in the vowel onglide of [w] are not flatter than those of [b]: indeed, a detailed calcula-

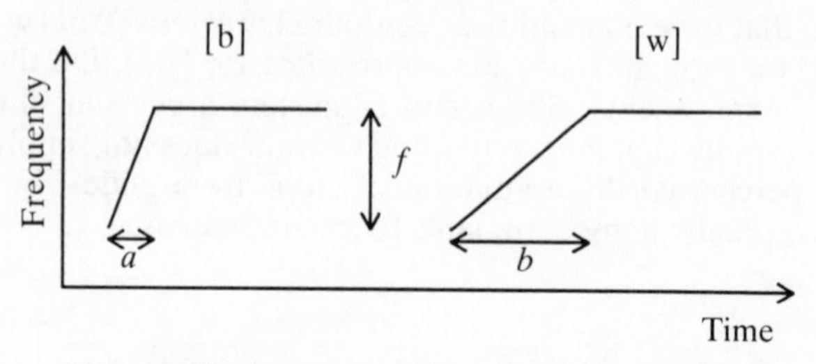


*Figure* 2.14. Schematic outline of the relationship between slope and transition duration in [b] and [w] initial syllables. For a constant onset frequency, the slope and transition duration (*f*/*a* and *f*/*b* for [b] and [w] respectively) necessarily covary. Some experiments have suggested that the transition duration is greater, and therefore the slope flatter, for [w] than for [b]. However, the simplified model assumes equal onset and offset frequencies for [b] and [w] initial syllables, which does not agree with the study by Mack and Blumstein (1983: see table 2.12).

| | bi | | be | | bɑ | | bo | | bu | | Mean | |
|---|---|---|---|---|---|---|---|---|---|---|---|---|
| | D | S | D | S | D | S | D | S | D | S | D | S |
| F1 | 13.1 | +0.38 | 25.4 | +1.96 | 39.9 | +2.43 | 26.6 | +2.52 | 12.1 | +1.82 | 23.4 | +1.8 |
| F2 | 47.2 | +4.26 | 34.2 | +3.16 | 22.2 | +0.63 | 11.8 | +1.27 | 15.1 | −1.19 | 26.1 | +2.3 |

| | wi | | we | | wɑ | | wo | | wu | | Mean | |
|---|---|---|---|---|---|---|---|---|---|---|---|---|
| | D | S | D | S | D | S | D | S | D | S | D | S |
| F1 | 29.0 | +0.72 | 33.6 | +3.18 | 63.7 | +2.86 | 50.6 | +1.50 | 30.3 | +0.42 | 41.5 | +1.7 |
| F2 | 76.6 | +17.4 | 95.2 | +8.96 | 91.4 | +1.92 | 66.7 | +1.89 | 46.3 | +1.58 | 75.3 | +6.3 |

*Table* 2.13. Duration (D, in ms) of vowel onglide in various [bV] and [wV] syllables and the slope of the onglide (S, in Hz/ms). The mean durations and slopes for the five vowels are shown on the far right of the display, and the F2 mean marked with an asterisk is based only on positive slopes (i.e. [bu] is excluded). The calculations of the slope are linear and do not, therefore, take account of the possibility that the first part of the onglide for [bV] syllables may be steeply rising.

tion of the slope of the formant transitions on Mack and Blumstein's data (table 2.13) shows that average F2 slope is considerably greater for [w] (6.35 Hz/ms) than [b] (2.33 Hz/ms). These results lend further support to the investigation of Sawusch *et al.*, which showed that only the duration and the extent of the transition (extent is the difference between offset and onset frequencies of the transition) but not the slope, are important perceptual cues in the /b/-/w/ distinction. It is emphasised that the slope calculations in table 2.13 assume linearity from onset frequencies to the vowel target: such calculations do not allow, therefore, for the possibility that the formant slopes may be steeper in the initial part of the transition for [b] than [w].

Thus, a low F1 and F2 in the steady-state of [w] are the prime acoustic cues that distinguish [w] from other sonorants. Three cues form the basis of the distinction of [b] from [w]: [b], unlike [w], is characterised by an abrupt change in the amplitude contour of the signal at the onset of periodicity; the onset frequencies of F1 and F2 of [w] are lower than those of [b]; and the duration of the formant transitions from the steady-state of [w] to the vowel target is greater than the acoustic vowel onglide in [bV] syllables.

*Summary*

A summary of the cues enabling identification of approximants from the acoustic waveform is shown in the discrimination tree in figure 2.15. At the first level of the tree, discrimination is based entirely on identifying the frequencies of the first three formants.

The identification of [w], [ɫ] and [ɫ̩] as a class depends on the following characteristics: F1 is less than 500 Hz; F2 is less than 900 Hz and F3 is in the range 2000–2400 Hz. These three features will probably also result in the automatic identification of the back vowels [u] and [ʊ] (and possibly even [o]). On the same level of the tree, [ɹ] is identified if F2 is greater than 900 Hz and if F3 is less than 1600 Hz; apart from sonorants produced with retroflexion due to the anticipatory and progressive coarticulatory effects of an adjacent /r/ segment, no other segment should be classified with [ɹ] on the basis of these features. For the identification of [j], F1 is less than 300 Hz, F2 is in the 1800–2400 Hz range and F3 is in the 2500–3100 Hz range. These three features will probably also result in the identification of the front vowels [i] and [ɪ]. Finally, for identification of clear [l], F1 is in the 300–500 Hz range; F2 is in the 900–1600 Hz range; and F3 is in the range 1800–2400 Hz (F3 in this figure is equivalent to Fant's use of F3 in table 2.5). These three features may also result in the detection of oral sonorants other

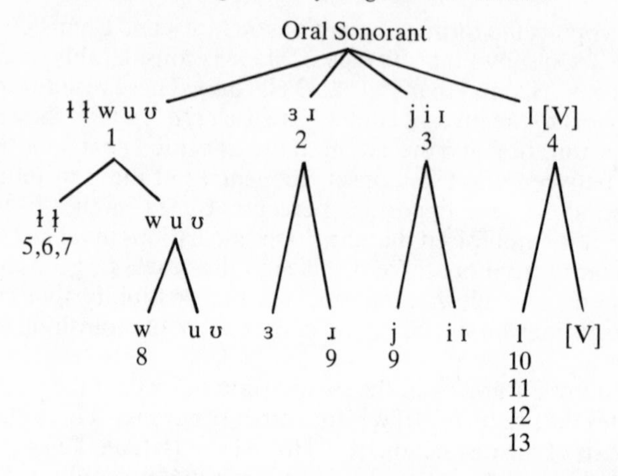


*Figure* 2.15. Discrimination tree for approximants. ɜ denotes vowels that have been 'r'-coloured due to the anticipatory and/or progressive coarticulatory effects of /r/. [V] stands for other sonorants. The relationship between numbers and features is as follows:

1 F1<500 Hz; F2<900 Hz; 2000<F3<2400 Hz
2 F2>900 Hz; F3<1600 Hz
3 F1<300 Hz; 1800<F2<2400 Hz; 2500<F3<3100 Hz
4 300<F1<500 Hz; 900<F2<1600 Hz; 1800<F3<2400 Hz
5 Intensity of F3 high
6 High density of formants in 2-5 kHz range
7 Energy above 2 kHz high
8 F1 is low in frequency and F2 is low in frequency
9 F1 is high in frequency and F3 is high in frequency
10 F1 is low in frequency and F3 is high in intensity
11 Anti-resonance in the 1800-2400 Hz range
12 High density of formants in the 2.5-5 kHz range
13 Energy above 2.5 kHz is high

than [l], although (in theory, at least) no other vowels should have formant frequency values in these ranges.

At the next level of the tree, a distinction is made between [ɫ] and [ɫ̩] on the one hand and [w], [u] and [ʊ] on the other based on three features (5, 6, 7): the intensity of F3 of [ɫ] and [ɫ̩] is greater than that of [w], [u] and [ʊ] (Lehiste 1964); the laterals have a higher density of formants in the 2–5 kHz range; the energy in the spectrum above 2 kHz should be greater for [ɫ] and [ɫ̩] than [w], [u] and [ʊ].

At the bottom of the tree, a low F1 and a low F2 separate [w] from back vowels (Lehiste 1964). 'R'-coloured vowels are separated from [ɹ] on the basis that F1 and F3 should both be higher for the former, as a result of less extreme stricture and less retroflexion (Nolan 1983, 90).

A low F1 and a high F3 separate [j] from [i] and [ɪ] (Lehiste 1964). Finally, the identification of [l] is based on four features (10, 11, 12, 13): F3 is weak; the presence of an anti-resonance in the 1800–2400 Hz range; high density of formants in the 2.5–5 kHz range; and the energy in the spectrum above 2.5 kHz should be greater for [l] than other vowels.

## Concluding Remarks

A set of acoustic features has been proposed for the progressive phonetic refinement of the broad categories *stop, fricative* and *sonorant* which have been hierarchically structured in the form of discrimination trees. The trees are, of course, only provisional and will have to be refined in the light of the performance of continuous speech recognisers.

The studies which have been reviewed show, perhaps, how little we know about the acoustic characteristics of some phonetic segments. Thus, while there are numerous studies dealing with the acoustic structure of oral stops in stressed CV syllables produced in citation form, very little is known about the acoustic cues that enable a within-class differentiation of nasal consonants (primarily because digital signal processing techniques in the last decade have been dominated by linear predictive coding techniques based on an all-pole model and have thus not been able to take anti-resonances into account), the allophones of /h/, non-sibilant fricatives and voiceless approximant consonants to mention but a few. It seems clear that many more detailed acoustic studies are necessary if substantial improvements are to be made in the performance of feature-based, continuous speech recognisers. At the same time, the task of constructing discrimination trees that encode the relationship between acoustic features and phonetic segments is a new and challenging area of investigation in Phonetics which will undoubtedly be enriched in the light of empirical results of automatic, continuous speech recognisers.

## Notes

1. The preparation of this chapter was completed as part of the Alvey Large-Scale Demonstrator Project on automatic speech recognition at CSTR, funded by the Science and Engineering Research Council.

2. All values were in mels in the study of Bladon and Al-Bamerni (1976); they have been converted to Hz for comparison with other studies reported in this chapter.

REFERENCES

Abramson, A. S., P. Nye, J. Henderson & C. Marshall (1981) Vowel height and the perception of consonantal nasality. *J. Acoustical Society of America 70,* 329-39.

Abbs, M. S. & F. D. Minifie (1969) Effects of acoustic cues in fricatives on perceptual confusions in preschool children. *J. Acoustical Society of America 46,* 1535-42.

Al-Bamerni, A. (1975) *An Instrumental Study of the Allophonic Variation of /l/ in* RP. MA dissertation, University College of North Wales, Bangor.

Bell-Berti, F. (1976) An electromyographic study of velopharyngeal function in speech. *J. Speech and Hearing Research 19,* 225-40.

Bladon, R. & A. Al-Bamerni (1976) Coarticulation resistance in English /l/. *J. Phonetics 4,* 137-50.

Bloomer, H. (1953) Observations on palato-pharyngeal movements in speech and deglutition. *J. Speech and Hearing Disorders 18,* 230-46.

Blumstein, S., E. Isaacs & J. Mertus (1982) The role of gross spectral shape as a perceptual cue to place of articulation in initial stop consonants. *J. Acoustical Society of America 72,* 43-50.

Blumstein, S. & K. Stevens (1979) Acoustic invariance in speech production: evidence from measurements of the spectral characteristics of stop consonants. *J. Acoustical Society of America 66,* 1001-17.

Blumstein, S. & K. Stevens (1980) Perceptual invariance and onset spectra for stop consonants in different vowel environments. *J. Acoustical Society of America 67,* 648-62.

Catford, J. C. (1977) *Fundamental Problems in Phonetics.* Edinburgh University Press: Edinburgh.

Chomsky, N. & M. Halle (1968) *The Sound Pattern of English.* Harper & Row: New York.

Cole, R. A. and W. E. Cooper (1975) Perception of voicing in English affricates and fricatives. *J. Acoustical Society of America 58,* 1280-87.

Cooper, F. S., P. C. Delattre, A. M. Liberman, J. M. Borst & L. J. Gerstman (1952) Some experiments on the perception of synthetic speech sounds. *J. Acoustical Society of America 24,* 597-606.

Curtis, J. F. (1942) *An Experimental Study of the Wave-Composition of Nasal Voice Quality.* PhD dissertation, University of Iowa.

Dalston, R. M. (1975) Acoustic characteristics of English /w, r, l/ spoken correctly by young children and adults. *J. Acoustical Society of America 57,* 462-9.

Darwin, C. J. & J. Seton (1983) Perceptual cues to the onset of voiced excitation in aspirated initial stops. *J. Acoustical Society of America 73,* 1126-35.

Delattre, P. C. (1951) The physiological interpretation of sound spectrograms. *PMLA 66,* 864-75.
—— (1954) Les attributs acoustiques de la nasalité vocalique et consonantique. *Studia Linguistica 8,* 103-9.
Delattre, P. C., A. M. Liberman & F. S. Cooper (1955) Acoustic loci and transitional cues for consonants. *J. Acoustical Society of America 27,* 769-73.
—— (1961) Formant transitions and loci as acoustic correlates of place of articulation in American fricatives. *Studia Linguistica 5,* 104-21.
Denes, P. (1955) Effect of duration on the perception of voicing. *J. Acoustical Society of America 27,* 761-4.
Derr, M. A. & D. W. Massaro (1980) The contribution of vowel duration, F0 contour, and frication duration as cues to the /juz/-/jus/ distinction. *Perception & Psychophysics 27,* 51-9.
Dickson, D. (1962) An acoustic study of nasality. *J. Speech and Hearing Research 5,* 103-11.
Diehl, R. (1976) Feature analysers for the phonetic dimension stop vs. continuant. *Perception & Psychophysics 19,* 267-72.
Dorman, M. F., L. J. Raphael & D. Isenberg (1980) Acoustic cues for a fricative-affricate contrast in word-final position. *J. Phonetics 8,* 397-405.
Dukiewicz, L. (1967) Polskie Gloski Nosowe. *Analiza Akustyczna.* Polska Akademia Nauk: Warsaw.
Edwards, T. (1981) Multiple feature analysis of intervocalic English plosives. *J. Acoustical Society of America 69,* 535-47.
Fant, G. (1968) Analysis and synthesis of speech processes, in *Manual of Phonetics* (ed. B. Malmberg). Amsterdam.
—— (1970) *Acoustic Theory of Speech Production.* 's-Gravenhage (2nd edition).
—— (1973) *Speech Sounds and Features.* MIT Press: Cambridge, Mass.
—— (1980) The relations between area functions and the acoustic signal. *Phonetica 37,* 55-86.
Fischer-Jørgensen, E. (1954) Acoustic analysis of stop consonants. *Miscellanea Phonetica II,* 42-59.
—— (1969) Voicing, tenseness and aspiration in stop consonants, with special reference to French and Danish. *Annual Report III,* Institute of Phonetics, University of Copenhagen.
Fischer-Jørgensen, E. & B. Hutters (1981) Aspirated stop consonants before low vowels, a problem of delimitation, its causes and consequences. *Annual Report,* Institute of Phonetics, University of Copenhagen.
Flege, J. & J. Hillenbrand (1986) Differential use of temporal cues to the /s/-/z/ contrast by native and non-native speakers of English. *J. Acoustical Society of America 79,* 508-17.
Fowler, C. A. (1984) Segmentation of coarticulated speech in perception. *Perception & Psychophysics 36,* 359-68.

Fujimura, O. (1962) Analysis of nasal consonants. *J. Acoustical Society of America 34*, 1865-987.

Fujimura, O. & J. Lindqvist (1971) Sweep-tone measurements of vocal-tract characteristics. *J. Acoustical Society of America 49*, 541-58.

Fujisaki, H. & O. Kunisaki (1976) Analysis, recognition, and perception of voiceless fricative consonants in Japanese. *Annual Bull. RILP 10*, 145-56.

Gerstman, L. (1957) *Cues for Distinguishing among Fricatives, Affricates, and Stop Consonants.* PhD dissertation, New York University.

Gimson, A. C. (1980) *An Introduction to the Pronunciation of English* (3rd edition). Edward Arnold: London.

Haggard, M. (1978) The devoicing of voiced fricatives. *J. Phonetics 6*, 95-102.

Haggard, M., S. Ambler & M. Callow (1970) Pitch as a voicing cue. *J. Acoustical Society of America 47*, 613-17.

Haggard, M., Q. Summerfield & M. Roberts (1981) Psychoacoustical and cultural determinants of phoneme boundaries: evidence from trading F0 cues in the voiced-voiceless distinction. *J. Phonetics 9*, 49-62.

Halle, M., W. Hughes & J. Radley (1957) Acoustic properties of stop consonants. *J. Acoustical Society of America 29*, 107-16.

Harrington, R. (1944) A study of the mechanism of velopharyngeal closure. *J. Speech Disorders 9*, 325-45.

Harris, K.S. (1958) Cues for the discrimination of American English fricatives in spoken syllables. *Language and Speech 1*, 1-7.

Hattori, S., K. Yamamoto & O. Fujimura (1958) Nasalisation of vowels in relation to nasals. *J. Acoustical Society of America 30*, 267-74.

Hawkins, S. & K. N. Stevens (1985) Acoustic and perceptual correlates of the non-nasal/nasal distinction for vowels. *J. Acoustical Society of America 77*, 1560-75.

Heinz, J. M. & K. N. Stevens (1961) On the properties of voiceless fricative consonants. *J. Acoustical Society of America 33*, 589-96.

Henderson, J. (1978) *On the Perception of Nasal Consonants.* Unpublished manuscript, Department of Linguistics, University of Connecticut, Storrs.

Hogan, J. T. & A. J. Rozsypal (1980) Evaluation of vowel duration as a cue for the voicing distinction in the following word-final consonant. *J. Acoustical Society of America 67*, 1764-71.

House, A. S. (1957) Analog studies of nasal consonants. *J. Speech and Hearing Disorders 22*, 190-204.

—— (1961) On vowel duration in English. *J. Acoustical Society of America 33*, 1174-8.

House, A. S. & G. Fairbanks (1953) The influence of consonant environment upon the secondary acoustical characteristics of vowels. *J. Acoustical Society of America 25*, 105-13.

House, A. S. & K. N. Stevens (1956) Analog studies of the nasalisation of vowels. *J. Speech and Hearing Disorders 21*, 218-31.

Howell, P. & S. Rosen (1983) Production and perception of rise time in the voiceless affricate/fricative distinction. *J. Acoustical Society of America 73,* 976-84.

Hughes, G. W. & M. Halle (1956) Spectral properties of fricative consonants. *J. Acoustical Society of America 28,* 303-10.

Isenberg, D. (1978) Effect of speaking rate on the relative duration of stop closure and fricative noise. *Haskins Laboratories Status Report on Speech Research SR-55/56,* 63-79.

Jassem, W. (1965) The formants of fricative consonants. *Language and Speech 8,* 1-16.

Jones, D. (1972) *An Outline of English Phonetics.* Heffer: Cambridge.

Joos, M. (1948) Acoustic phonetics. *Language 24,* 1-136.

Kewley-Port, D. (1982) Measurements of formant transitions in naturally produced stop consonant-vowel syllables. *J. Acoustical Society of America 72,* 379-89.

—— (1983) Time-varying features as correlates of place of articulation in stop consonants. *J. Acoustical Society of America 73,* 322-35.

Kewley-Port, D., D. Pisoni & M. Studdert-Kennedy (1983) Perception of static and dynamic acoustic cues to place of articulation in initial stop consonants. *J. Acoustical Society of America 73,* 1779-93.

Klatt, D. H. (1973) Interaction between two factors that influence vowel duration. *J. Acoustical Society of America 54,* 1102-4.

Kurowski, K. & S. Blumstein (1984) Perceptual integration of the murmur and formant transitions for place of articulation in nasal consonants. *J. Acoustical Society of America 76,* 383-90.

Lacerda, F. (1982) Acoustic perceptual study of the Portuguese voiceless fricatives. *J. Phonetics 10,* 11-22.

Ladefoged, P. (1971) *Preliminaries to Linguistic Phonetics.* University of Chicago Press: Chicago.

Lahiri, A., L. Gewirth & S. Blumstein (1984) A reconsideration of acoustic invariance for place of articulation in diffuse stop consonants: evidence from a cross-language study. *J. Acoustical Society of America 76,* 391-404.

Lariviere, C., H. Winitz & E. Herriman (1975) The distribution of perceptual cues in English prevocalic fricatives. *J. Speech and Hearing Research 18,* 613-22.

Lauttamus, T. (1984) *Distinctive Features and English Consonants: a Study of Five British English Fricatives.* University of Joensuu: Joensuu, Finland.

Laver, J. (1980) *The Phonetic Description of Voice Quality.* Cambridge University Press: Cambridge.

Lea, W. (1980) *Trends in Speech Recognition.* Prentice-Hall: New Jersey.

Lehiste, I. (1964) *Acoustical Characteristics of Selected English Consonants.* Mouton: The Hague.

Lehiste, I. & G. Peterson (1961) Transitions, glides and diphthongs. *J. Acoustical Society of America 33,* 268-77.

Liberman, A. M., P. C. Delattre, F. S. Cooper & L. J. Gerstman (1954) The role of consonant-vowel transitions in the perception of the stop and nasal consonants. *Psychological Monographs: General and Applied 68,* 1-13.

—— (1956) Tempo of frequency change as a cue for distinguishing classes of speech sounds. *J. Experimental Psychology 52,* 127-37.

Liberman, A. M., K. S. Harris, J. A. Kinney & H. Lane (1961) The discrimination of relative onset times of the components of certain speech and nonspeech patterns. *J. Experimental Psychology 61,* 379-88.

Lindblom, B. & K. Rapp (1973) Some temporal regularities of spoken Swedish. *Papers in Linguistics from the University of Stockholm 21,* 1-59.

Lindblom, B. & J. Sundberg (1971) Acoustical consequences of lip, tongue, jaw and larynx movement. *J. Acoustical Society of America 50,* 1166-79.

Lisker, L. (1957) Closure duration and the intervocalic voiced-voiceless distinction in English. *Language 33,* 42-9.

—— (1975) Is it VOT or a first formant detector? *J. Acoustical Society of America 57,* 1547-51.

—— (1978) Rapid vs. Rabid: a catalogue of acoustic features that may cue the distinction. *Haskins Laboratory Status Report 54,* 127-32.

Lisker, L. & A. S. Abramson (1964) A cross-language study of voicing in initial stops. *Word 20,* 384-422.

—— (1967) Some effects of context on voice onset time in English stops. *Language and Speech 10,* 1-28.

Løfqvist, A. (1975) Intrinsic and extrinsic F0 variations in Swedish tonal accents. *Phonetica 31,* 228-47.

Lubker, J. (1968) An electromyographic-cinefluorographic investigation of velar function during normal speech production. *Cleft Palate Journal 5,* 1-18.

Mack, M. & S. Blumstein (1983) Further evidence of acoustic invariance in speech production: the stop-glide contrast. *J. Acoustical Society of America 73,* 1739-50.

Magdics, K. (1969) Studies in the Acoustic Characteristics of Hungarian Speech Sounds. *Indiana University, Uraltic and Altaic Series 97.*

Malécot, A. (1956) Acoustic cues for nasal consonants: an experimental study involving a tape-splicing technique. *Language 32,* 274-84.

—— (1970) The lenis-fortis opposition: its physiological parameters. *J. Acoustical Society of America 47,* 1588-92.

Mann, V. A. & B. H. Repp (1980) Influence of vocalic context on perception of the [ʃ]-[s] distinction. *Perception & Psychophysics 28,* 213-28.

Manrique, A. & M. Massone (1981) Acoustic analysis and perception of Spanish fricative consonants. *J. Acoustical Society of America 69,* 1145-53.

Markel, J. & A. Gray (1976) *Linear Prediction of Speech.* Springer-Verlag: New York.

Massaro, D. W. & M. M. Cohen (1976) The contribution of fundamental frequency and voice onset time to the /zi/-/si/ distinction. *J. Acoustical Society of America 60,* 704-17.

McDonald, E. T. & H. K. Baker (1951) Cleft palate speech: an integration of research and clinical observation. *J. Speech and Hearing Disorders 16,* 9-20.

Mermelstein, P. (1977) On detecting nasals in continuous speech. *J. Acoustical Society of America 61,* 581-7.

Mohr, B. (1971) Intrinsic variations in the speech signal. *Phonetica 23,* 65-93.

Moll, K.L. (1962) Velopharyngeal closure on vowels. *J. Speech and Hearing Research 5,* 30-7.

Nakata, K. (1959) Synthesis and perception of nasal consonants. *J. Acoustical Society of America 31,* 661-6.

Nolan, F. J. (1978) The 'coarticulation' resistance model of articulatory control: solid evidence from English liquids? *Nottingham Linguistics Circular 7,* 28-51.

—— (1983) *The Phonetic Bases of Speaker Recognition.* Cambridge University Press: Cambridge.

Nusbaum, E., L. Foley & C. Wells (1935) Experimental studies of the firmness of velar-pharyngeal occlusion during the production of English vowels. *Speech Monographs 2,* 71-80.

Ochiai, Y., T. Fukimura & K. Nakatani (1957) Timbre study of nasalics, part II. *Memoirs of the Faculty of Engineering, Nagoya University 9,* 160-73.

O'Connor, J. L, L. Gerstman, A. M. Liberman, P. Delattre & F. S. Cooper (1957) Acoustic cues for the perception of initial /w, j, r, l/ in English. *Word 13,* 24-43.

Ohala, J. (1983) The origin of sound patterns in vocal tract constraints. in *The Production of Speech* (ed. P. F. Macneilage). Springer-Verlag: New York.

Ohde, R. & K. N. Stevens (1983) Effect of burst amplitude on the perception of stop consonant place of articulation. *J. Acoustical Society of America 74,* 706-14.

Öhman, S. E. G. (1966) Coarticulation in VCV utterances: spectrographic measurements. *J. Acoustical Society of America 39,* 151-68.

Parker, F. (1974) The coarticulation of vowels and stop consonants. *J. Phonetics 2,* 211-21.

Peterson, G. & I. Lehiste (1960) Duration of syllable nuclei in English. *J. Acoustical Society of America 32,* 693-703.

Polka, L. & E. W. Strange (1985) Perceptual equivalence of acoustic cues that differentiate /r/ and /l/. *J. Acoustical Society of America 78,* 1187-97.

Port, R. F. (1976) *The Influence of Speaking Tempo on the Duration of Stressed Vowel and Medial Stop in English Trochee Words.* PhD dissertation, University of Connecticut.

Port, R. F. (1979) The influence of tempo on stop closure duration as a cue for voicing and place. *J. Phonetics 7,* 45-56.

Potter, R. K., G. A. Kopp & P. Green (1966) *Visible Speech.* Dover Publications: New York (reprinted from 1947 Bell Telephone Labs. edition).

Raphael, L. J. (1970) *Preceding Vowel duration as a Cue to the Perception of Voicing of American English.* PhD dissertation, City University of New York.

Raphael, L. J. (1972) Preceding vowel duration as a cue to the perception of the voicing characteristics of word-final consonants in American English. *J. Acoustical Society of America 51,* 1296-1303.

Raphael, L. J., M. Dorman & A. Liberman (1980) On defining the vowel duration that cues voicing in final position. *Language and Speech 23,* 297-307.

Recansens, D. (1983) Place cues for nasal consonants with special reference to Catalan. *J. Acoustical Society of America 73,* 1346-53.

Repp, B. H. (1979) Relative amplitude of aspiration noise as a voicing cue for syllable-initial stop consonants. *Language and Speech, 27,* 173-89.

Repp, B. H., A. M. Liberman, T. Eccardt & M. Pesetsky (1978) Perceptual integration of acoustic cues for stop, fricative and affricate manner. *J. Experimental Psychology: Human Perception and Performance 4,* 621-37.

Revoile, S., J. M. Pickett & L. D. Holden (1982) Acoustic cues to final stop voicing for impaired and normal-hearing listeners. *J. Acoustical Society of America 72,* 1145-54.

Romportl, M. (1973) Zur akustischen Analyse und Klassifizierung der Nasale, in *Studies in Phonetics,* 78-83. Mouton: The Hague.

Schwab, E., J. Sawusch & H. Nusbaum (1981) The role of second formant transitions in the stop-semivowel distinction. *Perception & Psychophysics 29,* 121-8.

Schwartz, M. F. (1968) Identification of speaker sex from isolated voiceless fricatives. *J. Acoustical Society of America 43,* 1178-9.

Shinn, P. C. & S. Blumstein (1984) On the role of the amplitude envelope for the perception of [b] and [w]: further support for a theory of acoustic invariance. *J. Acoustical Society of America 75,* 1243-52.

Shinn, P. C., S. Blumstein & A. Jongman (1985) Limitations of context conditioned effects in the perception of [b] and [w]. *Perception & Psychophysics 38,* 397-407.

Slis, I. & A. Cohen (1969) On the complex regulating the voiced-voiceless distinction. *Language and Speech 12,* 80-102.

Smith, S. (1951) Vocalisation and added nasal resonance. *Folia Phoniatrica 3,* 165-9.

Soli, S. (1981) Second formants in fricatives: acoustic consequences of fricative-vowel coarticulation. *J. Acoustical Society of America 70,* 976-84.

—— (1982) Structure and duration of vowels together specify fricative voicing. *J. Acoustical Society of America 72,* 366-78.

Stevens, K. N. & S. Blumstein (1978) Invariant cues for place of articulation in stop consonants. *J. Acoustical Society of America 64,* 1358-68.

Stevens, K. N. & A. S. House (1956) Studies of formant transitions using a vocal tract analog. *J. Acoustical Society of America 28,* 578-85.

Strevens, P. (1960) Spectra of fricative noise in human speech. *Language and Speech 3,* 32-49.

Summerfield, C. (1987) Pole-zero analysis – an intuitive description of the theory. *Internal report,* Centre for Speech Technology Research, Edinburgh University, Scotland.

Summerfield, Q. & M. Haggard (1977) On the dissociation of spectral and temporal cues to the voicing distinction in initial stop consonants. *J. Acoustical Society of America 62,* 435-48.

Tarnóczy, T. (1948) Resonance data concerning nasals, laterals, and trills. *Word 4,* 71-7.

Umeda, N. (1981) Influence of segmental factors on fundamental frequency in fluent speech. *J. Acoustical Society of America 70,* 350-5.

Walley, A. & T. Carrell (1983) Onset spectra and formant transitions in the adult's and child's perception of place of articulation in stop consonants. *J. Acoustical Society of America 73,* 1011-22.

Wang, W. S-Y. (1959) Transitions and release as perceptual cues for final plosives. *J. Speech and Hearing Research 2,* 66-73.

Wardrip-Fruin, C. (1982) On the status of temporal cues to phonetic categories: preceding vowel duration as a cue to voicing in final stop consonants. *J. Acoustical Society of America 71,* 187-95.

Weinstein, C., S. McCandless, L. Mondshein & V. Zue (1975) A system for acoustic-phonetic analysis of continuous speech. *IEEE Transactions on Acoustics, Speech and Signal Processing 23,* 54-67.

Wolf, C. G. (1978) Voicing cues in English final stops. *J. Phonetics 6,* 299-309.

Yegnanarayana, P. (1981) Speech analysis by pole-zero decomposition of short-time spectra. *Signal Processing 3,* 5-17.

Yeni-Komshian, G. & S. Soli (1981) Recognition of vowels from information in fricatives: perceptual evidence of fricative-vowel coarticulation. *J. Acoustical Society of America 70,* 966-75.

Zimmerman, S. A. & S. M. Sapon (1958) Note on vowel duration seen cross-linguistically. *J. Acoustical Society of America 30,* 152-3.

Zue, V. (1976) *Acoustic Characteristics of Stop Consonants: a Controlled Study.* Indiana University Linguistics Club: Bloomington, Indiana.

THREE : G. DOCHERTY & L. SHOCKEY

# SPEECH SYNTHESIS

This chapter draws attention to the importance of incorporating large phonetic and linguistic knowledge bases into speech synthesis systems, in order to produce highly natural speech output for advanced man-machine communication systems. It is argued that the problems of producing natural speech synthesis are not entirely engineering problems, but are rooted in the failure, up to the present, to model adequately the systematic features of the speech waveform down to the finest detail. This aspect is crucial for improving the naturalness, and hence the acceptability, of synthetic speech output from machines.

The discussion begins with a non-technical, qualitative overview of speech synthesis strategies: abundant technical detail is available in recent reviews (Witten 1982, Bristow 1984, Linggard 1985, Fallside and Woods 1985). It is concluded that synthesis-by-rule is the most appropriate technique for complex applications which require flexible and natural speech output. The second section of this chapter presents a review of the phonetic characteristics of speech which should be incorporated into synthetic speech in order for high degrees of naturalness to be achieved. Many of these are areas which have fallen completely outside the scope of even state-of-the-art speech output systems. The suitability of various techniques for different types of applications is also assessed.

*Articulatory Synthesis*

Articulatory synthesis is the production of artificial speech by means of an analogue or digital simulation of the action of the human vocal apparatus. It involves devising a complex and comprehensive parametric model of the vocal organs (including the lungs, larynx, pharynx, tongue, jaw, lips, and the oral and nasal cavities), and of the aerodynamic and biomechanical processes which take place during speech production. A frequently-adopted approach is to model the vocal tract as a vector of cross-sectional area specifications corresponding to discrete points along its length. Complex models of glottal excitation have also been produced which incorporate important aerodynamic, mechanical and acoustic factors that are known to affect vocal fold activity in speech (Flanagan and Ishizaka 1978; Flanagan, Ishizaka and Shipley 1975; Cranen and Boves 1986). The parameter values can be manipulated to produce a range of possible vocal tract shapes and glottal configurations. These are then related to acoustic specifications according to standard acoustic and physical theories (Allwood and Scully 1982, Coker 1968, Mermelstein 1973, Mermelstein and Rubin 1977, Haggard 1979, Scully and Clark 1986).

This articulatory synthesis technique for producing artificial speech is primarily of interest for the experimental phonetician, who is thereby given the ability to perform controlled independent variations of key articulatory parameters in order to observe their importance for human speech production and perception. Used in this way, it is of potential use in the investigation and modelling of phonetic phenomena such as fundamental frequency declination, respiratory activity and inter-articulator timing relationships.

The cost for this high degree of flexibility is that the resultant articulatory model is highly complex. First, articulatory synthesis involves the specification and control of a very large number of parameters (far greater than in the acoustically based synthesis techniques to be discussed below). As well as making the generation of speech a computationally expensive process, this constraint requires that sufficient knowledge is available regarding the fine details of the physiology, timing and coordination of the vocal organs. Unfortunately, this is not the case, and even the most sophisticated articulatory approaches which are currently available involve over-simplified models of many aspects of human speech production which are as yet poorly understood.

However, in recent reviews of speech synthesis techniques, some writers (Witten 1982, Linggard 1985) have expressed the opinion that even though articulatory synthesis is an extremely complex procedure, it is potentially a good way to set about producing synthetic speech for all sorts of applications, not just those of the phonetics laboratory. The main premise on which this argument is based is the suggestion that many of the complexities involved in mapping a string of discrete linguistic units onto an acoustic signal are a function of regular, non-arbitrary, predictable properties of the vocal apparatus, and that, given a good model of these, the problems will be minimised. That is, if it were possible to devise a good enough model of human speech production, then the model would be capable of automatically providing all the intricacies and subtleties of the linguistic-to-acoustic mapping which takes place in speech production. All that would have to be input to the model would be a simple representation of the required utterance (e.g. a string of phonemes), and the result would be highly natural speech output, at a very low data rate.

This is quite an appealing idea. However, recent research in experimental phonetics suggests that the basic premise does not have firm foundations. Whilst it is undoubtedly possible to explain certain aspects of the speech signal by appeal to aerodynamic and biomechanical properties of the vocal organs, a number of studies (Bladon and Al-Bamerni 1976, Kelly and Local 1986) have shown that much of the fine-grained, systematic language- or accent-specific phonetic detail present in the speech waveform is quite arbitrary, and is far from being predictable from regular properties of the vocal apparatus. As suggested below, much of this detail is crucial for the provision of natural-sounding speech synthesis, and consequently, even with an extremely good simulation of the vocal apparatus it will not be possible to completely dispense with complex low level phonetic matters.

Articulatory synthesis is the only speech synthesis strategy that attempts to produce speech by modelling human speech production. Its main role is to be found in the phonetics laboratory rather than in a speech output device. The techniques to be described below involve modelling the spectral and/or time structure of the speech acoustic signal with no consideration being given to the articulatory/neuromuscular control processes which underly the production of that signal.

*Synthesis-by-Analysis*

All the techniques which belong in this category are essentially recording-playback techniques. They involve encoding speech in some form, such that the speech signal (i.e. the original token or utterance) can be regenerated at a later time as required. The differences between the techniques are based on the different sorts of coding used (i.e. time or frequency domain and degree of data reduction).

It may be worth noting that most of the recent reviews of synthesiser technology place great emphasis on the storage requirements and on the bit rate associated with various synthesis techniques. Increasingly, VLSI memory devices can overcome these constraints, and the bit rate is no longer such an important consideration for a number of applications (although it still is crucial when bandwidth conservation is an important factor; see Kingsbury (1985) for a summary of such applications). But as will be emphasised below, the ability to store large amounts of data does not automatically give the flexibility and generative capability required for a number of complex applications.

Storing a section of speech on magnetic tape for later retrieval and playback is obviously one of the original forms of synthesis by signal regeneration (Witten 1982). However, it does bring up the question as to what exactly constitutes speech synthesis. At best, tape storage and playback is a highly marginal example of speech synthesis, and should possibly be excluded from the present discussion since it involves relatively little data reduction. The point of including this form of speech storage in a discussion of speech synthesis techniques is to underline the fact that many methods of speech synthesis currently in use are functionally little more than highly sophisticated tape-recorders, and suffer from many of the same drawbacks, as will be pointed out below.

The next level of abstraction away from the original speech signal is that of a digitised speech signal. In digitising a speech signal, a sampling frequency in the region of 10 kHz is often used, thus preserving the perceptually significant portions of the spectrum below approximately 5 kHz. Transforming an analogue speech signal in this way and storing it for later regeneration is clearly very costly in terms of storage, and for this reason, alternative forms of digital coding are used to give a more compact and less redundant digital representation of the speech signal. Witten (1982) provides an accessible and comprehensive survey of techniques that can be used to encode speech in the time domain. These techniques differ primarily in the precise details of the quantization stage of analogue to digital conversion. Apart from straight digitisation, the following techniques have been used: log pulse code modulation (PCM); adaptively quantized PCM; differential PCM; differential PCM with either adaptive quantization, or adaptive prediction, or both; delta modulation; delta modulation with adaptive quantization. What these methods have in common is that they are all general-purpose signal-coding techniques, and do not take into account in any way the special features of the speech signal, or the manner in which the speech signal is produced.

However, even with these methods, the data rates are prohibitively high for many applications. As a consequence, the most-frequently used synthesis-by-analysis techniques involve a more complex parameterisation of the speech signal, and consequently a very substantial reduction in the data rate.

Waveform concatenation involves a complex time-domain encoding of speech, using different algorithms for voiced and voiceless waveforms. This representation is then used as the basis for regenerating synthetic utterances (an example of this technique is described by Mozer and Mozer 1985). One positive aspect of this technique is that it provides a very efficient encoding of the shape of the speech signal, storing it in a highly compressed form (1 kbit to 5 kbits per second of speech). An additional advantage of this technique over the techniques to be described below is that since the speech waveform is encoded in the time domain the utterance regeneration is relatively simple, because it is not necessary to convert frequency domain parameters into a time series. The drawbacks of this technique are those shared by all the synthesis-by-analysis techniques, centering on lack of flexibility and lack of generative ability.

Linear predictive coding (LPC) involves calculating for each frame of speech whether the excitation is voiced or voiceless, the fundamental frequency, and a set of predictor coefficients which represent the short-time vocal tract filtering properties (Atal and Hanauer 1971, Markel and Gray 1976). From this highly economic representation of the signal, it is possible to regenerate the speech waveform, and produce adequate output quality. In addition the form of coding used gives a certain amount of flexibility and since the source parameter is largely separated from the filter information, it is possible to regenerate an utterance with a different intonation contour from the original. LPC is now one of the most commonly used methods of coding speech for a range of purposes. It is a technique which is in current use in many military narrow band communication systems. The use of LPC coding in diphone synthesis will be discussed below.

LPC coding has become a popular technique for producing synthesised speech due to the speed and economy with which an utterance can be stored and re-synthesised. Its principal deficiencies lie in the all-pole model which LPC uses to represent the source and filter properties of the vocal tract. This involves an oversimplified approximation to the acoustic properties of speech on a number of counts: it assumes a single unbranched resonator, which leads to problems with modelling nasal sounds in which there is coupling of the oral and nasal cavities; it does not take into account the subglottal coupling which occurs during the open phase of the phonatory cycle which has important effects on the fine structure of the

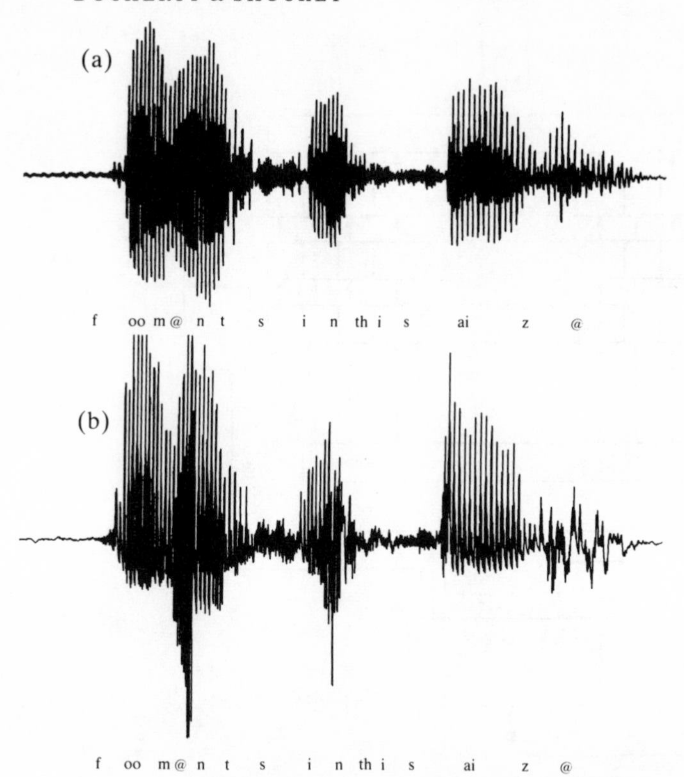


*Figure* 3.1. (a) Time waveform of the utterance 'formant synthesiser' produced by a male speaker. (b) Time waveform of the utterance 'formant synthesiser' regenerated from LPC coefficients.

acoustic signal (e.g. formant bandwidth); it is an all-pole model, and therefore it does not model zeros in the spectrum, which are important for producing natural-sounding nasals, laterals and fricatives. The nature of the LPC model means that it is able to produce satisfactory vowel sounds, since they are the class of sounds which correspond most closely to the all-pole model. Other sounds, and especially those which significantly violate the all-pole model, are less satisfactory (see figure 3.1).

Channel and formant vocoding represent the principal frequency-domain methods for providing an economic coding of the speech signal. Both channel and formant vocoders involve an analysis stage which extracts some representative information for

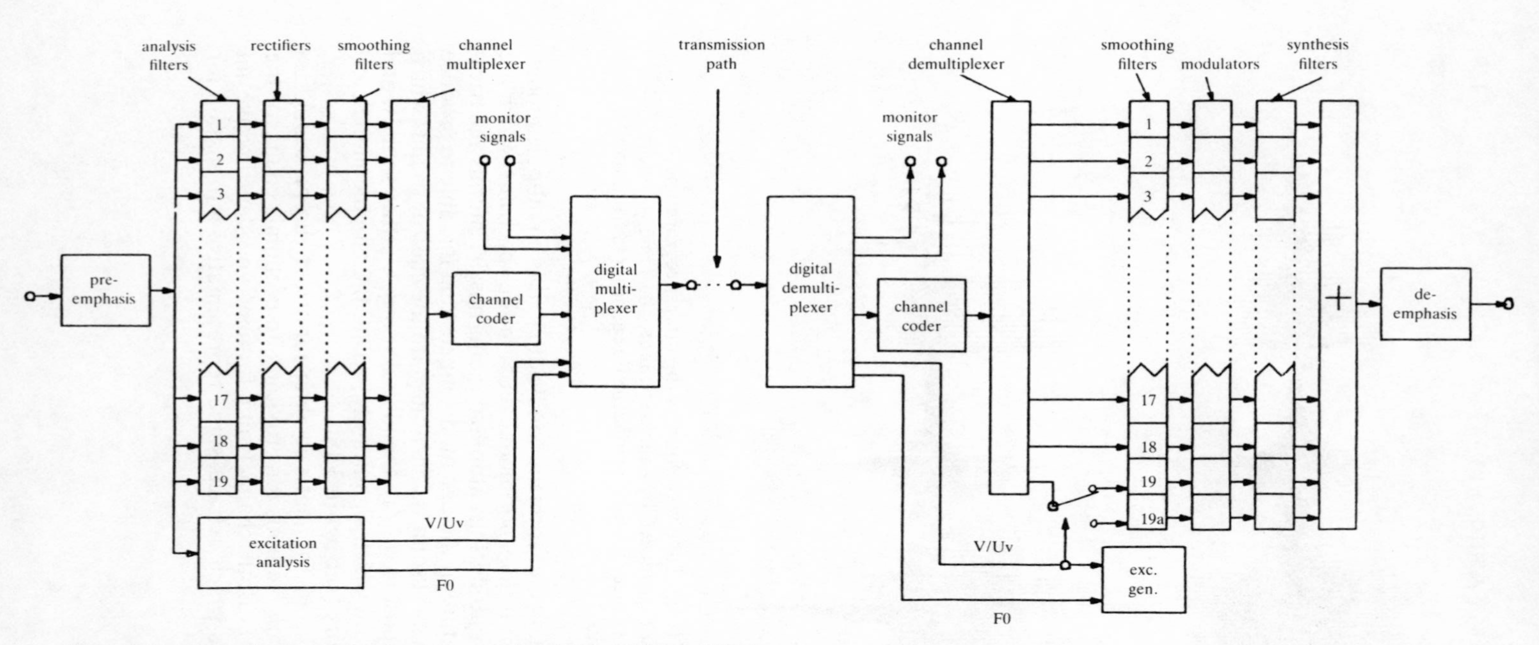


*Figure* 3.2. Block diagram of the JSRU vocoder (after Holmes 1980).

each frame of an input signal prior to storage and/or transmission, and a synthesis stage which uses these frequency parameters as its input, and regenerates the original speech.

The channel vocoder involves a filter-bank analysis of the input speech. The JSRU vocoder (figure 3.2) has a 19-channel filter bank and a 20 ms frame rate. It also transmits information on fundamental frequency, and on whether the frame is voiced or unvoiced. A vector containing information on the energy present in each band, together with the additional parameters, is transmitted to the synthesis stage, which uses this representation to reconstitute the speech signal. The channel vocoder is the standard technique used in NATO narrow band communication systems.

It can be seen that the channel vocoder makes no explicit use of the formant structure of speech. The formant vocoder (Chong 1978, Flanagan 1960, Coker 1965) was designed in an attempt to reduce the redundancy in speech coding (i.e. even the channel vocoder representation is redundant). The analysis component is computationally more complex, since it must pick out formant peaks from the spectrum, rather than just a value for the amount of energy in the band. The receiving end consists of separate glottal waveform and noise sources, and three or four variable formant filters. The values received drive all the synthesiser parameters and the speech signal is regenerated. It is the synthesiser end of the formant vocoder which has been widely used in synthesis-by-rule, as the so-called formant synthesiser. On the whole, formant vocoding gives a very economic representation of the speech signal, but the results of the output are far from ideal. The vowels are fairly good, but the consonants are less so. The main problems with this technique are that there is an insufficient number of acoustic parameters for adequately specifying the structure of the speech signal, and the frame rate does not allow modelling of some of the very rapid changes (e.g. in CV sequences). Formant synthesis, described below, has attempted to overcome these problems.

Whilst these techniques give very little insight into the nature of speech production/perception, they do lead to fairly good quality speech output (although see the comments above on formant vocoding). This is not surprising, since the representations used are either not far removed from the original speech signal, or of such a form that it is not difficult to reconstitute a version of the original speech signal.

*Synthesis-by-Rule and Diphone Synthesis*

Synthesis-by-rule and diphone synthesis are the two existing techniques which have the potential to provide natural and flexible synthetic speech for general-purpose speech output devices. The key to this lies in two important characteristics shared by both techniques. First, they allow utterances to be built up from a finite set of small units, and as a result they are capable of generating an infinite number of novel utterances. Secondly, both techniques permit the modeling of a wide range of the phonetic detail which is essential for the production of natural speech synthesis.

A third important merit of these techniques is that they exploit to varying degrees the large (but far from complete) knowledge base of the correspondence between linguistic units and the acoustic signal, which is the result of many years of multi-disciplinary research in experimental phonetics and speech acoustics. Synthesis-by-rule and diphone synthesis provide the best framework within which linguistic knowledge of the sort discussed below can be used.

*Synthesis-by-Rule*

Speech synthesis-by-rule is the only speech synthesis technique that does not involve regenerating pre-analysed and pre-stored speech waveforms. The hub of this technique is a formant synthesiser. A discrete input string (normally a string of phonemes) is converted into a continuous set of values for the acoustic parameters of the speech synthesiser, which are then sent to the synthesiser at regular intervals, normally of the order of 10 ms. The following description of a synthesis-by-rule system outlines the major features which are common to a number of different systems currently in use (e.g. Holmes, Mattingly and Shearme 1964; Klatt 1980; Clark 1981). Synthesis-by-rule is not a technique that has been devised very recently (the first important paper describing the technique was that of Holmes *et al.* 1964), but it has not changed a great deal since it was first used.

In the synthesiser, resonator circuits (analogue or digital) are used to simulate the filtering properties of the vocal tract, and must receive an input which specifies the formant frequencies, amplitudes, and bandwidths. Additional specification is required for determining which source to use (either voiced or noise), and if voiced, the fundamental frequency. The details of the design of formant synthesisers are well established. There are three main types of synthesiser in use (Holmes 1983, 1985; Klatt 1980; Summerfield 1986). They differ in the number of acoustic parameters

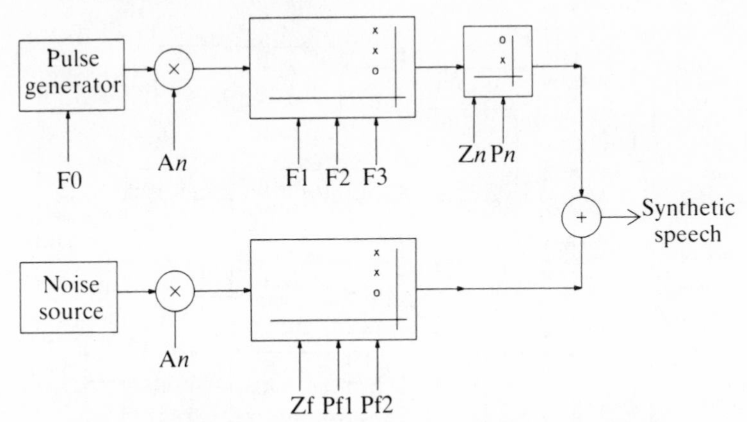


*Figure* 3.3. Serial formant synthesiser (after Flanagan, Coker and Bird 1962).

which they have, and therefore in the flexibility which they offer.

The design of the serial/cascade form of synthesiser resembles most closely the properties of the vocal tract (without nasal coupling), and is good for synthesising non-nasal sounds. Three to five resonators are needed for good spectral modeling. Each frame of data requires a specification of source type (periodic or noise), fundamental frequency, and frequencies/bandwidths of 3 to 5 resonators. Amplitude specifications are not necessary since they have been shown to be predictable from the bandwidths and formant frequencies. This design allows a fairly economic input representation, especially if fixed bandwidths and fixed higher formant frequency settings are used. Figure 3.3 shows a typical serial/cascade synthesiser.

The parallel synthesiser configuration gives greater flexibility for producing non-vowel sounds. It gives the user access to a greater number of parameters (e.g. formant amplitudes), at the cost of having to create a more complex input specification, and working with a model which is more removed from the human vocal apparatus (Quarmby and Holmes 1984, Clark *et al.* 1986). Figure 3.4 outlines a typical parallel synthesiser.

The hybrid synthesiser configuration incorporates the best of the previous two models (Klatt 1980). Non-nasal vowels, liquids and glides are produced serially, and the parallel configuration is used for nasals, stops, and fricatives.

Whilst there is a certain amount of disagreement over which is the optimum synthesiser configuration, it must be pointed out that

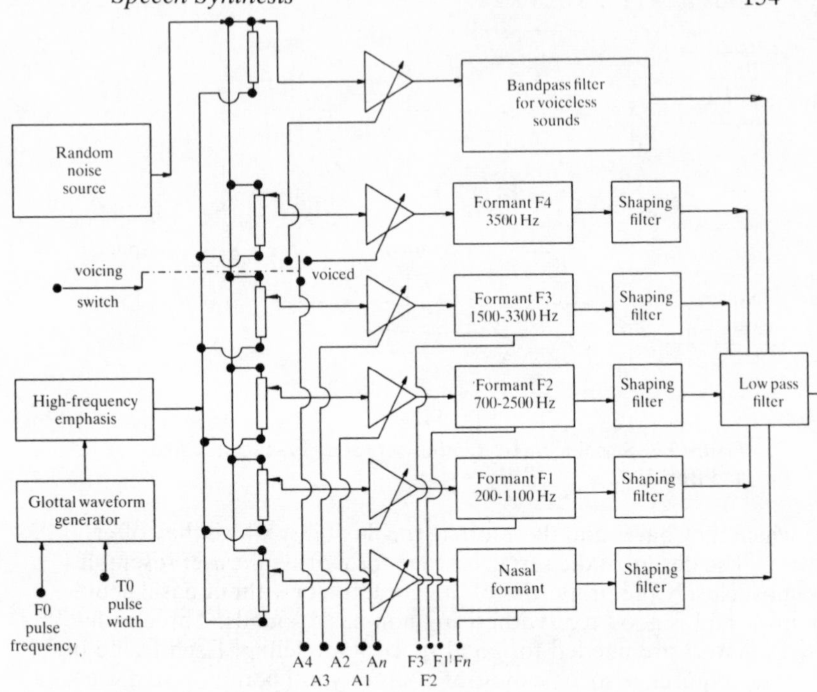


*Figure* 3.4. Parallel formant synthesiser (after Holmes 1972).

all three methods have been used to produce good synthetic speech. In fact, there is a strong consensus that the problems of synthesis-by-rule lie not in speech synthesiser design, but in the serious outstanding problems remaining to be solved in the conversion of a discrete phonemic input to the continuous string of parameter values sent on to the synthesiser, i.e. the rule component.

The rule component consists of a set of rules to map a string of parametrically coded, discrete linguistic units onto a stream of continuously updated acoustic parameter values appropriate for driving a speech synthesiser. In outline, it performs as follows. An annotated phonemic string is read. A look-up table of stored parametrically coded discrete segment targets is consulted, and the appropriate units are loaded into a buffer. The values associated with each of the units are then altered by context-sensitive rule application. A transition algorithm is applied to produce a continuous set of parameter values, which are then output to the synthesiser at the appropriate rate. Figure 3.5 is a diagrammatic representation of this process.

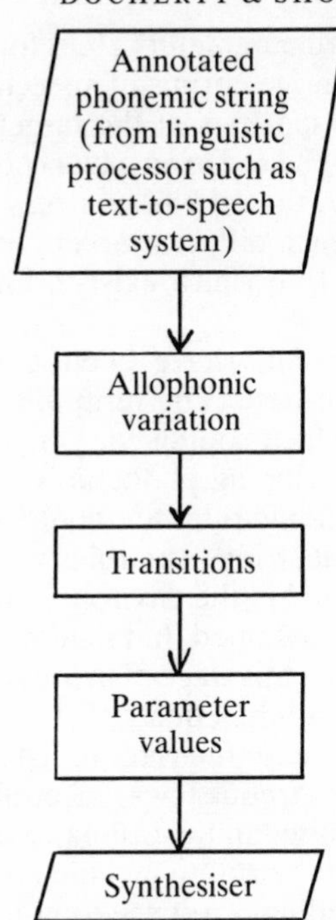


*Figure* 3.5. Typical algorithm for synthesis-by-rule.

The linguistic input to a synthesis-by-rule system is an annotated phonemic representation of the utterance to be produced. This may be produced as the output of the linguistic processor modules of a text-to-speech system (e.g. Allen, 1985, McAllister and Shockey 1986), or it may be entered manually at a keyboard. The annotation consists of markers for prosodic features such as stress, syntactic structure, and intonation tune type. Within the rule system, there is a table of synthesiser parameter values associated with each phoneme, which define that segment in terms of the acoustic parameters of the synthesiser, i.e. there are values for every parameter of the synthesiser. Also specified in the tables are values for parameters used in calculating the transitions between sounds. In addition, in some synthesis-by-rule systems, each table element which is called up from the look-up table is associated with a standard duration. These standard durations are then used as the input to a complex durational assignment routine, to be described below.

The construction of the tables of standard parameter values for a given language involves the analysis of a large corpus of speech from a single speaker, or from a group of speakers of the target language or accent. If synthesis-by-rule is to be used in an advanced multi-lingual, or multi-accent output device, then it will be necessary to have separate groups of tables for each different accent in each language, due to the substantial variation which exists both between and within speech communities.

Once the appropriate parametric elements have been obtained, the next call is to an allophony rule base. The term allophony is used to refer to the manner in which sounds vary as a function of a range of contextual factors, including phonetic context, and phonological structure. The allophonic rules are applied to the parametric representation of the input string. The rules will be of the general form 'if input unit *x* is located in the environment *y*, then change parameter *p* of *x* to *n*'. If it is assumed that minimal allophony is provided by the table values, then this stage of processing will have to deal with the assignment of all the effects of structural and phonetic environment on the phonetic realisation. This will include the effects of context on formant frequencies, spectral shape, intensity, voicing characteristics, timing and coordination, with context being interpreted in a very wide sense, to include not only immediate right/left context, but including reference to non-adjacent segments, and to structural properties of the utterance such as stress position, type of adjacent boundary, and position in syllable or sentence. Much of this contextual information is contained in the annotation which is input to the rule component at the same time as the phonemic string to be produced. The complexity involved in designing such a rule database cannot be underestimated, especially given that a good deal of the fundamental research which would underlie such a tool remains to be conducted. Up till now, in synthesis-by-rule devices, naturalness has been sacrificed to economy, and there is no synthesis-by-rule system available which approximates the level of systematic phonetic detail which is required for natural-sounding output, even though it has been recognised for some time (e.g. Umeda and Coker 1975, Allen 1984) that more natural output will be achieved by incorporating such detail.

An important part of the allophonic rule component is the assignment of segment durations. Segmental duration is dependent both on inherent properties of the input unit concerned and on a large number of phonetic and structural constraints imposed contextually. If standard durations have been obtained from the look-up tables, then these will be used as the input to the rules which will expand or contract the segments or leave them as they are. If standard durations are not available from the look-up tables, then it is necessary to adopt an alternative approach and work towards the assignment of segment durations by imposing a gross temporal structure on the utterance (e.g. by marking intonational tone group and rhythmic foot boundaries), and then to heuristically compute the durational values for each segment under those gross constraints. Both these approaches have been carried out in the past (Klatt 1979, 1980; Witten 1977; Witten and Smith 1977; Isard 1985). The former approach has been shown to perform highly satisfactorily, and has the advantage of demanding rules which have a similar structure to the rest of the allophony base (Carlson, Granstrom and Klatt, 1979).

Up to this point, the synthesis-by-rule system is manipulating a discrete representation of the input units. The final stage of the system involves transforming this discrete set of parameter values into a continuous stream of control signals for driving the speech synthesiser. This process requires the elaboration of a transition algorithm which interpolates in a complex way between the parameter specifications for each discrete element. This section is a very large problem in its own right. It is in the transition phase between segments that much of the temporal and articulatory complexity of the speech signal exists. It is very important that the transitions which are produced synthetically are an accurate reflection of those that are observed in real speech, since it has been recognised for over thirty years that transitions carry considerable perceptual importance, and that speakers are capable of distinguishing between sounds on the basis of transitional cues (Cooper *et al.* 1952, Liberman *et al.* 1954).

Previous transition algorithms, such as those proposed by Holmes *et al.* (1964), Rabiner (1968), Klatt (1983) or Quarmby and Holmes (1984), have aimed at producing linear or non-linear transitions between input units, and have met with some success in providing reasonably intelligible speech output. But if the resultant transitions are compared to those which can be observed in real speech (natural and synthetic segment transitions can be compared in the spectrograms shown in figure 3.6), it can be seen that the algorithms by which the synthetic transitions are created are far too simple. For example, it is not unusual for them to assume that transitions from a particular vowel into a consonant and those coming out of that consonant are identical, though this symmetry is rarely found in natural speech. In addition, in most systems, the same transition is used for each consonant-vowel (CV) combination regardless of the larger phonetic environment. It has long been recognised that this model does not reflect the truth (Lindblom 1963), and that in fact, for example, the transition from [t] to [a] is very different in the utterance [ita] from the corresponding transition in the utterance [ata].

The problem of finding an adequate algorithm for transitions is inherent in the choice of phoneme (or allophone) targets as the input, and represents one of the major disadvantages of working with elements of a segmental nature. It is, in fact, one of the major motivations for using the diphone synthesis technique.

*Synthesis-by-Rule: an Evaluation*

The fact that the utterance to be produced can be represented by the basic building blocks of language (i.e. the phonemes of the target language or accent) gives the complete generative capacity which was absent from the synthesis-by-analysis techniques and, in theory, it is possible to produce any utterance from a given language.

The flexibility offered by the synthesiser configuration (especially the parallel type, with its larger number of controllable parameters) and the look-up table of acoustic specifications is to be highly valued. It opens the way to synthesising different voices, different voice qualities, and the whole range of systematic segmental and prosodic features in speech, which is essential for increasing the quality and acceptability of speech output.

On the negative side, synthesis-by-rule essentially involves a steady-state and transition approach to modelling the speech waveform, so it is obviously flawed to a certain extent, since analysis of natural speech shows that the signal is almost continuously changing. The transitions in synthesis-by-rule output are very stylised approximations to the complex transition phenomena in real speech. This is a problem which is inherent in phoneme-based speech synthesis-by-rule. A possible solution is to construct the utterances from different units within which the transitions would be specified. However, this is only a part-solution and is associated with its own inherent difficulties.

The discrete-to-continuous conversion has in reality to be more complex than has been suggested in the outline description above. The acoustic properties of speech sounds are coloured by their phonetic and structural environment. So this context-sensitive variation must be incorporated in some way into the conversion process.

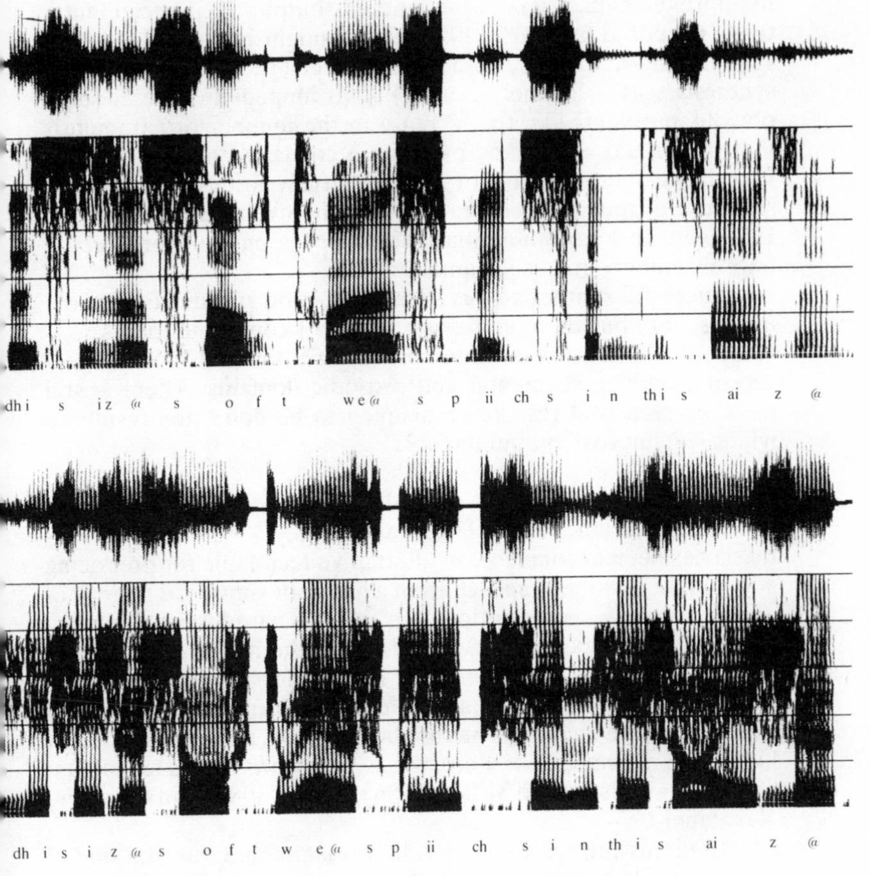


*Figure* 3.6. Spectrogram of the utterance ‘This is a software speech synthesiser’ (a) produced by a male speaker, (b) produced by a synthesis-by-rule system incorporating the Holmes, Mattingly and Shearme (1964) transition algorithm.

The formant-based approach concentrates on modelling the (linguistically) most significant peaks of energy in the spectrum, but pays little attention to spectral details such as the characteristics of the dips in the spectrum between the peaks, as can be seen in figure 3.6. If the principle is adopted that greater naturalness in speech output will be achieved by modelling as much as possible of all that is systematically present in the speech waveform, then it is clear that better specification of the overall spectral shape may lead to improved naturalness of the synthetic output. It has been tentatively proposed by Clark (1986) that it might be possible to base synthesis-by-rule on a channel synthesiser (i.e. analogous to the synthesiser of a channel vocoder). This immediately raises some obvious problems like the increase in the number of parameters that this would entail, and even more crucially, the fact that our knowledge of the acoustics of speech is largely based on the formant structure of the spectrum, rather than the overall spectral detail. However, this is a potential future development which could improve quality quite substantially.

Succesful synthesis-by-rule requires good specification of the input string, but there still exists a lack of comprehensive knowledge of phonetic realisation for English and other languages/accents, both in segmental and prosodic domains. There is still basic instrumental research remaining to be done, the results of which will improve output quality.

*Diphone Synthesis*

Since the late 1960s (Dixon and Maxey 1968) diphone synthesis has been available as an alternative technique for producing good quality synthetic speech from a linguistic input, for a range of applications. The basic notion on which it is based is a very attractive one, and appears to get over one of the major obstacles set by synthesis-by-rule. However, it does suffer from a number of problems, some of which are quite serious, and which probably explain why the technique has not been used as frequently as might have been expected. For more extensive reviews of this technique, see articles by Isard and Miller (1986), Stella (1985) and Olive and Nakatani (1974).

In discussing synthesis-by-rule it was pointed out that one of the drawbacks is the difficulty modelling the transitions between segments (see figure 3.7). The salient feature of diphone synthesis is that the transitions are already given in the system, and algorithms have to be provided to produce the steady-states between the transitions. The first step in devising a diphone system, is to store digitally (normally using LPC analysis) the transitions between all the different sound units which can occur together in the target language. This forms the database for the synthesis stage. When an utterance is generated, all the relevant transition specifications are read from the data base. For example, if the utterance to be produced was *computer,* it would be represented in the following way after reading the diphone database (the phonetic alphabet used throughout this paper is the Edinburgh University Machine-

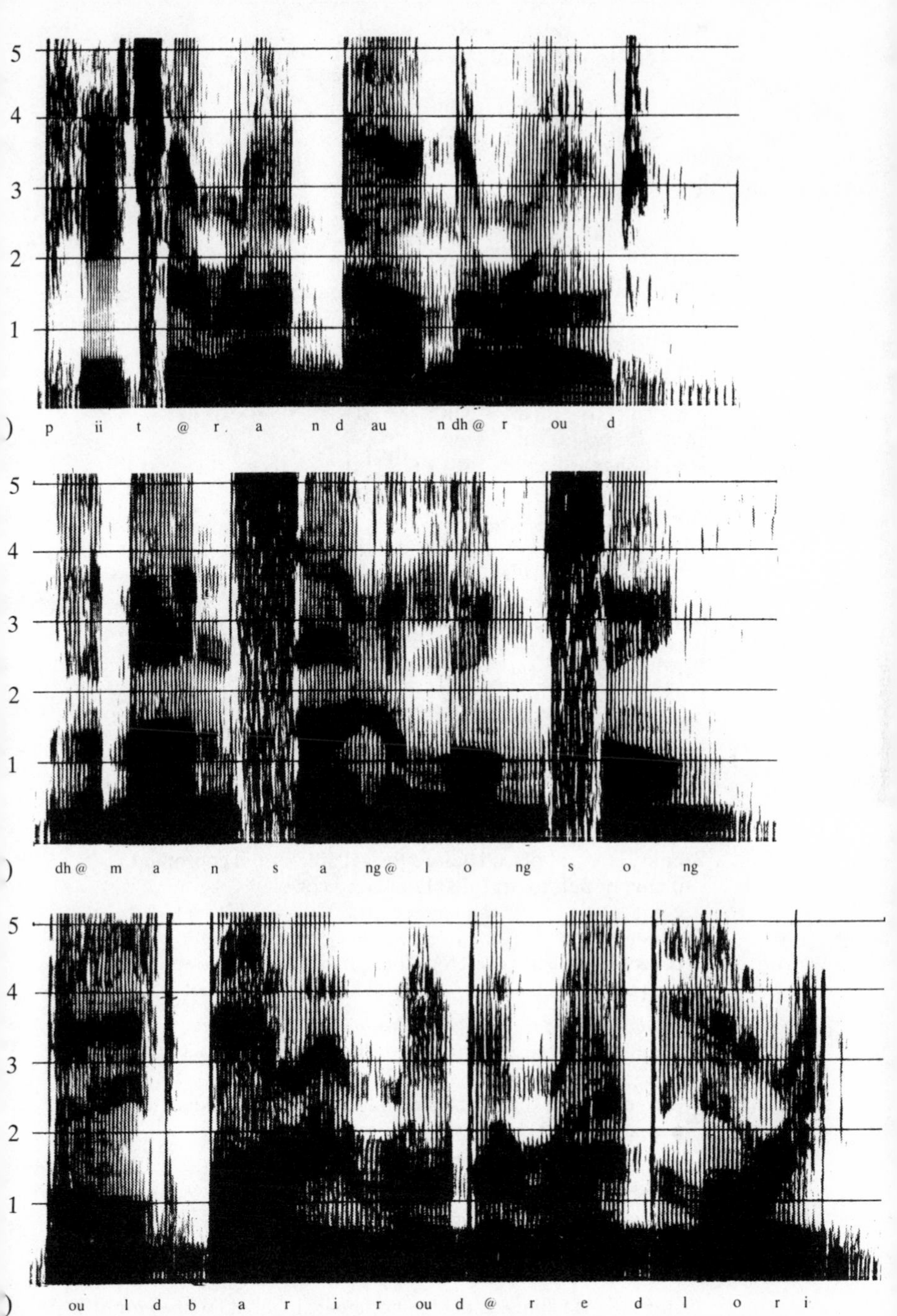


*Figure* 3.7. Examples of transitions between segments in conversational speech.

| Key-word | IPA | MRPA | Key-word | IPA | MRPA |
|---|---|---|---|---|---|
| Monophthongal | | | Consonants | | |
| Vowels | | | *p*ea | [p] | p |
| b*i*d | [ɪ] | i | *t*ea | [t] | t |
| b*ea*d | [i] | ii | *k*ey | [k] | k |
| b*e*d | [ɛ] | e | *b*ee | [b] | b |
| b*a*d | [a] | a | *d*ye | [d] | d |
| b*ar*d | [ɑ] | aa | *g*uy | [g] | g |
| b*u*d | [ʌ] | uh | *m*e | [m] | m |
| b*ir*d | [ɜ] | @@ | *kn*ee | [n] | n |
| *a*bout | [ə] | @ | si*ng* | [ŋ] | ng |
| p*o*t | [ɒ] | o | *th*in | [θ] | th |
| p*or*t | [ɔ] | oo | *th*en | [ð] | dh |
| p*u*t | [ʊ] | u | *f*an | [f] | f |
| b*oo*t | [u] | uu | *v*an | [v] | v |
| Diphthongal | | | *s*ea | [s] | s |
| Vowels | | | *z*ee | ]z] | z |
| d*ay* | [eɪ] | ei | *sh*e | [ʃ] | sh |
| g*o* | [oʊ] | ou | bei*g*e | [ʒ] | zh |
| c*ow* | [aʊ] | au | ea*ch* | [tʃ] | ch |
| *eye* | [aɪ] | ai | e*dg*e | [dʒ] | jh |
| b*oy* | [ɒɪ] | oi | *h*at | [h] | h |
| b*eer* | [ɪə] | i@ | *y*es | [j] | y |
| b*are* | [ɔə] | e@ | *w*ay | [w] | w |
| t*our* | [ʊə] | u@ | *r*ay | [r] | r |
| | | | *l*ay | [l] | l |
| | | | *wh*y | [ʍ] | hw |
| | | | lo*ch* | [x] | x |

*Stress Marking*

Primary stress: ″ placed before the syllable-initial consonant, if any, or before the syllable-nuclear vowel e.g. FISHING = /″f i sh i ng/ or RETREAT = /r i ″t r ii t/

Secondary stress: ′ placed before the syllable-initial consonant, if any, or before the syllable nuclear vowel e.g. PSYCHOLINGUISTIC = /′s ai k ou l i ng ″g w i s t i k/

*Utterance Boundary and Word Boundary Marking*

Utterance boundary: / (also stands for phonemic bracket) e.g. /dh @ * ″s t uh d i * @ v * f @ ″n e t i k s/

Word boundary: *, as in the example immediately above, with a space on each side

*Syllabic Consonant Marking*

A syllabic consonant can be indicated by placing the 'equals' sign (=) immediately after the relevant segmental symbol e.g. /dh @ * ″k i t n= * w @ z * i n * dh @ * ″g aa d n=/

*Figure* 3.8. Machine-readable phonemic alphabet (MRPA) for English (Received Pronunciation). With acknowledgement to J. Laver.

Readable Phonemic Alphabet, as listed in figure 3.8):

[#k ko om mp py yuu uut t@ @#]

in which each dyad or diphone represents the LPC-coded transition between the sounds which make it up. So, the first diphone contains the transition between a pause and [k], the second between [k] and [o], the third between [o] and [m], and so on. Unlike synthesis-by-rule where the phonemic steady states are concatenated, in diphone synthesis an utterance is generated by concatenation of the LPC-coded transitions prior to conversion of the LPC parameters back into an analogue signal for output.

*Diphone Synthesis: an Evaluation*

One of the major problems associated with synthesis-by-rule is that of context-sensitive modification of the acoustic properties of sounds, with the result that a highly complex allophonic rule base must be devised in order to model this. At first glance it would seem that diphone synthesis dispenses with this troublesome and complex requirement, since diphones are inherently context-sensitive because their acoustic properties are dependent on the the identity of the two sounds which they represent. However, there are two aspects of context-sensitivity which detract from what appear to be very good grounds indeed for adopting the diphone technique; the span of context-sensitivity, and the lack of constancy of context-sensitivity.

Diphone synthesis models only adjacent context effects, but there are a large number of effects which have been shown to occur across one or more intervening segments. One example noted above is vowel-consonant-vowel (VCV) sequences: the acoustic properties of the two vowels are interdependent, a fact that could not be taken into account if such a sequence has to be specified in diphone synthesis as /#V VC CV V#/. One solution to the problem of span of context-sensitivity is to use larger units of storage (triphones, demi-syllables, or full syllables). These will allow a greater amount of the long span effects to be captured in the output, with an appropriate increase in the output naturalness, but at the cost of a significant increase in the number of items that need to be stored in the system, and an increase in the amount of time required to analyse the items prior to synthesis. In practise, most diphone synthesis systems in use are, as their name implies, purely diphone based, and do not take these long span effects into consideration.

The problem of the lack of constancy in the transition between two sounds is even more problematic for diphone synthesis. If it were the case that the transitions in all the acoustic parameters between any given pair of phones were standard for all occurrences of that pair of phones, then it would be sufficient to store a single diphone specification for every permissible sequence of phones in the target language. However, large amounts of work in experimental phonetics have shown that the transition from one phone to another is not only a function of the articulatory and acoustic properties of the two phones, but also of a wide range of related phonetic and structural factors, such as whether adjacent vowels are stressed or not, rate of speech, the number of syllables in the word, the position of the word in a sentence, and the nature of the boundary that lies between the two sounds. This means that either rules must be written to alter the features of the basic diphone parameters accordingly (an almost identical task to that performed by the allophonic processing in synthesis-by-rule), or the diphone inventory must be greatly expanded so that there is a diphone (or what Isard and Miller (1986) call an *allodiphone*) covering all the different phonetic and structural environments in which a given pair of phones could occur. When all of this is taken into consideration, it is clear that diphone synthesis is far from being the solution to the problems which were encountered by synthesis-by-rule, and indeed that many of the problems are equally great in both techniques.

Apart from the questions of span and constancy of context-sensitivity, diphone synthesis has another major disadvantage when compared to synthesis-by-rule. Since the (allo)diphone inventory is coded in LPC parameters, it is essentially a signal regeneration technique, and it suffers from a number of the same drawbacks of that class of speech output, among them that it is speaker-dependent. Normally, the diphone-inventory is constructed from analysis of the speech of a single speaker, and since LPC is such a good form of storage and regeneration of speech signals, the resultant speech output has the vocal characteristics of that source speaker. In order for diphone synthesis to be used in a multi-speaker, multi-lingual speech output device, it would be necessary to have an enormous diphone corpus, at a very large cost in storage and in analysis effort.

It must be said, however, that diphone synthesis does have the advantage of allowing the creation of an adequate speech synthesis database in just a few weeks, most of this time being spent in analysing and generating the diphone database. By comparison, the construction of a synthesis-by-rule system requires a much larger period of time.

Diphone synthesis, then, suffers from a number of fundamental problems which counterbalance the gains that would be made by having segment-to-segment transitions wired-in to the system.

In summary whilst synthesis-by-analysis techniques are the most frequently used in commercial output devices, they all exhibit serious limitations. Synthesis-by-rule has very extensive, and hitherto largely untapped, potential as a technique for providing very natural speech output for speech output devices forming part of highly complex human/computer interactions (see the discussion of applications below). The realisation of this potential is less dependent on advances in electrical engineering or in computer architecture, and more on the building of a comprehensive phonetic knowledge base, which will allow synthetic speech to model as closely as possible all the systematic features of real speech, down to the finest detail. The following sections pursue this point, outlining the sort of knowledge which can be used to improve the quality of speech synthesis-by-rule.

*The Role of Phonetic Knowledge in Speech Synthesis*

It has been claimed above that the most promising strategy for producing unrestricted-domain, natural-sounding synthetic speech is to adopt a microscopic approach and build up the speech signal by modelling as accurately as possible all its systematic properties. The ability to carry this out depends on the availability of a very large knowledge source in which there are representations of the acoustic properties of speech, and the way they map onto units of linguistic analysis, such as the feature, segment, or syllable (see Hertz 1986, Hertz *et al.* 1985 and Huckvale 1985 for promising attempts to build such a knowledge base). In order to give an idea of the sort of information which will be stored in such a database this section reviews some of the major acoustic features of speech sounds, highlighting those areas which have not received a great deal of attention in the past from designers of speech synthesis systems, but which have a large contribution to make in improving the natural quality of speech output.

A distinction is made here between segmental units, which are speech sounds as normally represented by the alphabet-like symbols used by phoneticians (e.g. s, h, m, k) and suprasegmental features, which consist of pitch movements, loudness, length, and stress and are normally represented by diacritics in a phonetic transcription. Standard phonetic terminology is used in this section. This is defined exhaustively in the literature (Abercrombie 1967,

Ladefoged 1980, Laver forthcoming). A detailed account of the acoustic features of English consonants which are relevant for their recognition is also given in chapter 2 of this book, by Harrington.

*Segmental Units*

Stop consonants are usually thought of as consisting of three portions: a period of silence or near-silence (corresponding to oral closure), a burst (corresponding to the release of that closure), and another period of noise before the onset of a following vowel, if there is to be one. The third portion, which is sometimes referred to as voice-onset time (VOT), can be of short duration, but can last as long as 50 ms. Voiceless stops in English ([p, t, k]) generally display silence during the first portion, a strong release, and a long VOT which is filled with noise, the spectrum of which is largely determined by the quality of the following vowel. When a stop is followed by a perceptible VOT, it is said to be *aspirated.* English voiced stops ([b, d, g]) may show some low-energy, low-frequency periodic energy during the closure. The burst associated with a voiced stop is often much weaker than the burst of its voiceless counterpart. These characteristics vary depending on the place of articulation of the stop, the nature of the surrounding segments and the degree of stress on the word. The first and third portions have typically caused little problem for synthesisers, but the second can be problematic because it demands the reproduction of a short transient. This must be modelled meticulously in order that stops sound natural. A frame rate of less than 10 ms is necessary for this modelling to be achieved.

Nasal consonants in English ([m, n, ng]) are characterised by a band of low-frequency energy often termed the nasal murmur, which is accompanied by weaker components at higher frequencies. The higher components vary a great deal from speaker to speaker. Zeros are also present, the lowest of which is especially salient in attenuating frequencies in the spectrum. Both the low-frequency pole and the lowest-frequency zero are necessary for the synthesis of nasals as well as nasalised vowels. Most synthesis systems currently available can produce good-quality nasal consonants.

Like stops, fricatives vary noticeably in their voiceless and voiced realisations. Voiceless fricatives show random noise in a range of frequencies, with formant-like concentrations of energy giving them their distinctive characteristics. The fricatives [f] and [th] are normally much lower in amplitude than [s] and [sh]. Voiced fricatives ([v, dh, z, zh]) which have both periodic and aperiodic components, are less intense than their voiceless counterparts and usually show less high-frequency energy. This fact seems to have been overlooked in most formant synthesis control packages. Another feature of fricatives which is badly modelled in most systems is the amplitude envelope. Fricatives (not affricates) normally have a bell-shaped amplitude curve, beginning fairly attenuated and returning to that state before they end, while synthesised fricatives often display a uniform amplitude throughout. While many speech sounds can be adequately synthesised using frequencies lower than 5 kHz, most fricatives sound very much more natural if they include frequencies up to 12 kHz.

Liquids include the sounds [l] and [r] in English. They are represented acoustically by formants not unlike those seen in vowels, but typically of lower amplitude (given a similar degree of stress and position in utterance) and with less well-defined high frequency components. The liquid [l] generally offers little problem for speech synthesis, but [r] has been problematic with synthesisers with a fixed fourth formant. In natural speech, both the third and fourth formants are much lower for [r] than for any other sound in English.

The class of approximants or glides has only two members in English, [y] as at the beginning of *yet,* and [w], as in *wet.* These are the most vowel-like of the consonants and in fact must be seen in context to be distinguished from [i] and [u]. All else being equal, they differ from these two vowels in having slightly less intensity overall and especially in the higher formants. The main difficulty in synthesising these sounds is in achieving their characteristically slow rate of formant transition.

Vowels are characterised by well-defined formants. Naturally-produced vowels show a large number of formants on a spectral display, but typically only three are used to differentiate vowels in current synthetic implementations. The perceived quality of the vowel is based on the relation of these formants to each other. The actual synthesis of any particular vowel is not especially difficult in any of the strategies used today, and in fact recognisable vowels have been synthesised electrically since 1939 (Dudley *et al.* 1939). Flanagan (1972) comments on even earlier attempts using non-electrical means.

*Segmental Variation*

Speech sounds are sometimes considered as being constant because humans as language users have learned to perceive, as equivalent, sounds with very different characteristics. In fact, it is virtually impossible to make any two speech sounds which are absolutely identical, even in two repetitions of the same word. When used in fluent speech, sounds vary considerably depending on their phonetic and syntactic environments. This can be seen on one level in the variants of the sound /t/, which can be heavily aspirated at the beginning of a stressed syllable, as in *poTato,* almost completely unaspirated after /s/, as in *sTay,* and can be replaced by a glottal stop word-finally or before another consonant, as in *faTe* or *haTrack.* At a more detailed level, there is a great deal of variation in the degree of voicing during closure in English obstruents: the so-called voiced stops often show no voicing utterance-initially, complete or nearly-complete voicing intervocalically, and rapid fall-off of voicing before a voiceless segment or silence. These types will be discussed further below. The point to be made here is that most of the variation found in natural speech is conspicuouly absent in most present-day synthesis strategies, and that which is included does not reflect a consistent model of variation.

Vowels are especially likely to be affected by following segments which involve movement of a slow articulator such as the velum or the back of the tongue. High front vowels especially will show the influence of a following dark [l], all vowels will be somewhat influenced by a following [r], and all vowels will show some degree of nasalisation when followed by a nasal. Any given vowel is also likely to be strongly affected by an adjoining vowel, most strikingly by those most different in articulatory properties.

All segments (except those which are already so) can be rounded in anticipation of a following rounded segment. For example, in the word *skew* [s k y uu], the rounding can spread back from the [uu] all the way to the [s] at the beginning of the syllable. The consonants [r] and [sh], which are somewhat rounded in English, can also cause anticipatory rounding. Similarly, alveolar or palato-alveolar consonants which occur before or after [r] are likely to be more retroflex than normal. This will show as a lowering of the transition of the third and fourth formants of surrounding vowels.

The vowel-like consonants [w, r, y] and [l] can sometimes also be nasalised, usually via a nasalised vowel, as in *when* [w e n], *rum* [r uh m], *yam* [y a m] or *lamb* [l a m], and cases of nasalised [l] directly before a nasal as in *pulmonary* ["p uh l m @ n r i] are not infrequent.

Vowels influence the point of articulation of the consonants made before or after them if these consonants are made with the tongue. Two adjacent lingual consonants will also affect each other strongly, since they have to share an articulator and contrive to do so in the most economical way. The 'k' sound in *keel* is much farther forward in the mouth than the 'k' in *cool,* presumably because [ii] is a front vowel and [uu] is a back vowel. In much the same way, the [k] in *cocktail* is farther forward than the [k] in *backgammon,* since [t] is made farther forward than [g] and could be though of as pulling the [k] slightly forward. Another example is the [s] in *blister,* which is made farther forward then the [s] in *Ascot.* Consonants made at the lips or the very back of the vocal tract do not show these subtle shifts in point of articulation, but it is to be expected in the others.

One cause of segmental variation is, then, the influence of speech segments on each other, both in adjacency and at a distance. Much of this variation can be traced back to articulatory causes, but since the effects are not consistent from language to language, we assume that they are not simply inertial effects.

It is often argued that these phonetic effects occur more strongly word-internally than across word boundaries, so for example, the [ii] in *bean* could be expected to be more nasalised than the [ii] in *key nation*. More experimentation is needed to determine how abstractions like word boundaries affect the interrelations of adjacent segments. All of the effects discussed immediately above operate to some degree whenever the relevant segments are articulated continuously, regardless of higher structures. If higher-level structures do temper these effects, it is further evidence for the claim that cross-segmental effects are not caused entirely by inertia.

Another source of variation is even less mechanical, being related to the speaker's social habits and possibly to the rate of speech. One example might be the pronunciation of the [ng] at the end of words like *raining.* In formal speech, it is quite likely to be pronounced [ng], but in casual speech it very often appears as [n]. Another is devoicing of final consonants; the /z/ in *news* is very likely to be pronounced [s] in the sequence *news stand,* and there are a large number of similar cases. Using these variations is part of being a native speaker of English, and any attempt at modelling natural speech should include these effects, though no form of speech synthesis currently accommodates this fully.

In terms of voicing effects, a simple and commonly used approach to the synthesis of voicing in English is to introduce a robust periodic source whenever the segment in question is phonologically voiced, for the duration of the segment. This approach ignores the fact that voicing varies predictably in different environments, as hinted at above. In absolute initial position or after a voiceless segment, for example, English voiced stops, where *voiced* is being used as a phonological label, tend to show no phonetic voicing whatsoever. When there is a series of voiced consonants which impede airflow as in the word *charged* [ch aa jh d], voicing is very likely to begin strong but to slowly die out in the cluster. To maintain voicing, the air pressure below the vocal cords must be higher than that above the vocal cords. When there is closure or near-closure in the mouth, the pressure differential is reduced, and voicing diminishes. In absolute final position, voicing is very often not present in obstruent consonants and in fact is virtually never found in English in an utterance-final [z].

Another variation in voicing found in natural speech but normally not modelled in synthetic speech is the shape of the waveform during voiced stops. Because the articulatory correlate of the first period of a stop is a complete closure somewhere along the vocal tract, the source energy is highly attenuated during this period. The waveform is typically very low in intensity but also much more sinusoidal than normally seen, because the higher-frequency components are filtered out by the head.

*Suprasegmental Features*

Stress is a general term for a constellation of features which can cause a phonetic element to stand out perceptually with respect to surrounding elements. It has several acoustic correlates such as increase in intensity, duration, fundamental frequency, and variation in vowel formant frequencies. A stressed syllable may be perceptually louder, longer, or different in pitch from adjacent syllables, and its vowel may vary slightly in perceptual quality from the unstressed equivalent. Here, reference is not made to phonological reduction in English, but to the small differences found in the [d i sh] in *additional* said in isolation as opposed to the [d i sh] found normally in *an additional monetary comMITment,* which receives relatively less stress. The word *relatively* is especially important here: stress, like other suprasegmental features, is determined by perceptual comparison with adjacent elements.

It is not necessary that each of the intensity, duration, and fundamental frequency correlates of stress is present for a syllable to be perceived as stressed. In natural speech, one often finds only one or two of them operating, and it is difficult to predict which correlate will be present in a given case. In synthesis systems where all three are freely manipulable, stress can be convincingly simulated.

The speech sounds of English (or of any language) are not of arbitrary duration. Execution of adequate temporal patterns is crucial for intelligibility and incorporating accurate temporal patterns leads to major improvements in the naturalness of synthetic speech. Much research in the last decades has been geared towards discovering these patterns. There is a tremendous amount of variability in duration (see below), so that it is difficult to give a range of values to be expected for a given sound. One can only predict that in a given environment at a specified rate, a particular class of sounds will fall within some range. One can also predict that, all else being equal, a member of a given class will be longer or shorter than a member of another. These relationships have been explored thoroughly in Lehiste (1970).

It is also known that combination of sounds into a linguistic unit is not simply an additive process: the durational value of the [l] in *lay* cannot be used for the [l] in *play*. Phrases, words, and syllables have an internal durational structure which must be taken into account when synthesising speech.

There is also some evidence that there are fundamental frequency effects at the segmental level, apart from those discussed under voicing above. The primary effect among these is inherent pitch in vowels since in equivalent situations, high vowels such as [ii] will display a higher fundamental frequency than low vowels such as [aa]. The problem with using that information productively in synthesis is that it is very hard to define what equivalent situations are. Raising the fundamental frequency of high vowels by a few percent in all cases may appear to be a solution. But the human perceptual mechanism is extremely sensitive to small differences in fundamental frequency, which means that inappropriate fundamental frequency effects are very disruptive. Another well-known segmental effect is the instantaneous perturbation caused by the release of stops followed by vowels. After the release of a voiceless stop, the fundamental frequency will increase suddenly, then drop off suddenly before normalising itself for the vowel. After a voiced stop, there is a slight rise in fundamental frequency with a slower dropoff to the fundamental frequency of the vowel. In each case, the entire manoeuvre takes around 20 ms and is not perceived as a change of fundamental frequency at all. It has been suggested that this micro-perturbation is used (among the many other cues) to distinguish between voiced and voiceless stops, and its inclusion in a synthesis strategy will add to perceived naturalness. A very high data rate is needed to model these rapid changes.

English and most other languages use the tune of an utterance

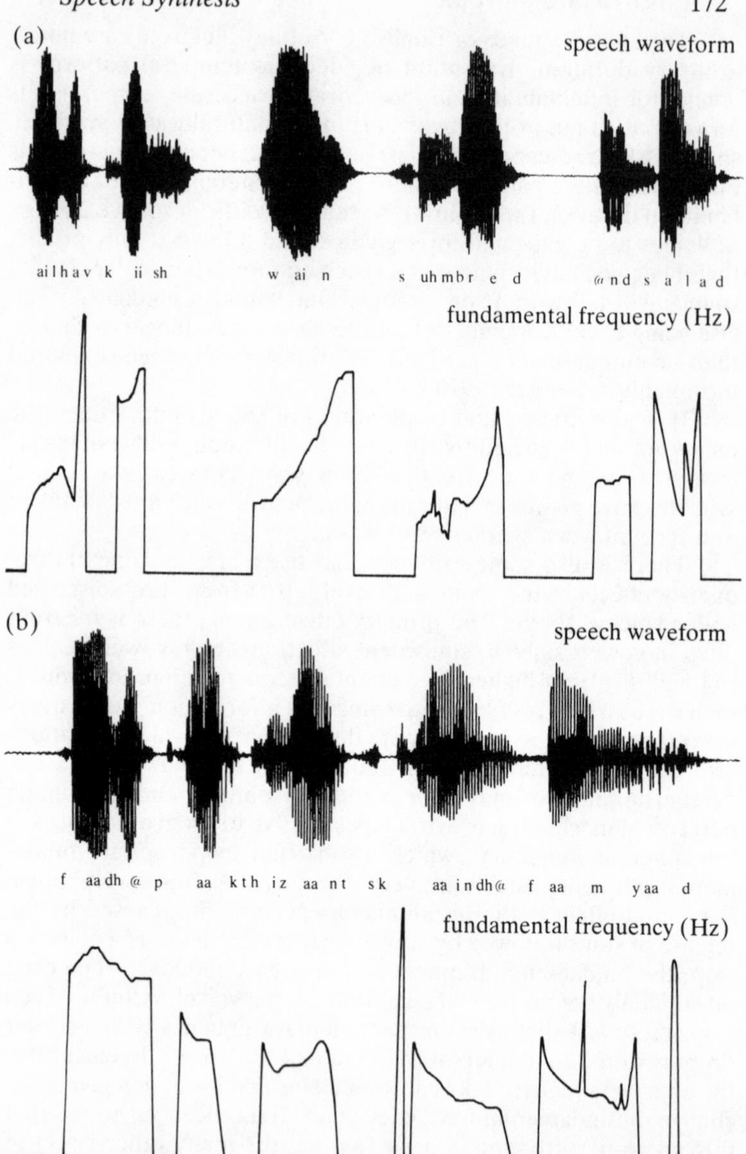


*Figure* 3.9. Speech waveform and F0 contour for (a) 'I'll have quiche, wine, some bread, and salad', and (b) 'Father parked his aunt's car in the farmyard'. (a) is an example of list intonation, and (b) shows F0 declination.

to convey information about grammatical type and grammatical structure as well as to convey the speaker's attitude towards what is being said. The spoken sentence *You have furry ears!* differs from *You have furry ears?* mainly in intonational contour, and the same can be said for the various realisations, *Yes!*, *Yes?*, and *Yes*. In the sentence, *I'll have quiche, wine, some bread, and salad,* intonation conveys the fact that a list is being produced (see figure 3.9). In the sentence, *Bill, who is the former manager, will present the trophy,* the embedded phrase (between the commas) is set off by intonational devices, to give cues about how to parse the sentence. In all of these cases, fundamental frequency must be manipulated to synthesise the sentences properly. Since most synthesis strategies include the ability to change fundamental frequency, they are potentially adequate in this respect. But, to produce natural-sounding intonation, it is necessary to know where to change fundamental frequency contours, in which direction to change them, and by how much. Unfortunately, while much of this knowledge is available in the literature, it has been taken little advantage of by designers of commercially-available synthesis-by-rule systems. One exception is a phenomenon known as declination, by which the fundamental frequency drops gradually within a grammatical phrase (see figure 3.9). However, existent systems use relatively unsophisticated algorithms which fail to capture the subtleties of fundamental frequency variation. Very promising work on modelling English intonation is being carried out by Pierrehumbert (1981) and Ladd (1987).

*Suprasegmental Variation*

Just as segmental features vary a great deal in connected speech, suprasegmental features can display a variety of realisations. Stress is one of the most variable. A syllable which is stressed in a word said in isolation may not be stressed noticeably when it is embedded in a sentence, or its stress may move onto another syllable. Consider the word *fifteen* when said alone and in the phrase *fifteen men*. New insights into how stress migrates around in higher-level constructions are currently being provided by modern phonologists, but their formalisation is difficult to use in synthesis-by-rule systems which do not take the syntactic structure of text into account.

The duration of linguistic units is very much affected by the environment in which they occur (Klatt 1976). It has already been mentioned that segmental duration changes with word structure. It is also true that in English, the overall duration of a larger unit can affect the duration of its components, so that, if said in the same linguistic environment, one would expect the *bag* of *bag check* to be longer than the one in *baggage compartment.* As noted above, stressed segments tend to be longer than unstressed. Syntax also has its effect on duration since units tend to be longer before major syntactic breaks and are very noticeably longer in sentence-final position. Speech rate must also be taken into account. Greater speed implies shorter segments, but not simply a proportional scaling-down of each of them. Vowels shorten relatively more than consonants as rate increases. Klatt (1980) attempts to construct an algorithm for prediction of duration for American English, taking a large number of these factors into account and to good effect for that dialect.

It is clear that connected speech is characterised by variability in virtually all of its components. Most devices which are used to produce synthetic speech can control all types of variability mentioned here in some way. Thanks to the robustness of the human speech perception mechanism, intelligible speech can be synthesised without taking much of this variability into account, but it must be included if the naturalness of synthetic speech is to increase.

*Speaker-Typing*

Different people have different voice qualities, and commercial synthesis systems have not ignored that fact. Several systems attempt to provide voices representing people of different age, sex, and size, with results which vary in acceptability. DECtalk, the text-to-speech system marketed by Digital Equipment Corporation, has nine built-in voices (which, incidentally, score differently on intelligibility tests; Pratt 1986), and one voice that is user-definable. The latter allows manipulation of sex, head size, breathiness, and several other parameters. Changing these parameter values causes effects which remain constant throughout resulting utterances. Laver and Hanson (1984) give a detailed analysis of properties other than gender which can contribute to speaker-specific voice characteristics. They divide these properties into two main classes: those governed by laryngeal settings, such as modal voice, whisper, and harsh voice (with a total of twenty possibilities) and those governed by supralaryngeal settings, such as raised larynx, lip-spreading, and protruded jaw position (with a total of twenty-two possibilities). In addition, they mention two overall muscular tension settings (tense and lax) which can function in conjunction with any combination of the above. Clearly, a very large number of combinations of these settings is possible, each combination providing a different perceptual impression. Such long-term settings are indeed one source of different voice types, but it must also be taken into account that the voice of a single speaker can vary a great deal under different circumstances. Nervousness, anger, sadness, tiredness, and a wide range of other factors can cause changes often not intended by the speaker, and voice quality can be manipulated intentionally to convey overtones of confidentiality, sarcasm, or ridicule. This variation must also be incorporated into a system which hopes to approximate the human facility. In order to do so, it is necessary to include a much more flexible model of long- and short-term speaker characteristics than has been previously used, allowing for a range of dynamically-adaptable laryngeal and supralaryngeal settings.

*Applications*

Some of the most likely applications for synthetic speech will now be considered, discussing what sort of speech quality is necessary to make the applications successful, and which of the techniques discussed above is most likely to provide the necessary quality. The chief applications for speech output fall into four major categories: monitoring/alarm systems, such as the patient-status monitoring system described by Peters (1986); database consultation systems, such as a speaking (railway) train timetable; interaction with a computer system, especially by remote users over a telephone channel; and reading machines, as found in aids for the blind or as educational tools.

These applications range from very limited ones using a small vocabulary in a restricted set of syntactic matrices to those exercising a potentially unlimited vocabulary in situations which change unpredictably. The speech synthesis strategy which is used is determined by the linguistic sophistication, degree of naturalness, and breadth of lexical coverage demanded in a given task. Figure 3.10 plots the relationship between the requirements of different types of speech output applications, and the suitability of different methods for generating synthetic speech.

On the limited end of the spectrum are located applications such as a speaking clock or an automatic banking machine. These systems are capable of only a few responses in very predictable situations and can be implemented using simple approaches. Waveform encoding techniques are more than adequate for such applications. Waveform encoding provides uniformly good-quality speech which is understandable in noisy environments and by inexpert users of the target language, as it certainly should, since the original

*Figure* 3.10. Applications.

| Task requirement | Example applications | Appropriate techniques | Amount of linguistic knowledge in system | Quality/flexibility tradeoff * |
|---|---|---|---|---|
| Very limited vocabulary : single speaker | Talking calculator<br><br>Motor vehicle warning system | Regeneration of preanalysed and coded utterances : insertion of single words into standard frames | nil | High quality, no flexibility |
| Moderate-sized or frequently updated vocabulary : single speaker, variable intonation | Directory assistance<br><br>Stock control system | Diphone synthesis | Specification of simple intonation contours | Acceptable quality, moderate flexiblity |
| Unrestricted vocabulary : multi-speaker : vocal affect | Reading machine for the blind | Synthesis-by-rule | Very extensive and wide-ranging linguistic knowledge required | High quality, high flexibility |

* Assessment of the flexibility of the system involves considerations such as 1) is the technique suitable for a single application, or can it be adapted to perform different tasks in different domains? Can it produce novel utterances or is it limited to a specific set? 2) Is it possible to specify different speaker types or is the output uniform with regard to speaker characteristics? 3) Is it possible to specify complex intonation contours similar to those observed in human speech? 4) Is it possible to modify the fine details of the acoustic parameters?

speech signal is essentially preserved intact.

However, despite the good results that can be obtained, these techniques are generally unsuitable for unrestricted domain speech synthesis (e.g. as in a general purpose text-to-speech system). Usually they involve storing words, phrases, or even sentences for subsequent regeneration. This means that they are domain-dependent; they cannot be used to generate novel utterances which fall outside the inventory of items that have been stored. As such they are suitable for applications such as the speaking sewing machine developed by Brother (Yoshimura and Noriyuki 1985) or the Bell Labs airline inquiry system described in Levinson and Shipley (1980). The smallest level of storage that is normally used with these techniques is the word, or possibly the syllable. This level of specification is incompatible with a high generative capacity due to problems of storage.

It is not just difficulties in storage and generating novel utterances which make synthesis by waveform regeneration an unattractive proposition. The problem of the inflexibility of these techniques is quite severe. It is largely impossible to model segmental/prosodic context-sensitivity in these coded representations, without a heavy replication of stored items. Similarly, the output is limited to the speech of a single speaker, the one who was the source of the original data, so these techniques cannot easily allow for a multi-speaker capability.

Thus, these techniques are good for applications which require good-quality, domain-dependent, limited-vocabulary speech output. The generative capacity which is essential for unlimited-vocabulary synthesis requires the storage of smaller, more basic units (e.g. phonemes or diphones).

A more elaborate but still simple application might be consultation of a larger, constantly updated database. One example might be a telephone-order warehouse, where user numbers and stock numbers, sizes, colours, and other data about required items are entered from a touchtone keypad, and all of the information keyed in is verified by synthetic voice. Such a system might deal with a vocabulary of several thousand words in simple, predictable syntactic configurations. Again, waveform-encoded words or larger units would not be unusable. If it were decided that the database had to be updated too frequently to allow for convenient encoding of key words, it would be feasible to use sizeable frames of encoded speech into which variable words were synthesised using a diphone technique. This would allow for preservation of voice quality (of a single speaker) and controllable intonation, while giving a wide latitude in the range of products to be handled. Other such enquiry systems include a speaking card catalogue in a library or an instruction program for equipment servicing. The water-supply control system with voice output described by Fallside and Young (1978) shows such an application already in place.

An example of an intermediate application is (telephone) directory assistance. In this case, a potentially unlimited vocabulary must be dealt with, since personal and place names can come from a wide variety of different language sources and population movements are unpredictable. The telephone directory will also be constantly updated for the same reason. Assuming that an automated directory assistance would repeat back the name, address, and phone number of the person in question, it would have to be very flexible in terms of pronunciation but we could assume a very simple stock of syntactic structures. For a system of this sort, diphone synthesis seems a good alternative, since some of the drawbacks associated with this type of synthesis revolve around the inability to handle the variation found in differing syntactic environments. A single speaker should be sufficient, so there would not be an untoward load on storage of diphones.

The most demanding application is a generalised text-to-speech system which is sufficiently adaptable for a variety of uses (see Nolan 1984 for a discussion of this and other advanced applications). One such use would be a text reader for the visually handicapped and others who are unable to read. It might be called upon to read newspapers, which would involve the use of different rates and intonations for different type faces, thus allowing the listener to distinguish between headlines and straight text. In reading a novel, it might assume different speech styles for different characters, or at least use different voices for different speakers within what could be recognised as discourse from the quotation marks and layout on the page. Scientific or legal texts might call for yet another, more measured, style, with frequent recourse to repeating or spelling out unfamiliar vocabulary items. Special-interest text such as cookery books would call not only for special routines but also for special intonations: the ingredients of a recipe would probably be read using list intonation, for example.

One market of non-readers which has not yet received much commercial interest is pre-school children. Cassette tapes of children's stories are very popular with those who have not yet learned to read, and the ability to listen to any favourite book at will would no doubt be even more so. Special intonation and voice quality are customarily used by adults reading to children, and these can be synthesised using the proposed generalised text-to-speech system. For an application such as this, synthesis by rule offers the needed generality, high quality, and flexibility.

*Concluding Remarks*

Society is entering an age in which human-machine communication by speech will become more and more common. At present, users are willing to accept fairly low-quality synthetic speech because it is still a novelty and because they are often in situations which give them extra motivation. But as applications and users become more sophisticated, natural-sounding synthetic speech will be expected. It must also be taken into account that some less highly-motivated potential users refuse to come to terms with synthetic speech which makes heavy demands on their attention and patience. This means that better-quality synthesis, developed within the context of a greater understanding of the human factors involved in the man-machine interface, is a priority if devices featuring speech output are to achieve full commercial success. While current speech synthesisers have yet to achieve a state of perfection, they seem capable of producing good output given a sufficiently well-specified input. An important task facing speech scientists is therefore the construction of control algorithms for speech synthesis which incorporate what is known by linguists about the acoustic structure of speech and the linguistic features which determine variability. Some degree of basic research is still called for, before the needed information is available, but much is known now which could be incorporated into synthetic speech to its benefit.

## REFERENCES

Abercrombie, D. (1967) *Elements of General Phonetics.* Edinburgh University Press.

Allen, J. (1984) Units in speech synthesis, *Proceedings of the 10th International Congress of Phonetic Sciences,* 151-5.

Allen, J. (1985) Speech synthesis from unrestricted text, in *Computer Speech Processing* (eds. F. Fallside & W. Woods) 461-77. Prentice-Hall.

Allwood, E. & C. Scully (1982) A composite model of speech production, *Proc. IEEE Conference on Acoustics, Speech, and Signal Processing, ICASSP-82,* 932-5.

Atal, B. & S. Hanauer (1971) Speech analysis and synthesis by linear prediction of the acoustic wave, *J. Acoustical Society of America 50,* 637-55.

Bladon, A. & A. Al-Bamerni (1976) Coarticulation resistance in English, *J. Phonetics 4,* 137-50.

Bristow, G., ed. (1984) *Electronic Speech Synthesis.* Granada.

Carlson, R., B. Granstrom & D. Klatt (1979) Some notes on the perception of temporal patterns in speech, in *Frontiers of Speech Communication Research* (eds B. Lindblom & S. Ohman) 233-43. Academic Press.

Chong, K. U. (1978) A low rate digital formant vocoder, *IEEE Transactions COM-23,* 1466-74.

Clark, J. (1981) A low level synthesis-by-rule system, *J. Phonetics 9,* 451-76.

—— (1986) Seminar paper on speech synthesis, presented at the Centre for Speech Technology Research, Edinburgh University, January 1986.

Clark, J., C. Summerfield & R. Mannell (1986) A high performance digital hardware synthesiser, *Proc. 1st Australian Conference on Speech Science and Technology,* 342-7.

Coker, C. (1965) Real-time formant vocoder using a filter bank, a general purpose computer, and an analog synthesiser, *J. Acoustical Society of America 38,* 940(a).

—— (1968) Speech synthesis with a parametric articulatory model, in *Speech Synthesis* (eds J. Flanagan & L. Rabiner) 135-9. Dowden, Hutchinson & Ross.

Cooper, F. S., P. C. Delattre, A. M. Liberman, J. M. Borst & L. J. Gerstman (1952) Some experiments on the perception of speech sounds, *J. Acoustical Society of America 24,* 597-606.

Cranen, B. & L. Boves (1986) A parametrical voice source model incorporating inter- and intra-speaker variation, *Proc. IEE Speech Input / Output Conference,* 94-8.

Dixon, R. & H. Maxey (1968) Terminal analog synthesis of continuous speech using the diphone method of segment assembly, *IEEE Transactions AU-16,* 40-50.

Dudley, H., R. Riesz & S. Watkins (1939) A synthetic speaker, *J. Franklin Institute 227,* 739-64.

Fallside, F. & W. Woods, eds (1985) *Computer Speech Processing.* Prentice-Hall.

Fallside, F. & S. Young (1978) Speech output from a computer-controlled water-supply network, *Proceedings IEE 125,* 157-61.

Flanagan, J. (1960) Resonance-vocoder and baseband complement, *IEEE Transactions AU-8,* 95-102.

—— (1972) Voices of men and machines, *J. Acoustical Society of America 51,* 1375-87.

Flanagan, J., C. Coker & C. Bird (1962) Computer simulation of a formant vocoder synthesiser, *J. Acoustical Society of America 35,* 2003(a).

Flanagan, J. & K. Ishizaka (1978) Computer model to characterise the air volume displaced by the vibrating vocal cords, *J. Acoustical Society of America 63,* 1559-65.

Flanagan, J., K. Ishizaka & K. Shipley (1975) Synthesis of speech from a dynamic model of the vocal cords and vocal tract, *Bell System Technical Journal 54*, 485-505.

Haggard, M. (1979) Experience and perspectives in articulatory synthesis, in *Frontiers of Speech Communication Research*, (eds B. Lindblom & S. Ohman) 259-74. Academic Press.

Hertz, S. (1986) English text-to-speech conversion with DELTA, *Proc. IEEE Conference on Acoustics, Speech, and Signal Processing ICASSP-86*, 2427-30.

Hertz, S., J. Kadin & J. Karplus (1985) The DELTA rule development system for speech synthesis from text, *IEEE Proceedings 73*, 1589-1601.

Holmes, J. (1980) The JSRU channel vocoder, *Proceedings IEE 127*, 53-60.

—— (1983) Formant synthesisers – cascade or parallel, *Speech Communication 2*, 251-73.

—— (1985) A parallel-formant synthesiser for machine voice output, in *Computer Speech Processing*, (eds F. Fallside & W. Woods) 163-87. Prentice-Hall.

Holmes, J., I. Mattingly & J. Shearme (1964) Speech synthesis-by-rule, *Language and Speech 7*, 127-43.

Huckvale, M. (1985) SP/PS speech-production production system. *Unpublished manuscript, Department of Phonetics and Linguistics*, University College, London.

Isard, S. (1985) Speech synthesis and the rhythm of English, in *Computer Speech Processing* (eds F. Fallside & W. Woods) 479-89. Prentice-Hall.

Isard, S. & D. Miller (1986) Diphone synthesis techniques, *Proc. IEE Speech Input / Output Conference*, 77-82.

Kelly, J. & J. Local (1986) Long-domain resonance patterns in English, *Proc. IEE Speech Input / Output Conference*, 304-9.

Kingsbury, N. (1985) Analysis and coding of speech at low data rates, in *Computer Speech Processing* (eds F. Fallside & W. Woods) 125-43. Prentice Hall.

Klatt, D. (1976) Linguistic uses of segmental duration in English: acoustic and perceptual evidence, *J. Acoustical Society of America 59*, 1208-21.

—— (1979) Synthesis-by-rule of segmental durations in english sentences, in *Frontiers of Speech Communication Research* (eds B. Lindblom & S. Ohman) 259-74. Academic Press.

—— (1980) Software for a cascade/parallel formant synthesiser, *J. Acoustical Society of America 67*, 971-95.

—— (1983) Speech synthesis-by-rule of consonant-vowel syllables, *MIT Speech Communication Group Working Papers 3*, 93-103.

Ladd, D. R. & A. Monaghan (1987) A model of intonational phonology for use in speech synthesis-by-rule, *Proc. 1st European Speech Technology Conference*, Edinburgh 1987.

Ladefoged, P. (1982) *A Course in Phonetics* (2nd edition). Harcourt, Brace, Jovanovich.

Laver, J. (forthcoming) *Introduction to Theoretical Phonetics.* Cambridge University Press.

Laver, J. & R. Hanson (1981) Describing the normal voice, in *Speech Evaluation In Psychiatry* (ed. J. Darby) 51-78.

Lehiste, I. (1970) *Suprasegmentals.* MIT Press.

Levinson, S. & K. Shipley (1980) A conversational mode airline information and reservation system using speech input and output, *Bell System Technical Journal 59,* 119-37.

Liberman, A. M., P. C. Delattre, F. S. Cooper & L. J. Gerstman (1954) The role of consonant-vowel transitions in the perception of stop and nasal consonants, *Psychological Monographs 8,* 1-13.

Lindblom, B. (1963) Spectrographic study of vowel reduction, *J. Acoustical Society of America 35,* 1173-81.

Linggard, R. (1985) *Electronic Synthesis of Speech.* Cambridge University Press.

Markel, J. & A. Gray (1976) *Linear Prediction of Speech.* Springer Verlag.

McAllister, J. & L. Shockey (1986) The Edinburgh University Centre for Speech Technology Research text-to-speech system, *Work in Progress, Linguistics Department, Edinburgh University, 19,* 36-44.

Mermelstein, P. (1973) Articulatory model for the study of speech perception, *J. Acoustical Society of America 53,* 1070-82.

Mermelstein, P. & P. Rubin (1977) Articulatory synthesis – a tool for the perceptual evaluation of articulatory gestures, *Haskins Labs Status Report on Speech Research SR53,* 1-11.

Mozer, F. & T. Mozer (1985) Algorithms drive time-domain speech synthesis, *Digital Design 11,* 70-6.

Nolan, F. (1984) Applying linguistics to synthesis, in *Electronic Speech Synthesis,* (ed. G. Bristow) 320-35. Granada.

Olive, J. & L. Nakatani (1974) Rule synthesis of speech by word concatenation: a first step, *J. Acoustical Society of America 55,* 660-6.

Peters, E. W. (1986) BST's text-to-speech product improves health care services, *Speech Technology 3.2,* 101-2.

Pierrehumbert, J. (1981) Synthesising intonation, *J. Acoustical Society of America 70,* 985-95.

Pratt, R. (1986) On the intelligibility of synthetic speech, *Proc. Inst. Acoustics 8.7,* 183-92.

Quarmby, D. & J. Holmes (1984) Implementation of a parallel formant speech synthesiser using a single chip programmable signal processor, *IEE Proceedings 131 (F),* 563-9.

Rabiner, L. (1968) Speech synthesis-by-rule: an acoustic domain approach, *Bell System Technical Journal 47,* 17-32.

Scully, C. & G. Clark (1986) Analysis of speech signal variation by articulatory synthesis, *Proc. IEE Conference on Speech Input/Output,* 83-7.

Stella, M. (1985) Speech synthesis, in *Computer Speech Processing* (eds F. Fallside & W. Woods) 421-60.

Summerfield, C. (1986) A review of VLSI structures for the implementation of formant speech synthesisers, *Proc. 1st Australian Conference on Speech Science and Technology*, 348-53.

Umeda, N. & C. Coker (1975) Subphonemic details in American English, in *Auditory Analysis and Perception of Speech* (eds G. Fant & M. Tatham) 539-64. Academic Press.

Witten, I. (1977) A flexible scheme for assigning timing and pitch to synthetic speech, *Language and Speech 20*, 240-60.

—— (1982) *Principles of Computer Speech*. Academic Press.

Witten, I. & A. Smith (1977) Synthesising English rhythm – a structured approach, *Work in Progress, Linguistics Department, Edinburgh University, 10*, 36-44.

Yoshimura, M. & Y. Noriyuki (1985) Application of voice I/O to home sewing machine, *Proc. SpeechTech '85 Conference*, 159.

FOUR : A. SUTHERLAND & M. JACK

# SPEAKER VERIFICATION

Use of biometric-based technologies for the identification or verification of an individual covers a wide spectrum of techniques and applications. This chapter explores aspects of speech technology in the area of automatic speaker verification where analysis of the acoustic (speech) waveform is used to authenticate an applicant's explicit claim to a particular identity.

Human listeners are skilled (in varying degrees) in the identification of an individual on the basis of some perceived aspects of their overall vocal performance. Such vocal attributes act as strong indicators of personal identity, but perceived vocal performance cannot, in general, offer unique speaker discriminability. Instead, such discriminability must be based on the measured contributions made by individual acoustic parameters to the vocal performance, such that the distribution of speakers on any acoustic parameter can be used singly, or in combination with other parameters, to characterise a population of speakers, and to discriminate individual speakers by means of those parameters. This discriminability (the inter-speaker distance) is conditioned by population distribution considerations, and also by considerations of long-term stability of the location of specific speakers within the relevant parameter space.

Instability factors which contribute to intra-speaker variability include such aspects as dependence on the details of the specific text of the utterances (text-dependence); random occasion-to-occasion variations of pronounciation by the speaker; ephemeral changes in the vocal performance brought about by fatigue, emotion or illness (such as a cold); and external acoustic conditions (multiple speakers or background noise) which affect the quality of the speech signals being analysed. The relationship between intra-speaker variability and inter-speaker distance is of crucial importance in biometric applications, and establishment of the degree of intra-speaker instability in relation to the magnitude of the typical inter-speaker distance on a given parameter for a given size of population is of prime importance in speaker verification system development.

For optimal system operation, each speaker requires to be characterised by a set of speech parameters which constitute a vector profile for the speaker. The parameters used in the vector profile should ideally be parameters that vary maximally between speakers and minimally within speakers. Further, the parameters should offer low susceptibility to external contamination.

Establishing the identity of an unknown speaker from the acoustic evidence of his voice, by any automatic means, can properly be called automatic speaker recognition. Within this definition however, it is useful to define two separate catagories: automatic speaker identification and automatic speaker verification. The primary difference between identification and verification lies in the certainty with which it is known that the speaker in question (the test speaker) is a member of the total population of speakers for whom the characterising acoustic vector-information has been collated (the reference set of speakers).

In automatic speaker identification, the test speaker may not necessarily be a member of the reference set of speakers. The fact that such membership is unknown, and is to be ascertained automatically by the system, means that the decision on identification relies on a comparison of the vector profile of the test speaker with each of the individual vector profiles of the speakers that make up the reference set. Identity of the test speaker is then based on a statistical similarity measure (or distance) between the input (test) vector profile and one of the reference vector profiles. The adequacy of the actual distance measure used is dependent upon the absolute size of the reference population set, the statistics of inter-speaker distances in the reference set, and the statistics of worst-case intra-speaker drift. The likelihood of error in a speaker identification system may be described by considering a population of $N$ speakers. In operation, the speaker whose reference vector profile is closest to that of the input vector profile of the test speaker is determined, and identified as the originator of the input vector. The overall relationship between the likelihood of error and the size of population, $N$, exhibits monotonicity in this case.

In automatic speaker verification, the test speaker is assumed to be represented within the reference set. This assumption is held even when an imposter is making a false claim by using the identity of a valid member of the reference set. Here again, appeal is made to a distance measure between the input vector profile and a reference vector profile (belonging to the claimed individual) for verification to be achieved. Typically, the size of the potential reference set will be significantly smaller in automatic speaker verification systems than in automatic speaker identification systems. In a speaker verification system the (authenticated) reference vector profile of the claimed individual is compared with the input vector profile of the claimant and if the distance measure of the similarity between the two vector profiles is below a predefined threshold then positive verification is made. In this case (assuming a representative population distribution) the likelihood of system error is generally independent of population size.

Whereas the two types of systems are identified by clear differences in the form of user input, their implementations are achieved in very similar manners. Indeed, the published accounts of many systems developed primarily for automatic speaker identification have included specific evaluations of their performance in automatic speaker verification mode. In the remainder of this present discussion, having developed some basic concepts, comment will be focused explicitly in the area of speaker verification.

An automatic speaker verification system may be regarded as a process involving a time variant analogue (speech) input signal and a binary (occasionally ternary to allow for a no-decision output) output response. The system can be decomposed to three separate sections: the pre-processor, to condition the received acoustic signal into a form suited to further analysis; the parameter extractor, to derive the speech parameters used in the vector profile; and the classifier, to carry out the verification statistics (figure 4.1).

The input to the pre-processor is the speech signal, and the pre-processor must perform tasks such as gain control and (word) endpoint detection as required. In the parameter extractor, signal transformations take place to one of several domains, such as temporal, spectral or cepstral domains. Transformations may take place in either analogue or (most commonly) digital form. The output of the parameter extractor is thus a representation of the speech signal, in an appropriate domain. Since the information contained in the acoustic waveform is dependent on several variables in addition to those introduced by the originator of the speech, it is important to extract from the speech waveform those parameters which are maximally tolerant of degradations of the signal introduced by the prevailing acoustic conditions under which the

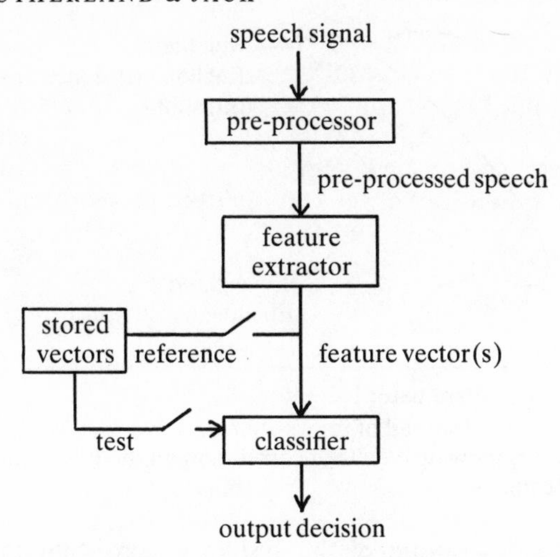


*Figure* 4.1. Typical speaker verification system.

sound was made; and minimally dependent upon the text of the phrase being spoken. This process, of extracting parameters which are ideally dependent only on the identity of the speaker, forms the heart of all automatic speaker recognition systems.

Three major methods of parameter extraction can be identified. The first of these reduces the text-dependence of parameters by averaging them over different speech segments. Such methods are known simply as statistical methods. The other two methods are termed segmentation and contour (Jesorsky 1978). With the segmentation method, a particular phonetic event is identified and isolated for detailed analysis. This will often result in some restriction on the text to be used as input. The contour approach results from the direct use of the temporal variation of the input speech to form time functions of one or more parameters. The end result of each of these processes is the vector profile of the speaker.

Having obtained the input vector profile, it now rests with the classifier to verify the identity of the speaker by comparison with the (authenticated) vector profile of the claimed speaker. The authenticated vector (reference) profile is obtained during a training phase, and can be accessed for comparison on some prompt (such as a personal identity number (PIN) or some other code). The output of the classifier, and hence the output of the system, is then given in binary form, or ternary if the no-decision condition is permitted.

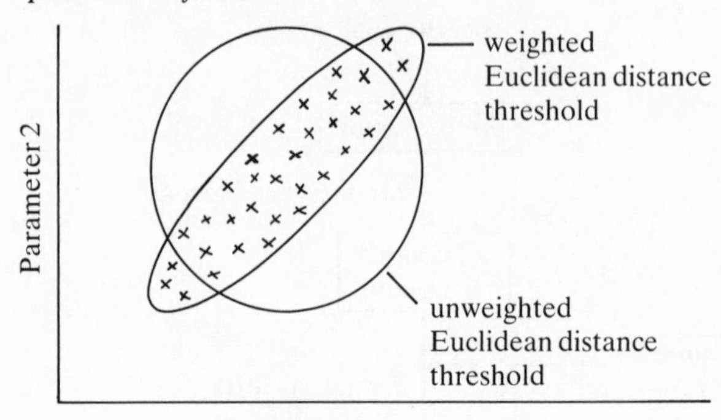


*Figure* 4.2. Typical spread of utterance feature vectors for a given speaker, showing effects of correlation on the Euclidean distance metric.

The above description of the system is consistent over the range of speaker verification systems which have been studied. The options available within each block however, and in particular the parameter extraction process, are manifold and form the basis of this chapter.

*Measures of Similarity*

A variety of methods can be used to determine the similarity between an input parameter vector (or vectors) and a pre-stored reference vector, (obtained during an earlier training session), belonging to the claimed individual.

Formulae for measuring the similarity between a test vector and a reference vector are usually arranged in such a way as to produce a value of zero in the event of identical vectors, and increasingly larger positive values for lesser similarities. For these reasons, such measures are known as distances. To further enhance the notion of distance, both the reference and input parameter vectors may conveniently be considered as points in $M$-dimensional space, where $M$ is the dimensionality of the vectors.

The most basic distance metric is known as the simple Euclidean distance. This is the straight-line distance between the two points in $M$-dimensional space. The problems of simple Euclidean distance measuring can be shown by example. If the example of a two-dimensional parameter vector is adopted, use of several training vectors may result in the diagram shown in figure 4.2. Notice that, if the mean of these vectors is assumed as the reference vector, then an unweighted Euclidean distance measure will result in the threshold being represented by a multi-dimensional hyper-sphere (or circle in this case) centred on that mean. This is inadequate, however, since no allowance is made for any correlation of the individual parameters. To overcome this, a weighted distance can be used where the values adopted by the weighting matrix effectively alter the overall distance metric. Thus a more useful metric would be one whereby points lying on an ellipse around the mean point would be represented as being of equal distance from the mean and the effects of inter-variable correlation could thus be removed (figure 4.2). Such an elliptical distance measure is dependent on the covariance matrix of individual speakers. The covariance matrix contains information on the variance of each parameter for a speaker and also the correlations between individual parameters for the speaker. Non-singularity is an essential feature of the covariance matrix, implying that, in general, the number of reference utterances made by each speaker must be greater than the dimensionality of the parameter vector. A further restriction on the use of such a weighted distance metric is that data storage and computation time required may become prohibitive, since a covariance matrix must be held for every member of the population. One possible extension to the above measure is to pool the individual covariance matrices thus forming a pooled, within-speaker covariance matrix. While this makes the (generally reasonable) assumption that most speakers have similar covariance matrices, it has the advantage of offering a more easily met restriction of ensuring the non-singularity of the covariance matrix: the dimensionality of the parameter vector must, in general, be less that the total number of known reference-utterances from all speakers, minus the number of speakers. This distance metric is known as the Mahalanobis distance (Mahalanobis 1936). A further, similar, measure of distance can be obtained through the use of discriminant analysis. However, the main advantage of the discriminant analysis approach is that it may be used to reduce the dimensionality of the parameter vector, as discussed further below. It is significant that the above distance metrics assume the underlying probabilty density function of the parameter vector to be multivariate Gaussian.

One commonly used non-parametric decision method is known as the nearest-neighbour technique (Cover and Hart 1967). In this case, a test vector will be classified (in an identification system) as belonging to the same speaker as its nearest neighbour. Alternatively, the test vector may be classified as belonging to the speaker who is most heavily represented within the *k* nearest neighbours. Such techniques require that every vector obtained during a speaker's training session is retained (and not simply the mean point), and that a distance metric is used to determine what exactly constitutes a near neighbour. While a Euclidean distance metric may be used, other metrics have been designed specifically for use with the nearest neighbour method (Short and Fukunaga 1981).

The correlation coefficient between two vectors may also be used as an effective distance metric, (Bunge 1977a). While this has the property that, should both the test and reference vectors be multiplied by some constant, the measure will be unaffected, it has been found to be less effective than other metrics (Shridhar, Mohankrishnan and Sid-Ahmed 1983).

Alternative techniques of similarity measurement which have been developed are probabilistic methods (Schwartz, Roucos and Berouti 1982; Wolf *et al.* 1983), which offer scope for retention of the actual characteristics of the parameter distributions as an alternative to applying an overall assumption of Gaussian distribution to the parameter. Research in this area has developed the use of probability density functions (PDF) to identify speakers (Schwartz, Roucos and Berouti 1982) as a method which can offer an improved decision technique in comparison to the standard Mahalanobis distance measure.

Thus far, the problem of temporally ordered sequences of parameter vectors has not been specifically addressed. Such sequences occur in speaker verification systems which require the speaker to utter specific words or sentences and constitute the contour analysis system described above. The first difficulty encountered in attempting to compare two such contours is that utterances are rarely spoken at the same rate on different occasions, even by the same speaker. Some type of time normalisation is thus required. The most simple solution to this problem is to linearly normalise the duration of the test utterance to suit that of the reference. However, the difference in duration between two utterances is seldom uniform, but is more likely to be confined to certain segments. It is thus of importance that some type of non-linear warping technique be used to facilitate accurate time registration of the contours. This is accomplished using dynamic time warping (DTW) (Myers *et al.* 1980). The use of DTW assumes that the input test utterance has been transformed into a temporally sequential list of parameter vectors (or frames) in the same way as the reference utterance. Each frame of the test utterance is then compared, using a simple distance metric, with several frames of the reference utterance. (The frames presented for comparison are dictated by predefined local and global contraints.) The frame which results in the lowest accumulated distance is matched to the input frame and the process continued. The result of this process is both a time-aligned utterance pair, and a final accumulated distance. This final distance describes the similarity between the test utterance and the reference utterance and may thus be compared against a pre-determined threshold to accomplish the speaker verification task.

*Parameter Selection*

Parameter selection and extraction is the heart of any speaker verification system design and three major constraints will affect selection of suitable parameters (Doddington 1976). These are variation in speech from the same speaker; similarity in speech from different speakers; and problems in accurate parameter measurement. Thus a desirable parameter will be one in which errors due to these constraints are reduced to a reasonable minimum. In a similar vein, it has been suggested (Wolf 1972) that desirable parameters must occur naturally and frequently in normal speech; be easily measurable; have small intra-speaker and large inter-speaker variance; must not change with time or be affected by the speaker's health; must not be affected by any reasonable background noise and must not depend on specific transmission characteristics.

One possible method of identifying such parameters is to determine those which contribute most to the perceptual bases of speaker recognition. While studies have been carried out to this end (Voiers 1964), more objective methods of determining an effective parameter exist.

The most common measurement method for the effectiveness of a particular parameter is known as the *F*-ratio and may be described as being the ratio of inter-speaker variation to intra-speaker variation, or, slightly more precisely, as the ratio of the variance of speaker means to the mean of speaker variances (Pruzansky and Mathews 1964).

This measure however, falls short of the goal of comprehensive parameter evaluation in that it takes no account of the inherent redundancy in a selected parameter set should there be any inter-parameter correlation. This problem has been tackled by Wolf (1972) by providing estimates of pairwise inter-parameter dependence by measuring the overlap of the parameter distributions for individual speakers. Another method is to consider the entire selected parameter vector and enter it into a multi-dimensional version of the *F*-ratio. The resulting figure from such a calculation is the divergence (Marill and Green 1963), and has been shown to be related to the accuracy of a speaker recognition system (Atal 1976). It should be noted however, that the divergence depends wholly on linear relationships (correlations) between parameters, whereas the method using pairwise comparisons could, in circumstances of strong non-linear dependence, be considered superior.

Finally, the most simple method of determining the effectiveness of a parameter is to employ it in a full speaker verification system and thus obtain a final system error-rate. Unfortunately this is also the most time-consuming method.

In contour analysis systems the possibility of weighting parameters according to their effectiveness can produce appreciable gains in system performance (Ney and Gierloff 1982, Tohkura 1986). The restriction of parameter weights to 1 or 0 can simplify the problem to that of parameter selection.

As mentioned above, inter-parameter correlation may reduce the efficiency of the parameter set though redundancy. Furthermore, it has been found (Markel and Davis 1979) that the inclusion of too large a number of parameters in the set can be detrimental to the performance of the resulting system. The investigation of methods whereby an optimal subset of some full parameter set may be found has thus been an area of intensive study. The simplest method of finding the optimal subset, the exhaustive search technique, will in most cases be prohibitively time-consuming. This technique requires all possible subsets (of size $R$) from the total parameter set (size $N$) to be evaluated in an operational speaker verification system, and the final results examined to find the optimal subset. This is impractical for parameter sets of all but the smallest size. A more efficient scheme, known as the knock-out technique (Sambur 1975a), reduces the search space considerably. This algorithm assumes the original number of candidates in the parameter set to be $N$. The effectiveness of the $N$ subsets of size $N-1$ is then determined and the parameter not evident in the most effective of these sets is discarded. The size of the set is thus reduced by one and the process repeated until all the parameters have been discarded. The reverse ordering of the discarded parameters is a rank ordering of their efficacy. Although this is an efficient technique, the resulting parameter set may still not be optimal. An improved method of selection is the adoption of dynamic programming (Nelson and Levy 1968, Chang 1973). This has been shown to be capable of producing parameter sets which, for a given dimensionality, are more effective than those produced by the knock-out method (Cheung and Eisenstein 1978).

If a high level of parameter redundancy is prevalent, it may be possible to remove parameters which are, on average, highly correlated by using the average correlation coefficient method of parameter set selection (Mucciardi and Gose 1971). Finally, the well-known technique of discriminant analysis (Wilks 1962, James 1985) has been used successfully in speaker recognition systems (Mohn 1971, Bricker *et al.* 1971) . This technique uses the *F*-ratio as a method of parameter evaluation and, employing eigenvector analysis, produces a new parameter space, the axes of which are linear combinations of the old parameter axes, and orthogonal to one another. Furthermore, it is often the case that only a subset of the orthogonal parameter set accounts for almost all the variance (among speakers). In such cases a small subset of orthogonal parameters will suffice.

*System Evaluation*

The methods used to evaluate the performance of speaker verification and speaker identification systems are very simple, but must be explained prior to any investigation of existing techniques. Two separate error measures are used in the description of speaker verification system performance. The first of these relates to rejection of a claimed identity, when in fact the claimed identity was true (false reject). This is commonly known as a Type I error. The Type II error is dependent upon the number of times a speaker is erroneously verified as a claimed individual, when in fact the identity claimed was false (false accept). Both Type I and Type II errors are expressed as percentages of a total number of attempted verifications. A more compact way of describing the performance of a system is simply to average the above two error figures, as the average error rate. The ratio of Type I to Type II errors is dependent upon the distance threshold chosen to mark the boundary of acceptance/rejection. For this reason well-defined thresholds are usually given in order that the error rate of a system is adequately described.

Another possible measure, termed the minimal error rate (MER), uses the intersection of the probability density functions produced by the within-speaker and between-speaker distances (Fakotakis, Dermatas and Kokkinakis 1986). This effectively minimises the sum of the false reject and false accept rates, thus providing a threshold at which the average error rate will be at a minimum.

The most common measure of error, however, is the equal error rate (EER). For each claimant, the overall distance from the input vector to the claimed reference vector can be calculated. A plot may then be made, with the overall distance as a variable along

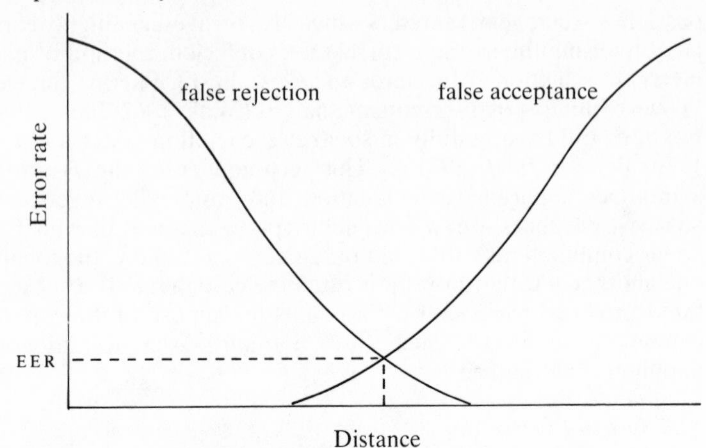


*Figure* 4.3. Typical false acceptance and false rejection error rates against threshold.

the abscissa. For a given abscissa value (overall distance) assumed to be the threshold distance, the percentage of genuine claimants whose distances were greater than this threshold value is plotted, together with the percentage of imposters whose distances were less than the same threshold. This results in a cumulative distribution function with respect to the overall distance (threshold). The intersection of the false reject (Type I) and false accept (Type II) curves gives the equal error rate and the corresponding (*a posteriori*) distance threshold (figure 4.3).

The nature of the *a posteriori* EER measure has been firmly challenged as being unrealistic when related to the problem of selecting a real-life decision threshold, and two methods for determining *a priori* threshold values have been suggested (Furui 1981a). The first of these involves setting an experimentally determined, fixed-value threshold to remain constant for all claimants. For the second, the optimum threshold is estimated for each claimant, based on his reference vector and a set of utterances from other speakers. The results produced using such thresholds show significant deviation from those produced using the EER. The EER, however, remains the most common form of performance measurement for automatic speaker verification systems, and will be used as a basis of comparison here.

In addition to the actual error statistic obtained, it is important to note the conditions under which the measurement was carried out. In the case of automatic speaker verification, major considerations include the type of population being used, e.g. casual imposter or trained mimic, and the relationship between the number of utterances made and the dimensionality of the parameter vector. Also, as in any pattern-recognition problem, the selection of the test and reference data sets must be considered.

While the majority of speaker verification systems are evaluated using casual imposters, the use of professional mimics provides a much more stringent criterion. The term casual imposter refers to an individual who attempts to gain false acceptance by speaking the correct text, but in his own natural voice. In one test (Lummis and Rosenberg 1972), four intensively trained mimics achieved a false accept rate of 27 %, compared with 1.2 % for casual imposters. The problem of identical twins appears to affect listener performance in (human) speaker verification tests to a greater extent than it does in automatic speaker verification system performance. In one study (Rosenberg 1973), listeners falsely accepted a twin as his brother 96 % of the time, whereas the automatic speaker verification system accurately discriminated between the two on all occasions. It remains true, however, that the ability of a speaker to disguise his voice interferes with accurate speaker verification both by human listeners (Reich and Duke 1979) and by automatic speaker verification systems (Hollien, Majewski and Hollien 1974).

A somewhat less obvious cause of difficulty in the evaluation of speaker verification systems is the consideration of sample size in relation to the dimensionality of the parameter vector (Sarma and Venugopal 1977). It has been suggested (Foley 1972) that under certain conditions the ratio of the number of utterances from each speaker ($S$) to the dimensionality of the parameter vector ($M$) should be at least 3 in order to avoid a large optimistic bias on the error rate. This ratio has not always been achieved in reported evaluations of speaker verification systems.

The relationship between the utterances used to train a system (reference) and those used to evaluate it (test) is of great importance. Several methods of separating the collected data have been suggested (Toussaint 1974). The most optimistic results are obtained when the same data is used both to train and test the system. A more common method, the holdout method, divides the data into two separate sets – one for reference and one for test. This method is inefficient, however, and other methods exist whereby the reference and test data are shuffled to make maximum use of the database. The effects of a time difference between reference and test data collection sessions is also important when evaluating an automatic speaker verification system. In general, the longer the hiatus between the collection of each set, the worse the error rate can be expected to be (Furui 1981b). These effects can be minimised however, by using reference data collected over several sessions or by adaptation strategies (Rosenberg and Shipley 1985).

*Classification of Systems*

The first system classification offered here is based on the dependency of the systems on the text of the input utterance. Systems for true text-independent speaker verification will allow any utterance to be employed (with no linguistic constraints of any sort) and perform successful speaker verification on that utterance. Such techniques are almost always based on statistical methods (time averaging, etc.).

Other candidates for the label of text-independent operation include systems which segment the speech in order to extract certain phonetic events or words for further analysis. This usually results in the restriction of the input speech to a small subset of words. These words are carefully chosen so as to include the particular phonetic event being used for speaker verification.

The earliest text-dependent speaker verification techniques were restricted to fixed (usually all voiced) utterances. These most frequently constituted the previously described contour matching systems.

While the primary reason for the development of text-independent systems was for their use in speaker identification system situations involving an unco-operative or unaware speaker, use of such techniques in speaker verification systems could simplify the entire operation. In general, text-independent systems require less processing time than do their text-dependent counterparts, which can be partly attributed to the removal of the need for time registration, and whole contour dissimilarity measurements (Foil and Johnson 1983). The simplest form of text-independent operation involves the averaging of each parameter within a vector over a considerable duration of input speech. This is done in the hope that the linguistic content of the signal will be averaged out leaving an envelope of speaker-dependent information. The quantities produced, however, cannot include temporal information, and thus some amount of speaker-dependent information is unavoidably lost.

Although the above classification boundary is of importance, it is of more relevance when considering existing systems which are based on a variety of speech parameterisations, to classify these systems on the basis of the main parameters being extracted, since these provide the most basic reason for differences in performance amongst systems and thus offer an improved classification of techniques.

The first systems examined here revolve around the pitch, or fundamental frequency of the speech. A large number of systems have been developed using this parameter (Wolf 1972; Das and Mohn 1971; Rosenberg and Sambur 1975; Bunge 1977b; Hunt, Yates and Bridle 1977; Atal 1972; Lummis 1973; Luck 1969; Markel, Oshika and Gray 1977; Markel and Davis 1979).

The second parameter commonly used in verification systems (particularly earlier ones) is the gain or intensity of the signal. Although a number of systems have incorporated this parameter (Das and Mohn 1971, Rosenberg and Sambur 1975, Lummis 1973, Markel and Davis 1979), it is used in all cases in conjunction with other parameters.

The short-time spectrum of speech can itself be used as a parameter and its time average and correlation characteristics, over a number of frames, are also a commonly used parameters (Li, Dammann and Chapman 1966; Li, Hughes and House 1969; Luck 1969; Das and Mohn 1971; Li and Hughes 1974; Hollien, Majewski and Hollien 1974; Doddington 1976; Bunge 1977b).

The resonant frequencies of the vocal tract and nasal cavities (formants) along with their bandwidths have, despite their relative difficulty in measurement, been used to a large extent (Das and Mohn 1971, Lummis 1973).

Finally, linear prediction coefficients are not only a parameter in their own right, but also a starting point for many other parameters and have become very popular in speaker recognition systems (Atal 1974; Rosenberg and Sambur 1975; Sambur 1976; Wohlford, Wrench and Landell 1980; Mohankrishnan, Shridhar and Sid-Ahmed 1982; Li and Wrench 1983; Soong *et al.* 1985). Cepstral coefficients are experiencing similar popularity (Bunge 1977b; Luck 1969; Furni 1981a; Gish *et al.* 1985).

Other parameters, less frequently adopted, are nasal co-articulation (Su, Li and Fu 1974), reflection coefficients and the relative timing of phonetic events (Sambur 1975a).

### *Pitch and Gain*

Before use is made of either pitch or gain they must be removed from the speech signal and this, in the case of pitch, is a difficult task. Pitch-detection algorithms are manifold and may be broadly classed as being either time-domain or short-term analysis (Hess 1983). Major methods in the time domain include first harmonic extraction and temporal structure investigation (Gold and Rabiner 1969; Sutherland, Jack and Laver 1986). Short-term analysis techniques produce, in general, a pitch estimate formed by averaging over a short interval (or frame). Autocorrelation is one such technique, relying on the enhancement of the signal periodicity (Rabiner 1977), and frequency-domain techniques analyse the glottal harmonics to determine the basic pitch (again on a short-term basis; Seneff 1978). In multiple spectrum analysis (cepstrum) the peak corresponding to the glottal harmonics gives a relatively reliable pitch estimate (Noll 1967). Accurate evaluation of the performance of pitch detection algorithms is still a difficult task (Rabiner *et al.* 1976) (although the laryngograph device has aided such study considerably; Hess and Indefrey 1984) and their performance under conditions of noise remains variable (Oh and Un 1984).

The gain, or intensity, of a speech signal is basically a measure of the energy contained in the speech waveform and is thus relatively easy to measure. In a major study (Lummis 1973) the intensity was found by summing the the squares of the first 60 points of a fast Fourier transform (FFT). Each output was of bandwidth 10 Hz, and thus the gain corresponded to the 600 Hz intensity pattern of the speech.

In one published study (Sambur 1975a) the entire pitch contour of the sentence *cash this bond please* was examined. A simple zero-crossing time-domain technique was employed for pitch extraction. The resulting contour was split into four sections and both the slope and average value of each section examined. A great intra-speaker variation of pitch across different recording sessions was found, although such variations always remained within certain boundaries. It was also noted that average values of the fundamental frequency tended to give better results than slope parameters; a fact also noted in more recent research (Hunt, Yates and Bridle 1977).

An early study of automatic speaker recognition based on pitch contours alone (Atal 1972) provided proof that pitch contours were adequate for recognition purposes. In this study, the short-time correlation analysis of the (cubed and low-pass filtered) speech input was used to determine the basic pitch. Each contour was linearly time-normalised to 2 seconds and represented by 40 samples uniformly spaced across the utterance. Owing to the high correlation between neighbouring pitch samples, a data-compression technique was adopted. On sixty attempts, by ten speakers, the identification error rate was found to be 3%.

In a further study (Luck 1969), pitch measures gained by

cepstral methods were used in a speaker verification system. The method of segmentation prior to extraction was adopted and the phrase *my code is* was split up. Various parameters, including pitch, were extracted from the word *my*. It was found that a vector consisting of 32 cepstral coefficients, combined with the pitch and duration of the word, was the most effective combination (compared with all cepstral coefficients and spectral information, etc.). In further tests using this combination, it was found that EERs of between 6% and 13% were possible – depending on the claimed speaker.

The use of an entire gain contour to time-register other contours, including pitch, has been examined (Lummis 1973). In this study the pitch period, the gain and first, second and third formants were extracted by FFT cepstral methods (Schafer and Rabiner 1970). In the cepstral domain, the fundamental frequency component was measured and removed, and a return to the frequency domain was made in order to determine the formant frequencies.

Using five different types of distance measure on the pitch and gain contours, an EER of 0.5% was produced. An EER of 8.4% was achieved using 3 formant contours. However, as only 152 independent utterances were used in this study, its statistical significance is in doubt. Although the author is bold enough to claim that formant data may be of little use in verification systems, a study using professional mimics (Lummis and Rosenberg 1972) found that formants are of use in combating trained imposter acceptance.

There is further evidence of the successful marriage of pitch and gain (Rosenberg and Sambur 1975, Rosenberg 1976). In this case the pitch was extracted using the parallel processing method (Gold and Rabiner 1969). The time required by this system for pitch and gain analysis alone was 25–30 times the length of the utterance. Using selected distance measures, this system achieved 1% EER on pitch and gain contours. A mimic acceptance of 15.6%, however, was reported.The above EER was achieved on a database of only thirty-two speakers, and a more representative seventy-seven-speaker database gave 2.5% as the EER.

The use of pitch and gain has not been restricted to text dependent systems: one study (Markel, Oshika and Gray 1977) used the mean and standard deviation of the pitch and gain variation (along with reflection coefficients). The pitch was determined by cepstral methods. Here the gain was found to be the least effective of these parameters, the pitch the second most effective, and the spectrally based parameters the most effective. In further experiments (Markel and Davis 1979), again using pitch, gain and reflection coefficients (their means and standard deviations), an EER of 7.2% was achieved using about 40 seconds of completely unconstrained text. When the gain parameters were removed, however, the EER improved to 4.25%.

The pitch of a speech signal is therefore heavily dependent upon speaker identity. Its relative ease of extraction and its immunity to the frequency characteristics of the transmission medium has made it a suitable parameter for many speaker verification systems. The stability of its mean value (over several seconds) has allowed its use in text-independent systems. It is however, susceptible to both mimicry and day to day variation with the mood of the speaker. The use of the gain of a speech segment has been less successful. Initially, in text-dependent form, it proved an adequate contender, but has seen less use in recent systems.

*Spectral Parameters*

Spectral transformations on successive frames of input speech represent one of the most commonly used parameter extraction methods in speaker verification systems. This parameter matrix may be used in text-dependent systems by subjecting it to a dynamic time-warping routine or, for text-independent systems, the spectra may by averaged over the time axis such that linguistic information is averaged out and only speaker-dependent information remains. The applicability of the long-term averaged spectrum to the speaker verification problem has been studied several times.

A major experimental automatic speaker recognition system, AUROS (Bunge 1977b), used the long term averaged spectra of 11 seconds of speech (from fifty speakers) to attempt text-independent speaker identification. It succeeded with an error rate of about 1%. Another study (Furui, Itakura and Saito 1972) used cepstral coefficients which were derived from long-term average spectra. However when employed in a text-independent system (Wohlford, Wrench and Landell 1980), an identification error rate of 25% (17 speakers) was achieved using 13 seconds of speech. When the speech duration was extended to 34 seconds, however, the error rate improved to 5%. Further studies (Li and Hughes 1974; Hollien, Majewski and Hollien 1974) have used the speech spectrum formed from 30 seconds of speech.

One major successful speaker verification system (Doddington 1974, 1976, 1985) making use of spectral parameters originally used a 16-channel analogue filter bank as its pre-processor, although later versions use a 14-channel digital version. This system requires the input text to be from a small group of monosyllabic words. The vowel-portion of each word is analysed and stored as the reference vector (template). The input word of a claimant is then scanned against this template and the point of minimum distance compared with a threshold. This system has been operational as an access control mechanism for over a decade now (Doddington 1985), and has allowed some particularly interesting figures to be measured. For example, only 25% of the user population exhibit rejection rates greater than the average rejection rate, implying that the population consists of *sheep* (the section of the population who have little difficulty in being correctly accepted) and *goats* (the section of the population which suffers a high false rejection rate) (Cameron and Millar 1986). Overall false rejection rate is reported as being 0.9% for a false acceptance (of casual imposters) of 0.7%.

A similar system, although employing LPC-derived spectral coefficients, employs a two-tier verification algorithm (Naik and Doddington 1986). Analysis consists of a phrase being reduced to 10 principal spectral components every 20 ms, which are then passed to a dynamic time-warp routine. The user must first successfully verify (within three attempts) using a fixed text utterance. A secondary verification is then carried out using a variable 5-digit sequence spoken in a continuous fashion. Within-speaker variability due to coarticulation is limited by ensuring the digit pairs employed in each sequence have distinctive boundaries. In use, on a forty-speaker database, a false reject rate of 0.75% for a false accept rate of 0.04% has been reported. Later evaluation on a 200-speaker database (Naik and Doddington 1987) produced similar results.

Further reported results showed the use of spectral coefficients gained by a 16-channel filter bank to give an EER of 2.99% (Mohankrishnan, Shridhar and Sid-Ahmed 1982); and a false rejection rate of 0.6% for a false acceptance of 0.3% again using spectral features (Feix and DeGeorge 1985). In the latter example, a speaker database of only eighteen persons was used, and the results were obtained under supervision.

Thus, the spectrum, be it long-term averaged or short term, appears to be excellent parameter for automatic speaker verification and may be described as being easy to measure, having a good level of discrimination and being difficult to mimic. It is, however, susceptible to the varying frequency characteristics of any input channel, unlike parameters such as pitch. It may thus be considered as suitable only for systems where strict control of the transmission medium quality can be maintained.

*Formants*

Formants may be described as the resonances of the vocal tract and the nasal cavities. They vary in frequency, relative amplitude, and bandwidth according to the speech sound being made and the speaker making it. The latter fact is of interest in automatic speaker verification systems. Extraction and measurement of formant frequencies poses a difficult problem in signal processing. Their presence in the spectrum as peaks may be masked by the harmonics of the excitation signal, and thus some smoothing is required prior to the use of peak picking algorithms. Two major spectrum smoothing methods are used.

If the cepstrum of the speech signal is formed, its high-time peaks relate principally to the harmonics of the excitation signal. It is thus a simple step to remove this peak, thus achieving homomorphic deconvolution, whereupon a return to the spectral domain will allow the spectral peaks due to formants to be identified more simply (Schafer and Rabiner 1970).

By applying a linear predictive filter to the speech signal, the inverse transfer function of a model of the vocal tract may be obtained and transformed to the spectral domain. As the representation of each peak in the spectrum requires a complex conjugate-pole pair, the order of the linear predictive filter imposes a strict constraint on the smoothness of the resulting model. The order of the filter must thus be carefully chosen to both remove the effects of excitation harmonics, yet retain sufficient resolution of the formant peaks. Alternatively, the direct factorisation of the predictor polynomial may be used to find the spectral poles and hence the formants. Linear predictive analysis allows accurate measurements of formants to take place providing the number of poles required to model the speech is known in advance. However, the spectral zeros produced during nasal sounds cannot be modelled by the linear predictive technique. The use of cepstral smoothing allows complete freedom in the number of peaks which may be present in the spectral model at one time.

Having determined the position of the peaks in the smoothed spectrum, and measured their amplitude and bandwidth, the results may be employed in automatic speaker verification to good effect.

Nasal consonants (/m/, /n/) are reported (Glenn and Kleiner 1968) as being particularly appropriate for the speaker verification task. However, the measurement of the poles of the transfer function during nasals is particularly difficult due to the complication of the oral cavity acting as an anti-resonance. One study (Wolf

1972), attempted to use individual filter values in the spectral regions where formant peaks could be expected, and normalised the values obtained by subtracting the energy function of the middle of the following vowel. Using this method, it was shown that the spectral regions exhibiting the highest *F*-ratio correspond to features such as the formants at 1, 2 and 3 kHz in /n/. Another study (Sambur 1975a) measured precisely the same nasal by means of linear predictive analysis and similar findings were made. Although nasal formants are amongst the best parameters available for speaker verification, their sensitivity to physiological changes is high.

The use of formants measured directly from the unsmoothed speech spectrum has also been attempted (Das and Mohn 1971). No figures are available for the system operating on formant data only, but when combined with pitch and filter averages a figure of 1% average error, with 10% no decision, was obtained.

The technique of cepstral smoothing to determine the frequencies of the first three formants (F1, F2, F3) has been used (Lummis 1973) in a text-dependent speaker verification system. The values of the formant frequencies were traced as a function of time, for a given sentence, thus forming contours. An EER of between 8.4% and 10% was obtained for various combinations of the formants on an all-male database of forty speakers.

Finally, a 12-pole model linear prediction analysis has been used (Rosenberg and Sambur 1975) to represent formants on the basis that there was no *a priori* reason for using formant contours rather than the contours of the the LPC filter coefficients from which they were derived. Results showed that the best formant contours (F1, F2 and F4) were slightly less informative than the contours of the best three of the filter coefficients. Only four speakers were included in this test however, thus no firm conclusions could be reached. Nevertheless, the 50% reduction in processing time resulting from the use of filter coefficients rather than formants suggests strongly the use of linear prediction coefficients as a possible parameter set.

### *Linear Predictive Coefficients*

One relatively fast and certainly accurate method of speech signal analysis is linear predictive coding (Atal and Hanauer 1971, Schroeder 1985, Makhoul 1975). The theory of this technique is closely linked to models of the vocal tract and relies upon the fact that any particular speech sample may be predicted by a linear combination of several previous samples plus a predictor residual term. The weights applied to each of the previous speech samples in the combinations are known as the linear prediction coefficients (LPC), and are calculated in such a way as to minimise the predictor residual term. Several methods of carrying out this task exist, and are explained in Rabiner and Schafer (1978).

The usefulness of linear predictive coefficients can be extended by their tranformation into other parameters. Cepstral coefficients are obtained by use of a simple relationship with the LPC (Schroeder 1981), while the partial correlation (parcor) coefficients may be obtained from a set of linear prediction coefficients and vice versa. The notion of the vocal tract as being a set of connected lossless tubes can also produce the parameters of log area ratio. These are equal to the log of the ratio of the areas of adjacent tubes in such a model, and can be derived from the parcor coefficients.

The first major study into the use of LPC with relation to automatic speaker recognition (Atal 1974) was carried out using data from 60 utterances (six repetitions of the same sentence by ten different speakers). Each utterance was divided into 40 uniformly spaced time frames, and 12 predictor coefficients produced for each frame. An identification error rate of 36.2% on a single segment of length 50 ms was achieved.

In a later text-dependent system (Rosenberg and Sambur 1975), a 12-coefficient predictor analysis of the sentence *we were away a year ago* was carried out. A high correlation between adjacent coefficients was noted and confirmed experimentally by the fact that results did not improve appreciably using all 12 coefficients over using only the best set of 3. The effectiveness of the LPC contours was clearly exhibited when, on choosing two coefficient contours and combining them with pitch and gain, an EER of 0.9% was achieved, compared with 2.5% for pitch and gain alone.

Linear prediction coefficents exhibit significant redundancy between adjacent contours. While in the above experiment, the solution was simply to use only two (clearly different) contours, a more efficient method is to orthogonalise (Sambur 1975b) the parameters prior to their use. This approach has been taken (Sambur 1976) and it was noted that only a small subset of the resulting orthogonal parameters exhibited any significant variation over the duration of an utterance; the remaining parameters being adequately specified simply by their means. Here, a single fixed-text utterance was subjected to a 12th-order linear predictive analysis. The closely related parcor and log area ratio coefficients were also produced. The orthogonalisation of the above parameters was followed by the calculation of their mean values. In identification tests, on a speaker database of twenty-one male speakers, an error rate of 3.2% was obtained through the use of the LPC orthogonalised parameters (providing that more than six were used in the comparative stages). The use of orthogonal parcor coefficents gave an improved error figure of 0.8%, as did the use of the orthogonal log area coefficients. In verification mode, EERs of 4.8%, 0.8% and 0.8% were obtained from the orthogonalised LPC, parcor and log area coefficients respectively.

The success of the above experiments prompted attempts at text-independent identification using the same arrangement. In this case, the speakers recorded additional sentences (of different text) from those recorded for the initial experiment. Identification error rates (no verification tests were conducted) were reported as being 6% using orthogonal parcor coefficients. A more recent study of the orthogonalisation of linear prediction coefficients (Bogner 1981), used 14 reflection coefficients, but did not produce such impressive results as its predecessor. Nevertheless, an EER of 5% was obtained.

As was mentioned briefly at the start of this section, various other parameters are closely related to the coefficients of linear predictive analysis. The most useful of these has frequently proved to be the cepstral coefficients.

*Cepstrum Coefficients*

The cepstrum of a digitised input signal may be obtained in one of two ways. The inverse discrete Fourier transform (IDFT) of the log magnitude of the spectral representation of the original signal is one method, and the manipulation of linear predictive coefficients is the other. In both cases, the process results in a representation of the speech signal within which relatively easy deconvolution of the vocal tract from the excitation signal can take place.

The method involving the IDFT was used in an early study (Luck 1969) in which 16 cepstral coefficients were measured during production of the word *my* by twenty-three male imposters. When compared, using a nearest-neighbour classification technique, to the stored coefficients of a genuine speaker, an EER of approximately 6% resulted.

The use of LPC-derived cepstral coefficients was explored (Atal 1974) as a continuation of the study used to determine the effectiveness of the linear predictive coefficients themselves. The cepstral coefficients actually proved to be the more effective of the two sets. An average verification error rate of 10% was obtained for 200 ms of speech, but by allowing an average to be taken over 500 ms a reduction of the error rate to 5 % was achieved. By separating each of 60 utterances into 40 equal segments, and then recombining the 40 segments in random manner, a constrained text-independent test was carried out. A set of 12 cepstral coefficients were gained for each segment. Averaged over 500 ms, an identification error rate (on ten speakers) of 28 % was noted, and improved to 7 % when averaging was carried out over 2 seconds of speech (compared with better than 2 % error for the text-dependent case).

A recent, and detailed, survey (Furui 1981a) of the capabilities of cepstral coefficients relied upon the production of cepstral coefficients from LPC. A fully phrase-dependent, contour-based, time-warping system structure was adopted. The conversion from LPC to cepstral coefficients was carried out on a frame-by-frame basis and the technique of subtracting the time average of each coefficient (over the duration of the input utterance) from its value in each segment could be used for channel normalisation if deemed necessary. The time contours obtained from the above conversions were then expanded by an orthogonal polynomial representation over short time segments. The three polynomial coefficents produced – PC0, PC1, PC2 – represent mean value, slope and curvature of the contour (in each segment) respectively. The selection of the cepstral or polynomial coefficients to be used was made on the basis of a statistical analysis of the effectiveness of each component. Using the sentence *we were away a year ago*, spoken by ten genuine female users and forty female imposters over a standard telephone line, a preselected distance threshold resulted in a false rejection rate of 0.29 %, and a false acceptance rate of 0.43 %, for LPC-cepstral coefficients. In order to carry out a comparison, a system was developed in which the cepstral coefficients were derived by means of an FFT process. The equivalent error rates obtained were 0.29 % and 0.33 %. In view of this result, it is suggested that use of the FFT-based cepstral analysis is unnecessary, due to the fact that it takes twice as long to derive as the LPC method. It should be noted that in a later publication (Rosenberg and Shipley 1985) other workers claim that a further, unpublished, evaluation of this (LPC-based) system has taken place, and that verification error rates of 4 % were achieved.

The cepstral coefficients of a speech signal may thus be seen to hold important speaker-dependent information. Also of interest is their apparent independence from the pitch and gain of the speech. Correlation coefficients between the error rate distributions of a major system (Furui 1981a) using cepstral coefficients and using pitch and gain have been calculated and found to be in the order of

0.3. This suggests that any system employing cepstral parameters may well benefit from the addition of pitch or gain parameters.

*Nasal Co-Articulation*

A detailed study of the use of nasal co-articulation with relation to automatic speaker recognition systems has been carried out (Su, Li and Fu 1974). Nasal consonants are produced with the oral cavity closed, and energy radiated mainly from the nasal cavity. Nasal co-articulation can occur when a nasal consonant is followed by a vowel. In this case, the tongue anticipates the following vowel segments and moves to the vowel position during nasal phonation (Moll and Daniloff 1971).

In an experiment to determine the usefulness of the above measurments with relation to speaker dependence, it was determined that the nasal /m/ provides better measurements, because it produces greater co-articulation. (The tongue is completely free to move to the following vowel position, since oral closure is at the lips.) A nasal coarticulation measure was used in an experiment involving ten speakers. This, it was claimed, produced zero errors. However, a similar experiment using thirty-three speakers found the identification error rate to be 15%.

*Vector Quantisation*

The technique of vector quantisation (VQ) has been used extensively in signal processing (Gray 1984). Its essential feature, that of data compression, has prompted its adoption in speech coding systems (Buzo *et al.* 1980), while its similarity measures have provided reason for its application to speaker verification (Soong *et al.* 1985; Soong and Rosenberg 1986; Buck, Burton and Shore 1985; Rosenberg and Soong 1986; Higgins and Wohlford 1986).

Vector quantisation, as the term suggests, quantises the parameter space occupied by the parameter vectors gained from a speaker (eg. $M$-dimensional spectral coefficient vectors) into several partitions. The centroid vector of each partition may then be used to represent any vectors which fall within its catchment area. The set of centroid vectors is known as a codebook. The formation of a codebook, by determining the partitions of the parameter space, is a complex task and depends upon the minimisation of some distortion figure. The average distortion over a reference set of data may be found by calculating the costs of representing an input vector by the centroid vector of its partition. The cost may be determined using a distance metric such as the Euclidean distance.

In a speaker verification system, codebooks are formed for each speaker using the reference utterances. In the text-independent case, codebooks may be formed from widely varying input text, in the hope that most phonemic elements will be represented. For text-dependent systems, however, a code-book may be formed using multiple repetitions of one utterance. The actual task of verification is carried out by calculation of the average distortion of an input (test) utterance against the claimed speaker's codebook. The average distortion figure may then be compared against some threshold value to authenticate the speaker's identity.

The effectiveness of the above technique was demonstrated in one study (Soong *et al.* 1985), which used a 100-speaker database. A constrained vocabulary was chosen (10 digits) and the data gathered over the standard telephone lines. The analysis technique employed resulted in an 8-dimensional LPC parameter vector. Although the verification paradigm was not specifically investigated, identification trials using a codebook of 64 vectors and a test token of 10 digits gave an error rate of only 1.5%. The number of vectors used in the codebook has considerable effect upon the accuracy of the system. In the above experiment, a codebook of only two vectors resulted in an error rate of 34%.

In a later study (Rosenberg and Soong 1986) the performance of a similar system was examined in the context of speaker verification in both text-dependent and text-independent modes. Again, a 100-speaker data base was employed. In the text-independent case, a 7-digit long test token resulted in an EER of 2.2%. For text-dependent operation, this figure improved to become 0.3%. In both the above experiments, adaptation was employed in order to mitigate the effects of the natural variation in speaker behaviour with time.

*Practical Implementation Problems*

Two major practical issues are the effects of the user interface upon system performance, and the use of speaker verification systems over public telephone lines.

In tests with a voice-controlled access system (Doddington 1976) entry was controlled by the use of a large booth with electrically controlled exit and entrance doors. An acoustic prompt was used for such instructions as *louder please* and *verified thank-you,* as well as providing the actual text to be spoken. The microphone was placed in such a way as to require the user to stand while verifying his identity.

A detailed study (Feix and DeGeorge 1985) of the user interface issue found that wall-mounted microphones were not optimal, in that users had difficulty in positioning themselves comfortably, and exhibited nervousness during their use. The use of telephone handsets was found to be much more effective, although adaptations to standard units were of course required. The use of visual prompting, compared with that of acoustic, was found to be most successful. This study suggests the visual solution is to be preferred, because it requires less attention of the user. The effects of the user imitating an acoustic prompt were also eliminated. The importance of user interface issues in automatic speaker verification systems is undeniable, and ideal solutions are yet to be found.

The use of automatic speaker recognition over normal dial-up telephone lines has received considerable attention from many experimenters (Sambur 1976; Furui 1981a; Federico, Ibba and Paoloni 1987). One study (Sambur 1976), using a database of twenty-one speakers, with utterances transmitted across telephone lines, and a 12th-order LPC analysis, produced 16.7% as the best possible error rate. Increasing the LPC analysis to 14th-order, in the hope that 4 poles would model the telephone media line, and 10 the speech signal, and by modifying the distance metric slightly, an error rate of 4% was achieved. This study was text-dependent.

Several other studies have been made on the use of speaker recognition systems over telephone and other communication lines (Hunt, Yates and Bridle 1977; Hunt 1983; Krasner *et al.* 1984; Gish *et al.* 1985, 1986).

### *Concluding Remarks*

When drawing a discussion such as this to a close, final comments must be carefully balanced between outright speculation, and simple repetition of the factual content already given. For this reason, a brief study of what has been, and what may be, done on a hypothetical system will be made.

The signal processor has in the past varied from analogue filter banks to array processor fast Fourier transforms. Both of these devices remain options for inclusion in present-day systems, and the ever increasing speed of digital signal-processing (DSP) devices ensures their usefulness in small, real-time speaker verification systems.

The engineering problems of parameter extraction, and the phonetic problem of identifying reliable speaker-dependent parameters are linked. The perfect compromise of an easy-to-extract parameter which is heavily speaker-dependent has yet to be found, although some parameters are proving to be worthy candidates (Bogner 1981, Furui 1981a). The use of more and more parameters may appear as an option for increased accuracy (in view of the fact that future DSP chips will have increased capacity for such an approach), but must be examined with care.

The use of reference profile adaptation to combat long-term intra-speaker variance has been used (Doddington 1976), where a new reference profile was formed from a weighted average of the input vector and old reference on each successful attempt. No figures are available regarding the effect of this upon error rates, but it was stated that the memory requirement more than doubled over that which was required for no adaptation. A study of a similar adaptation routine (Rosenberg and Shipley 1985) showed the average self-distance of a speaker to increase by approximately 14% of its original value over a given time interval (ten trials) when no profile adaptation was carried out, and by only approximately 4% of its original value when adaptation was used.

The decision between text-dependent and text-independent operation must, at present, depend upon the application environment of the final system. The use of text-dependent systems will allow the user to become accustomed to a phrase, and the speech period need be no longer than 2 seconds. For text-independent systems, however, present systems require more speech input in order that the speaker dependent parameters can stabilise. Moreover, their accuracy does not achieve that which is required for forensic applications.

Automatic speaker verification is now a rapidly developing field. Its use in commerce, industry and perhaps domestic environments in our future society seems likely to become increasingly common.

## REFERENCES

Atal B. S. (1972) Automatic speaker recognition based on pitch contours. *J. Acoustical Society of America 52 (6)*, 1687-97.

—— (1974) Effectiveness of linear prediction characteristics of the speech wave for automatic speaker identification and verification. *J. Acoustical Society of America 55 (6)*, 1304-12.

—— (1976) Automatic recognition of speakers from their voices. *Proc. IEEE 64 (4)*, 460-75.

Atal, B. S. & S. L. Hanauer (1971) Speech analysis and synthesis by linear prediction of the speech wave. *J. Acoustical Society of America 50 (2)*, 637-55.

Bogner, R. E. (1981) On talker verification via orthogonal parameters. *IEEE Trans. Acoustics, Speech and Signal Processing 29 (1)*, 1-12.

Bricker, P. D., R. Gnanadesikan, M. V. Mathews, S. Pruzansky, P. A. Tukey, K. W. Wachter & J. L. Warner (1971) Statistical techniques for talker identification. *Bell System Technical J. 50(4)*, 1427-54.

Buck J. T., D. K. Burton & J. E. Shore (1985) Text dependent speaker recognition using vector quantization. *Proc. IEEE Conf. Acoustics, Speech and Signal Processing ICASSP-85*, 391-4.

Bunge, E. (1977a) Speaker recognition by computer. *Philips Technical Review 37(8)*, 207-19.

Bunge, E. (1977b) Automatic speaker recognition system AUROS for security systems and forensic voice identification. *Proc. 1977 International Conference on Crime Countermeasures*, 1-7.

Buzo, A., A. H. Gray, R. M. Gray & J. D. Markel (1980) Speech coding based upon vector quantization. *IEEE Trans. ASSP-28*, 562-74.

Cameron, I. R. & P. C. Millar (1986) Speaker recognition, fact or fiction? *Digest of IEE Colloquium on Man-Machine Interfaces in Computer Security*, 80.

Chang, C. (1973) Dynamic programming as applied to feature subset selection in a pattern recognition system. *IEEE Trans. SMC-3(2)*, 166-71.

Cheung, R. S. & B. A. Eisenstein (1978) Feature selection via dynamic programming for text independent speaker identification. *IEEE Trans. ASSP-26(5)*, 397-403.

Cover, T. M. & P. E. Hart (1967) Nearest neighbor pattern classification. *IEEE. Trans. IT-13(1)*, 21-7.

Das, S. K. & W. S. Mohn (1971) A scheme for speech processing in automatic speaker verification. *IEEE Trans. AU-19*, 32-43.

Doddington, G. R. (1974) Speaker verification. *Final Report Rome Air Development Center Griffiss AFB NY, Tech. Rep RADC*, TR-74-179.

—— (1976) Personal identity verification using voice. *Proc. Electro-76*, 22-4.

—— (1985) Speaker recognition – identifying people by their voices. *Proc. IEEE 73(11)*, 1651-64.

Fakotakis, N., E. Dermatas & G. Kokkinakis (1986) Optimal decision threshold for speaker verification. *Proc. EUSIPCO 1986*, 585-7.

Federico, A., G. Ibba & A. Paoloni (1987) A new automated method for reliable speaker identification and verification over telephone channels. *Proc. IEEE Conf. Acoustics Speech and Signal Processing ICASSP-87*, 1457-60.

Feix, W. & M. DeGeorge (1985) A speaker verification system for access-control. *Proc. IEEE Conf. Acoustics, Speech and Signal Processing, ICASSP-85*, 399-402.

Foil, J. T. & D. H. Johnson (1983) Text independent speaker recognition. *IEEE Communications Magazine 21 (9)*, 22-5.

Foley, D. H. (1972) Considerations of sample and feature size. *IEEE Trans. IT-18(5)*, 618-26.

Furui, S. (1981a) Cepstral analysis technique for automatic speaker verification. *IEEE Trans. ASSP-29(2)*, 254-71.

—— (1981b) Comparison of speaker recognition methods using statistical features and dynamic features. *IEEE Trans. ASSP-29 (3)*, 342-50.

Furui, S., F. Itakura & S. Saito (1972) Talker recognition by long-term average speech spectrum. *Electronics and Communications in Japan 55A (10)*, 54-62.

Gish, H., K. Karnofsky, M. Krasner, S. Roucos, R. Schwartz & J. Wolf (1985) Investigation of text independent speaker identification over telephone channels. *Proc. IEEE Conf. Acoustics, Speech and Signal Processing ICASSP-85*, 379-82.

Gish, H., M. Krasner, W. Russell & J. Wolf (1986) Methods and experiments for text-independent speaker recognition over telephone channels. *Proc. IEEE Conf. Acoustics, Speech and Signal Processing ICASSP-86*, 865-8.

Glenn, J. W. & N. Kleiner (1968) Speaker identification based on nasal phonation. *J. Acoustical Society of America 43 (2)*, 368-72.

Gold, B. & L. Rabiner (1969) Parallel processing techniques for estimating pitch periods in the time domain. *J. Acoustical Society of America 46 (2)*, 442-8.

Gray, R. M. (1984) Vector quantization. *IEEE ASSP Magazine 1 (2)*, 4-29.

Hess, W. (1983) *Pitch Determination of Speech Signals, Algorithms and Devices*. Springer-Verlag.

Hess, W. & H. Indefrey (1984) Accurate pitch determination of speech signals by means of a laryngograph. *Proc. IEEE Conf. Acoustics, Speech and Signal Processing ICASSP-84* 18 (B), 1-4.

Higgins, A. L. & R. E. Wohlford (1986) A new method of text-independent speaker recognition. *Proceedings IEEE Conf. Acoustics, Speech and Signal Processing ICASSP-86*, 869-72.

Hollien, H., W. Majewski & P. Hollien (1974) Speaker identification by long-term spectra under normal, stress and disguise conditions. *J. Acoustical Society of America 55*, S-20.

Hunt, M. J. (1983) Further experiments in text-independent speaker recognition over communications channels. *Proc. IEEE Conf. Acoustics, Speech and Signal Processing ICASSP-83*, 563-6.

Hunt, M. J., J. W. Yates & J. S. Bridle (1977) Automatic speaker recognition for use over communication channels. *Proc. IEEE Conf. Acoustics Speech and Signal Processing ICASSP-77*, 764-7.

James, M. (1985) *Classification Algorithms*. Collins, London.

Jesorsky, P. (1978) Principles of automatic speaker-recognition, in *Speech Communication with Computers*. L. Bok (ed.), Carl Hanser Verlag/MacMillan.

Krasner, M., J. Wolf, K. Karnofsky, R. Schwartz, S. Roucos & H. Gish (1984) Investigation of text-independent speaker identification techniques under conditions of variable data. *Proc. IEEE Conf. Acoustics, Speech and Signal Processing ICASSP-84 18 (B)*, 51-4.

Li, K. P. & G. W. Hughes (1974) Talker differences as they appear in correlation matrices of continuous speech spectra. *J. Acoustical Society of America 55*, 833-7.

Li, K. P. & E. H. Wrench (1983) An approach to text-independent speaker recognition with short utterances. *Proc. IEEE Conf. Acoustics, Speech and Signal Processing ICASSP-83*, 555-8.

Li, K. P., J. E. Dammann & W. D. Chapman (1966) Experimental studies in speaker verification using an adaptive system. *J. Acoustical Society of America 40 (5)*, 966-78.

Li, K. P., G. W. Hughes & A. S. House (1969) Correlation characteristics and dimensionality of speech spectra. *J. Acoustical Society of America 46*, 1019-25.

Luck, J. E. (1969) Automatic speaker verification using cepstral measurements. *J. Acoustical Society of America 46 (4)*, 1026-32.

Lummis, R. C. (1973) Speaker verification by computer using speech intensity for temporal registration. *IEEE Trans. ASSP-21 (2)*, 80-8.

Lummis, R. C. & A. E. Rosenberg (1972) Test of an automatic speaker verification method with intensively trained professional mimics. *J. Acoustical Society of America 51*, 131.

Mahalanobis, P. C. (1936) On the generalised distance in statistics. *Proc. National Inst. of Science Institute, India, 12*, 49-55.

Makhoul, J. (1975) Linear prediction: a tutorial review. *Proc. IEEE 63 (4)*, 561-80.

Marill, T. & D. M. Green (1963) On the effectiveness of receptors in recognition systems. *IEEE Trans. IT-9*, 11-17.

Markel, J. D., B. T. Oshika & A. H. Gray (1977) Long-term feature averaging for speaker recognition. *IEEE Trans. ASSP-25 (4)*, 330-7.

Markel, J. D. & S. B. Davis (1979) Text-independent speaker recognition from a large linguistically unconstrained time-spaced data base. *IEEE Trans. ASSP-27 (1)*, 74-82.

Mohankrishnan, N., M. Shridhar & M. A. Sid-Ahmed (1982) A composite scheme for text-independent speaker recognition. *Proc. IEEE Conf. Acoustics, Speech and Signal Processing ICASSP-82*, 1653-6.

Mohn, W. S. (1971) Two statistical feature evaluation techniques applied to speaker identification. *IEEE Trans. C-20 (9)*, 979-87.

Moll, K. L. & R. S. Daniloff (1971) Investigation of the timing of velar movements during speech. *J. Acoustical Society of America 50 (2)*, 678-84.

Mucciardi, A. N. & E. E. Gose (1971) A comparison of seven techniques for choosing subsets of pattern recognition properties. *IEEE Trans. C-20 (9)*, 1023-31.

Myers, C., L. R. Rabiner & A. E. Rosenberg (1980) Performance tradeoffs in dynamic time warping algorithms for isolated word recognition. *IEEE Trans. ASSP-28*, 623-35.

Naik, J. M. & G. R. Doddington (1986) High performance speaker verification using principal spectral components. *Proc. IEEE Conf. Acoustics Speech and Signal Processing ICASSP-86*, 881-4.

—— (1987) Evaluation of a high performance speaker verification system for access control. *Proc. IEEE Conf. Acoustics, Speech and Signal Processing ICASSP-87*, 2392-5.

Nelson, G. D. & D. M. Levy (1968) A dynamic programming approach to the selection of pattern features. *IEEE Trans. SSC-4 (2)*, 145-51.

Ney, H. & R. Gierloff (1982) Speaker recognition using a feature weighting technique. *Proc. IEEE Conf. Acoustics, Speech and Signal Processing ICASSP-82*, 1645-8.

Noll, A. M. (1967) Cepstrum pitch determination. *J. Acoustical Society of America 41*, 293-309.

Oh, K. A. & C. K. Un (1984) A performance comparison of pitch extraction algorithms for noisy speech. *Proc. IEEE Conf. Acoustics, Speech and Signal Processing ICASSP-84 18 (B) 4*, 1-4.

Pruzansky, S. & M. V. Mathews (1964) Talker recognition based on analysis of variance. *J. Acoustical Society of America 36*, 2041-7.

Rabiner, L. R. (1977) On the use of autocorrelation analysis for pitch detection. *IEEE Trans. ASSP-25 (1)*, 24-8.

Rabiner, L. R. & R. W. Schafer (1978) in *Digital Processing of Speech Signals*. Prentice-Hall.

Rabiner, L. R., M. J. Cheng, A. E. Rosenberg & C. A. McGonegal (1976) A comparative study of several pitch determination algorithms. *IEEE Trans. ASSP-24 (5)*, 399-418.

Reich, A. R. & J. E. Duke (1979) Effects of selected vocal disguises upon speaker identification by listening. *J. Acoustical Society of America 66 (4)*, 1023-8.

Rosenberg, E. (1973) Listener performance in speaker verification tasks. *IEEE Trans. AU-21 (3)*, 221-5.

—— (1976) Automatic speaker verification: a review. *Proc. IEEE 64 (4)*, 475-87.

Rosenberg, A. E. & M. R. Sambur (1975) New techniques for automatic speaker verification. *IEEE Trans. 23 (2)*, 169-75.

Rosenberg, A. E. & K. L. Shipley (1985) Talker recognition in tandem with talker independent isolated word recognition. *IEEE Trans. ASSP-33 (3)*, 574-86.

Rosenberg, A. E. & F. K. Soong (1986) Evaluation of a vector quantization talker recognition system in text independent and text dependent modes. *Proc. IEEE Conf. Acoustics Speech and Signal Processing ICASSP-86*, 873-6.

Sambur, M. R. (1975a) Selection of acoustic features for speaker identification. *IEEE Trans. ASSP-23 (2)*, 176-82.

—— (1975b) An efficient linear prediction vocoder. *Bell System Technical J. 54 (2)*, 1693-1723.

—— (1976) Speaker recognition using orthogonal linear prediction. *IEEE Trans. ASSP-24 (4)*, 283-9.

Sarma, V. V. S. & D. Venugopal (1977) Performance evaluation of automatic speaker verification systems. *IEEE Trans. ASSP-25*, 264-6.

Schafer, R. W. & L. R. Rabiner (1970) System for automatic formant analysis of voiced speech. *J. Acoustical Society of America 47(2)*, 634-48.

Schroeder, M. R. (1981) Direct (non-recursive) relations between cepstrum and predictor coefficients. *IEEE Trans. ASSP-29(2)*, 297-301.

—— (1985) Linear predictive coding of speech: review and current directions. *IEEE Communications Magazine 23(8)*, 54-61.

Schwartz, R., S. Roucos & M. Berouti (1982) The application of probability density estimation to text-independent speaker identification. *Proc. IEEE Conf. Acoustics, Speech and Signal Processing ICASSP-82*, 1649-52.

Seneff, S. (1978) Real time harmonic pitch detector. *IEEE Trans. ASSP-26*, 358-64.

Short, R. D. & K. Fukunaga (1981) The optimal distance measure for nearest neighbor classification. *IEEE Trans. IT-27(5)*, 622-7.

Shridhar, M., N. Mohankrishnan & M. A. Sid-Ahmed (1983) A comparison of distance measures for text-independent speaker identification. *Proc. IEEE Conf. Acoustics, Speech and Signal Processing ICASSP-83*, 559-62.

Soong, F. K. & A. E. Rosenberg (1986) On the use of instantaneous and transitional spectral information in speaker recognition. *Proc. IEEE Conf. Acoustics, Speech and Signal Processing ICASSP-86*, 877-80.

Soong, F. K., A. E. Rosenberg, L. R. Rabiner & B. H. Juang (1985) A vector quantization approach to speaker recognition. *Proc. IEEE Conf. Acoustics, Speech and Signal Processing ICASSP-85*, 387-90.

Su, L. S., K. P. Li & K. S. Fu (1974) Identification of speakers by use of nasal coarticulation. *J. Acoustical Society of America 56*, 1876-82.

Sutherland, A. M., M. A. Jack & J. Laver (1986) A pitch detection algorithm optimised for microperturbation analysis. *Proc. IOA Autumn Meeting*, 331-8.

Tohkura, Y. (1986) A weighted cepstral distance measure for speech recognition. *Proc. IEEE Conf. Acoustics, Speech and Signal Processing ICASSP-86*, 761-4.

Toussaint, G. T. (1974) Bibliography on estimation of misclassification. *IEEE Trans. IT-20(4)*, 472-9.

Voiers, W. D. (1964) Perceptual bases of speaker identity. *J. Acoustical Society of America 36(6)*, 1065-73.

Wilks, S. S. (1962) in *Mathematical Statistics*. Wiley, New York.

Wohlford, R. E., E. H. Wrench & B. P. Landell (1980) A Comparison of four techniques for automatic speaker recognition. *Proc. IEEE Conf. Acoustics, Speech and Signal Processing ICASSP-80*, 908-11.

Wolf, J. J. (1972) Efficient acoustic parameters for speaker recognition. *J. Acoustical Society of America 51(6)*, 2044-56.

Wolf, J., M. Krasner, K. Karnofsky, R. Schwartz & S. Roucos (1983) Further investigation of probabalistic methods for text-independent speaker identification. *Proc. IEEE Conf. Acoustics Speech and Signal Processing ICASSP-83*, 551-4.

FIVE : G. DUNCAN & M. JACK

# THE HELIUM SPEECH EFFECT

The most primitive attempts of man to enter into the ocean environment can be traced to native divers, who would take natural sponges to the surface, wring them dry, and then, after covering them with oil, dive to depths of some 40–50 feet carrying the sponges in their mouths, breathing the air trapped within the capillary channels of the sponge in order to prolong their stay under water to some five minutes.

The earliest recorded reference (Schott 1664) to a man-made contrivance to prolong the working duration of man underwater relates to an experiment which took place at Toledo, Spain, in the year 1538, before Emperor Carlos v and several thousand spectators. In his collection of monographs on *'Technical Curiosities, including the Wonders of Art'*, Gaspar Schott recounts the details of the experiment as witnessed by one Jean Taisner, born in Hainault, France, in 1509. Two Greek divers were placed inside an 'underwater cooking-pot' (see figure 5.1), which was immersed totally in the River Tajo with the open end downwards. Much to the amazement of the onlookers, when the diving bell was lifted back out of the river, the two occupants were completely dry, and moreover, the candle they had taken down with them still remained lit.

The first attempts to allow man more freedom of movement underwater were provided by the swim bladder of Giovanni Borelli (1685), first proposed in 1679. This first true diving suit consisted of an air-filled bladder which completely surrounded the diver's head, onto which two pipes were fitted, the one taking air supplied from bellows to the bladder, the other venting the diver's expired breath into the sea. Such a device was impracticable, however, beyond a depth of three fathoms, the main problem being the sewn goat-leather suit to which the bladder was attached, which was apt to spring leaks at such depths, despite a tight neck-collar, much to the

*Figure* 5.1. *Cacabus aquaticus,* the 'underwater cooking pot', showing (I) entry into the water and (II) the internal frame and harness arrangement of the earliest known diving bell (from Schott 1664).

consternation of the diver involved.

The diving bell, so-called because of its conical shape, was closed at the top (apex) end, but open to the sea at the bottom end in order to allow the diver to carry out his work. The technology of the diving bell was scarcely improved at all until the late seventeenth century. Describing the primitive version of the bell Sir Edmund Halley (1716) wrote '. . . and if the Cavity of the Vessel may contain a Tun of Water, a single Man may remain therein as least half an Hour, without much inconvenience, at 5 or 6 Fathoms Deep.'

Sir Edmund, however, does indeed enumerate several inconveniences. At six fathoms, the bell is half full of water due to the ambient pressure; the diver cannot bear the coldness of the water in the bell for very long, and lastly, the main inconvenience is due to the accumulation of the diver's expired, tepid air, which '. . . in passing the Lungs, loses its Vivifying Spirit, and is rendered effete, not unlike the Medium of Damps, which is present Death to those that breath it.'

This problem of serious accumulation of carbon dioxide, which rendered the whole atmosphere in the bell rapidly unfit for respiration, had been previously identified by other researchers in the field but had not been solved (Bacon 1620). Sir Edmund's contribution to the science of diving was a bell which, first, had a glass panel inserted at the top to improve lighting. Most importantly, however, the air inside the bell could be replenished by a system of two barrels, each holding 36 gallons of air, which were alternately lowered by a pulley down to the diver, who then emptied the fresh air into the bell. There was also a stop-cock near the top of the bell so that tepid air could be vented before it made the atmosphere of the bell intolerable. This technique, first used in 1691, both prolonged the useful stay of the diver under the sea and increased his working depth, and was to predominate for almost a century until the development of pumps capable of producing compressed air in 1788. Extolling the virtues of his device, Sir Edmund wrote:

> There was such a plentiful supply of air, that I myself have been One of the Five who have been together at the Bottom, in nine or ten Fathoms of Water, for above an Hour and Half at a time, without any sort of ill consequence: and I might have continued there as long as I pleased, for any thing that appeared to the contrary. Besides, the whole Cavity of the Bell was kept entirely free from Water, so that I sat on a Bench, which was diametrically placed near the Bottom, wholly drest with all my Cloaths on. I only observed, that it was necessary

> to be let down gradually at first, as about 12 foot at a time; and then to stop and drive out the Water that entred, by receiving three or four Barrels of fresh Air, before I descended further.

It is indeed a paradox that, some two centuries later, it was to be shown beyond question that it was the ascent to the surface, and not the descent to the ocean floor, which was to prove the worst problem. Further, the 'Vivifying Spirit' which Sir Edmund so painstakingly sought to restore to the respiratory mixture, and which was later identified in 1774 by Joseph Priestley to be oxygen, was to prove to be one of the most toxic elements known to man when respired in a hyperbaric atmosphere.

The larger version of the diving bell, termed a caisson due to its box shape, can hold some 50 workers, and was first used in the construction of the support pillars for the Kehl bridge, near Strasbourg, France, in 1858–61. A few years later, the Frenchman Dr A. Jaminet made observations of Irish immigrant caisson labourers during the construction of the Eads bridge over the Mississippi at St Louis, USA. The men had spent four hours working on the bridge substructure at a pressure of four atmospheres (i.e. at a depth of approximately 100 feet). It was recorded that, on their return to the surface, several of the men suffered severe convulsions and bleeding from the nose, and indeed there was a series of fatalities (Jaminet 1871). In fact, of the 352 diving labourers employed in the project, some 30 or so became severely and permanently paralysed, whilst 12 suffered sudden death. Many others experienced violent convulsions, and pains in the muscles and joints coupled with temporary paralysis, especially in the lower limbs.

In 1876 it was another Frenchman, Professor Paul Bert, who identified for the first time the root causes of decompression sickness, or 'the bends' as it came to be colloquially known. It was shown that in saturation diving, in which the diver can be considered as saturated with gas for a specific depth (pressure), the crucial factor determining the diver's susceptibility to the bends was not, as believed, the rate of compression or descent in the water, which could in fact be carried out at any desired rate, but rather depended on the rate of decompression or ascent (Bert 1876). At depth, the diver's bloodstream can be considered to be saturated with the component gases of air for the corresponding ambient pressure. If decompression occurs too rapidly, that is the pressure is relieved too quickly, then the gas appearing out of solution from the bloodstream does not have enough time to diffuse through the skin and muscle tissue, but instead accumulates in the veins, acting as an air-lock, thereby, at the very worst, halting circulation and causing death (Behnke, Thompson and Motley 1935; Clark 1982). The most troublesome gas in this respect has been shown to be nitrogen, which constitutes some 78% by volume of air. The released oxygen has been shown to recombine with the blood almost as quickly as it appears.

However, Professor Bert was able to demonstrate that it is the oxygen in high-pressure air respired at depth which is in fact toxic to man and is the cause of violent convulsions. Further experimentation additionally found that breathing a gaseous mixture whose partial pressure of oxygen exceeds 0.8 bar for long periods in hyperbaric conditions will produce violent convulsions and may even result in death.

With the advance of technology, deeper dives were becoming possible, and with them came the discovery that not only could oxygen promote convulsions, but nitrogen too was troublesome. When breathed for prolonged periods at depth, nitrogen produces a state of narcosis in the diver which worsens with increasing depth or pressure (Case and Haldane 1941). Investigations have shown that the first symptoms of nitrogen narcosis are headaches, vertigo and vomiting, leading to the rapid onset of delirium tremens, during which phase it appears that the most common hallucination is for the diver to visualise someone once very close to him but now dead walking across the sea bed towards him, and actually start a conversation with him (Webster 1978). A second distressing condition which can be directly attributed to high concentrations of high-pressure nitrogen is dyspnoea, described as being like trying to breathe air through a straw. This ailment arises through breathing a very dense gas mixture under pressure. Despite both of these distressing indications, it in fact takes a very experienced diver to recognise them himself, and in most cases it is the surface crew who first realise that the diver is undergoing nitrogen narcosis.

In order to avoid these dangerous physiological conditions, and in addition increase the maximum possible diving depth, a lightweight gas was sought to replace nitrogen and the other heavy gases present in air. The most convenient replacement so far has proved to be helium gas. It has various advantages. It is inert, non-toxic, non-flammable, colourless, odourless and tasteless, and of course is very light. Note, however, that the use of a lighter respiratory mixture does not affect the diver's susceptibility to decompression sickness, nor does it significantly reduce the time required for safe decompression during ascent to the surface.

The US Navy was the pioneer in the use of helium-oxygen (heliox) respiratory mixtures in the early 1930s, and also produced tables of the required decompression times according to working depth and duration of the dive. For example, diving to 250 ft for 20 minutes requires two-and-a-half hours' decompression, whereas diving to 200 ft for 1 hour requires three hours' decompression. Working depths of 500 ft are now commonly attained using a helium-oxygen mixture, and require twelve hours or more of decompression.

Although the percentages of helium in the respiratory mixture may appear surprisingly high, typically around 97.6% helium and 2.4% oxygen at 500 ft, it should be recalled that it is the partial pressure of oxygen that is critical. Thus, although the volume of oxygen is low, its partial pressure is sufficient to sustain life.

One of the drawbacks of using a lighter-than-air respiratory mixture with such a small average molecular weight is that it tends to find escape routes from equipment in a way that heavier gases would not, and therefore causes problems in maintaining hermetic seals in breathing equipment. However, the principal disadvantage of using a heliox mixture relates to its effect upon voice communications. The heliox mixture, with its increased velocity of sound and different acoustic impedance with respect to air, alters the speech uttered by the diver to such an extent that what reaches the surface is not an intelligible acoustic waveform, but an 'incoherent squeaky jumble', which none but the extremely experienced can hope to comprehend: even then, serious and occasionally fatal mistakes in comprehension can occur (Hunter 1968; Baume, Godden and Hipwell 1982).

With the realisation that oil drilled from the ocean floors is a viable source of national wealth, there has been a dramatic increase in deep-diving operations since the early 1960s, with the attendant obligatory use of heliox respiratory mixtures. To assist the increasingly frequent diving operations that involve long-term endurance of a heliox environment, a range of electronic signal processing systems has been developed to enhance the intelligibility of the divers' squeaky voice emissions, known as 'helium speech'. This chapter reviews the development of these signal processing systems, termed 'helium speech unscramblers'.

As with all problems in signal processing, it is imperative to know as much as possible about the entity to be processed. In the following sections we consider the factors that are known to have a direct bearing on speech intelligibility under normal everyday atmospheric conditions. These are then related to the known acoustic events of helium speech in terms of how changes in various acoustic phenomena affect speech intelligibility with respect to the listener. Here, it is also shown that the diver himself can adapt his own speech while breathing heliox, presumably in an attempt to render it more intelligible as he himself judges. The chapter explores the architectures of both experimental and currently in-service unscrambler systems in relation to the presently accepted acoustic phenomena of helium speech.

*The Speech Mechanism*

In normal air environments, the human interpretation of the acoustic speech signal depends upon a complex interplay of both temporal and acoustic events, the most salient of which are described below in terms of method of acoustic waveform production by the talker and aural perception by the listener. The effects of pressure and gas mixture on the acoustic waveform are explored here and related to the perceptual factors directly affecting the intelligibility of helium speech.

The human vocal tract comprises a non-uniform tube from the vocal cords to the lips and includes the nasal cavity (see figure 5.2a). In voiced speech, the vocal tract can be likened to a pulse-excited causal, minimum-phase filter whose effects are represented by an acoustic tube of non-uniform cross-sectional area. Generally, the tube is approximated by a series of short concatenated sections each having uniform cross-sectional area, with the vocal cords, through which the pulse excitation enters the system, at one end and the lips at the other (Fant 1960, Pinson 1963), from which the pressure waveform carrying the acoustic information travels out into the surrounding medium (see figure 5.2b). As a pressure wave enters the system through the vocal cords, it undergoes various transmissions and reflections at the boundaries of each acoustic tube section. The system as a whole can be regarded as a series of resonating cavities, with each cavity contributing to the overall waveform shape and the characteristic distribution of the power spectral density, which, for voiced sounds, exhibits distinct areas of resonant energy, or formants, centered on certain frequencies.

In air at normal temperatures and pressures, the tissue of the vocal tract walls and articulators, or moveable elements, such as the palate, tongue, teeth and lips, can be essentially considered to be perfect acoustic reflectors, thereby forming lossless boundaries. The vocal tract filter is not, however, time-invariant in the long term, but can in most cases of voiced speech be so considered over periods of up to 30 ms (Rabiner and Schafer 1975).

Changing the shape of the various cavities of the acoustic tube alters their respective contributions to the overall waveform shape

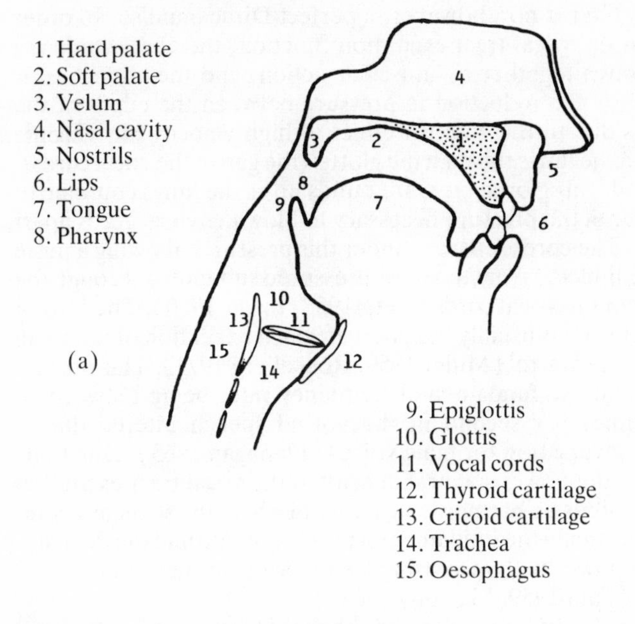


from glottis

*l*

(b)

to lips

*Figure* 5.2. (a) Section through the human vocal apparatus.
(b) Concatenated acoustic tube model of the vocal tract.
Each resonant section is of length *l*.

(Wood 1977), thereby modifying the filtering characteristic of the vocal tract to produce changes in the power spectral density distribution of the resulting speech signal. The excitation source for the vocal tract filter is not, however, a perfect Dirac impulse. In order to produce the vocal tract excitation function, the vocal cords are initially drawn together by muscular action, and the adduction is completed by the reduction in pressure between the edges of the vocal folds due to the Venturi effect of high-velocity air which is blown from the lungs through the glottis (the gap in the vocal cords). Once closed, sub-glottal pressure builds up as the lungs continue to expel air, until the pressure necessary to blow the vocal cords apart is reached. The cords separate under this pressure, allowing a pulse of air to be injected from the over-pressured sub-glottis through the gap between the vocal cords (Berg 1958, Laver 1980). The instant of glottal closure is usually the point at which excitation of the vocal tract is most powerful (Miller 1959, Rosenberg 1971). This process is repeated at the fundamental frequency rate, being between 50 and 250 times per second in the voiced speech uttered during ordinary conversation for male voices (Flanagan 1965). Due to its periodicity, the power spectral density of the vocal tract excitation function exhibits an harmonic structure in which the strongest component is normally the fundamental repetition rate and the decrease in spectral power with increasing frequency is of the order of −12 dB/octave (Fant 1959, Martony 1964a).

The combined effect, then, of the vocal tract filter function and the vocal cord excitation source is to produce a pressure waveform emanating from the lips. The power spectral density of this acoustic waveform will exhibit the periodicity of the excitation source and, in addition, will contain concentrations of energy at the formant frequencies, denoted F1, F2, F3, etc., corresponding to the filtering action of the vocal tract. The lips themselves typically act as an acoustic horn radiator, thereby imparting an emphasis of +6 dB/octave to the overall power spectral density (Morse 1948). An example of the power spectral density for the waveform corresponding to the vowel 'ee' is shown in figure 5.3a.

The quality of speech sounds as perceived by the listener (phonemes), depends on several acoustic attributes of the received acoustic waveform. Speech intonation, for example, is a function of the variation in time of the fundamental frequency of excitation of the vocal tract, whereas the pitch of the perceived speech signal is, strictly speaking, the psychological impression of the tone of voice, and is a function of both the fundamental frequency and overall intensity of the voiced waveform (Denes and Pinson 1963). However, the term pitch is often used to signify only fundamental frequency, and indeed the two terms will be used interchangeably throughout this chapter.

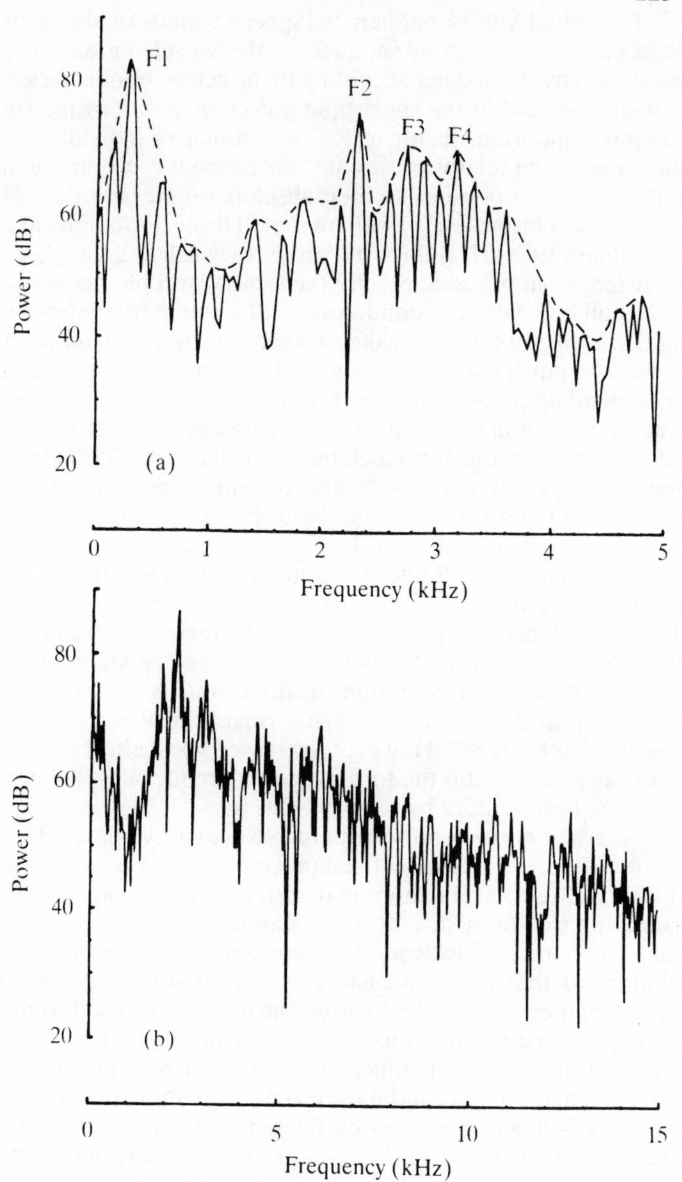


*Figure* 5.3. (a) Power spectral density of a 25.6 ms segment from the voiced vowel 'ee'. (b) Power spectrum from the unvoiced fricative 'sh'.

Individual voiced phonemes (speech sounds made when the vocal cords are in vibration, such as the vowels /a/ and /e/) are recognised by the listener according to the centre frequencies of the formants present in the short-time power spectrum of the signal. The most important factor in the recognition of individual vowel phonemes is the relative ratios of these formant centre frequencies to each other, as opposed to their absolute frequency values (Fant 1959, 1960; Pinson 1963; Fairbanks and Grubb 1961; Fujisaki and Kawashima 1968). It is this ratio property which helps explain why the listener can perceive exactly the same vowel phoneme spoken by a small boy, whose absolute values of formant frequency will be high, and an adult male speaker, whose absolute formant frequencies will be much lower (Peterson and Barney 1952). Although the corresponding power spectrum for any particular phoneme will in general contain many formants, it is generally accepted that only the first three or four formants, between 50 Hz and 3500 Hz in the short-time spectrum of speech, play the most crucial role in speech perception (Peterson 1952; Fant 1959; Fairbanks and Grubb 1961; Holbrook and Fairbanks 1962).

The relative amplitudes of formants have some bearing upon the perceived quality of the particular phoneme (Williams and Stevens 1972, Scherer 1981). For example, increasing the energy in the acoustic waveform to make the phoneme appear louder is reported to affect the intensities of the formants from the second formant upwards, whereas the first formant remains relatively unaffected (Miller 1959). However, increased vocal effort in this way is not expected to alter the formant centre frequencies to any great extent; at best perhaps by some 10–20 Hz.

Another most important perceptual quality, particularly in relation to the problems of helium speech, is that of nasality. Theoretically, a nasal quality may result from the coupling of any resonating side-branch, that is, an acoustic shunt element, to the main vocal tract. However, the main side-branch resonator contributing to the nasalised quality of voiced sounds is, in air at normal temperatures and pressures, the nasal cavity itself. It may at first appear that rotating the velum down and away from the back pharynx wall is in itself sufficient to allow air-flow into the nasal cavity and hence induce nasal resonance. In reality, however, nasal resonance is a complex function related to nasal chamber dimensions, and in particular to the ratio of the nasal port openings and the area of the velic opening in the vocal tract (Bjork 1961, Lintz and Sherman 1961, Laver 1980).

The acoustic correlates of resonance in the nasal cavity are, first, that nasal resonances and associated antiresonances are introduced into the vocal tract frequency response, leading to a modified power spectrum. The most commonly reported nasal formant frequencies are between 200–300 Hz and around 1 kHz (House and Stevens 1956), with another around 2.5 kHz, their associated bandwidths being of the order of 300 Hz for the lowest formant increasing to 1 kHz for those near to 2.5 kHz (House 1957). Nasal antiresonances are each paired with a nasal formant, and values of between 500–700 Hz have been reported for the lowest (Fujimura 1962), with another between 900 Hz and 1.8 kHz (House and Stevens 1956).

Secondly, as regards the acoustic waveform overall, the most general effect associated with nasality is an overall loss of power, which can be directly attributed to the introduction of the nasal antiresonances, which absorb acoustic energy, especially in the higher frequencies. A further effect of nasality is the detuning of existing vocal tract formants such that their bandwidths increase, with an attendant decrease in formant peak amplitude (Martony 1964b). This general attenuation is considered to be responsible for the drop in intelligibility of nasal voices (Moser, Dreher and Adler 1955; Diehl and McDonald 1956).

In unvoiced fricative speech production, the speech sounds are produced due to turbulent air flow at some constriction in the vocal tract, e.g. between the tongue and the soft palate. The turbulent air flow acts as a white noise excitation source for the vocal tract, since the vocal cords are not in vibration (i.e. the sound is unvoiced), but are held slightly apart by muscular tension to allow free passage of air from the lungs into the vocal tract towards the point of constriction. In the case of unvoiced frication, the vocal tract can still be approximated by a non-uniform acoustic tube; in this instance, however, the excitation source, rather than being a train of quasi-periodic pulses, is best approximated by a random train of impulses or a white noise source. In consequence, for any particular vocal tract configuration corresponding to a particular vowel, the power spectrum would in this case, exhibit no line structure, and formant amplitude would be expected to decrease whilst formant bandwidth would increase (Heinz and Stevens 1961). However, the mechanism of production of the excitation source itself produces other changes in the overall power spectrum. Specifically, the excitation source is no longer situated at the vocal cords, which are held apart by muscular tension to permit the free passage of air, but rather the source is located at a point of constriction within the vocal tract, such as by the bringing together of the teeth and lower lip, as in /f/, or by the action of the tongue, as in /ʃ/ (Flanagan 1965). The action of forcing air at high velocity through the constriction produces turbulent flow in the vicinity of the constricting passage and possibly also at the teeth. Acoustic noise is generated as a result of the turbulent flow, and it is this noise which acts as the excitation source for the frontal cavities between the constriction and the external atmosphere (Fant 1959). In the case of the fricative /ʃ/, for example, the constriction is formed by the tongue and the front of the hard palate. The power spectral density corresponding to /ʃ/ is shown in figure 5.3b.

Since the power spectral density exhibits a formant structure for fricatives, perception of individual fricative phonemes is based to some extent on the relative centre frequencies and bandwidths of formants, particularly in the case of fricatives whose relative power is high, such as /s/ and /ʃ/ (Denes and Pinson 1963).

However, for certain of the lower-intensity weak fricatives, such as /f/ and /θ/ (th as in *think*), there is an important contribution to perception from the segmental environment of the fricative consonant. Thus, in the case of a weak fricative preceded or followed by a (voiced) vowel, perception of the fricative phoneme is based not only on spectral attributes of the fricative itself, but also on formant transitions into the vowel, on the direction of these transitions, and on other secondary acoustic characteristics such as vowel duration and relative ratio of power from vowel to fricative (Lisker *et al.* 1960, Martony 1962). This perceptual mechanism is also important to the perception of the next class of unvoiced speech sounds considered here, plosive or stop consonants.

In plosive or stop consonant speech production, there is a build-up of pressure at some location due to closure of the vocal tract by, for example, the teeth or lips. The subsequent, sudden release of pressure causes a transient excitation of the vocal tract which results in the sudden onset of sound. The characteristic power spectral density for any particular plosive phoneme, such as /p/ and /t/, is considered to be concentrated within some particular region of the power spectrum (Lindqvist and Lubker 1970; Carlson, Granstrom and Pauli 1972; Carlson and Granstrom 1977). For the plosive /t/, for example, the main concentration of spectral power is in the region of 1.8 kHz.

It is generally considered that the spectral concentration of energy associated with the plosive burst is not, in itself, sufficient to allow identification of a particular plosive phoneme. Experiments have shown, for example, that a plosive burst centered at one frequency may be heard as /k/ when associated with one vowel, but

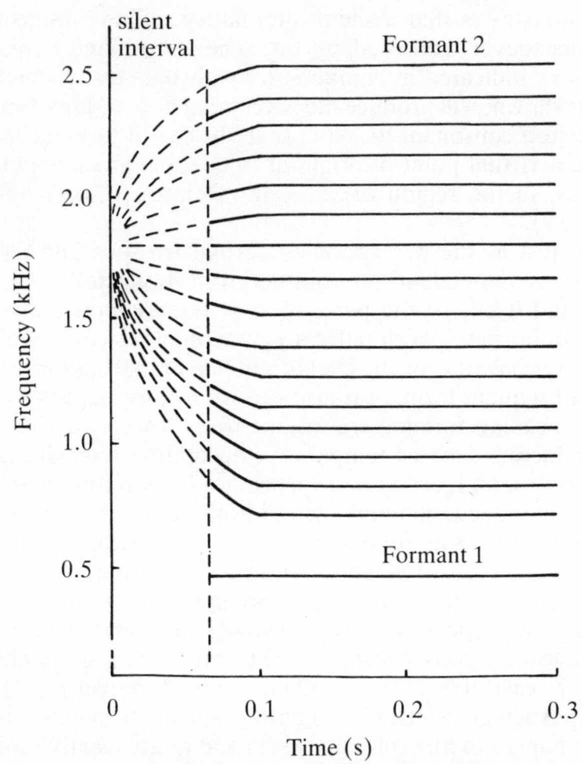


*Figure* 5.4. Second formant transitions all perceived as the stop consonant /t/ (after Denes 1963).

as a /p/ when associated with another vowel, implying that identification of a particular plosive is dependent to some extent on the nature of the following vowel (Delattre, Liberman and Cooper 1955; Stevens and House 1956; Stevens and Klatt 1974). Further experimentation has since shown that the most important attribute of the following vowel in this respect relates to the direction in frequency and duration in time of the second formant transition. Specifically, in the absence of a distinct plosive burst itself, all second formant transitions perceived as one particular plosive phoneme have a virtual point of origin within the same frequency region (Harris *et al.* 1958; Borovickova, Malac and Pauli 1970). For example, for a /p/ plosive the virtual origin of the second formant transition has been found to be at 700 Hz, but for a /k/ plosive is at around 3 kHz. This principle is illustrated in figure 5.4, which demonstrates that a silent interval, containing no acoustic energy whatsoever, followed by the second formant frequency transitions as indicated in combination with the steady-state first formant as shown, will produce the perceptual impression of having heard the stop consonant /t/. Note that all second formant transitions have a virtual point of origin in frequency, as shown by the dotted arcs, in the region of 1.8 kHz (Delattre, Liberman and Cooper 1955).

Thus, just as the perception of certain fricative and plosive consonants is dependent on formant transitions following the speech sound itself, so the perception of the composite acoustic waveform of normal speech reflects a perceptual system based on a complex interplay of acoustic events and perceptual cues involving not only subsequent formant transitions from consonant-to-vowel, but also preceding formant transitions from vowel-to-consonant, and other factors such as temporally- fluctuating intensity ratios from sound to sound, changes of speech production from voiced to unvoiced and indeed, combinations of both of the latter classes of speech production as in, for example, the voiced fricative /z/ (Fairbanks, House and Stevens 1950; Fairbanks and House 1953; Fant 1967; Scott 1976). Differentiation between the consonants /b/ and /p/, for example, depends not only on vowel formant transitions but also on the low intensity voicing present up to the instant of plosive release in the case of /b/ (Denes and Pinson 1963; Stevens and Klatt 1974). In particular, whilst voiced nasal consonants may be identified from changes in formant bandwidth and relative amplitude, as mentioned earlier, their perception depends also on transitions of the formants of the following vowel (Öhman 1966). Refering to figure 5.4, if the silent interval before the commencement of the second formant transition is replaced instead by a low intensity buzz, then the same formant trajectories now give rise to the perception of a nasal consonant; that is, those transitions with a virtual origin at 700 Hz are perceived as /m/, whereas those with a virtual origin of 1.8 kHz are perceived as /n/ (Liberman *et al.* 1954).

The importance of perceptual cues in the identification of specific phonemes is highlighted in experiments to find difference limens for vowel formant centre frequencies. It has been found that, for example, the shift in second formant frequency needed to produce a just-noticeable difference in vowel identity for vowels spoken within a consonant-vowel-consonant framework is much higher than that for vowels uttered in isolation (Flanagan 1955, Mermelstein 1978). This property is exploited by speakers in everyday conversation, since it is rare in any case that a speaker achieves exactly the same formant frequencies for a given vowel from instance to instance; rather the formant frequencies are placed into approximate frequency areas with phonemic identification aided by one or several of the factors outlined above (Lindblom and Floren 1965).

A further parameter affecting perception and intelligibility, which has particular reference to helium speech, relates to the effects of variations of the relative power ratios of consonant to vowel. It has been shown that a small consonant-to-vowel power ratio inherently favours intelligibility, whereas an increased power ratio has an adverse effect upon intelligibility (Fairbanks and Miron 1957). Experiments have demonstrated, for example, that in the presence of speech signal distortion, decreasing the consonant-to-vowel intensity ratio will aid intelligibility, and that in such cases, the effects of peak-clipping and amplitude compression on the signal are in fact advantageous (Licklider and Pollack 1948), whereas increasing the speaker's overall vocal effort tends to increase the vowel-consonant intensity ratio and therefore has a detrimental effect upon intelligibility.

*Speech in Deep Submergent Atmospheres*

Since the transmission medium for the speech signal is the atmosphere surrounding the talker, and since the signal itself is an acoustic pressure wave propagating through the medium, then the pressure waveform is dependent upon the physical properties of the medium. From acoustic theory, the speed of sound, $c$, in a gas is given by:

$$c = \sqrt{(\gamma P/\rho)} \qquad (5.1)$$

where $\gamma$ is the adiabatic constant (ratio of specific heats), $P$ is the gas pressure and $\rho$ the density of the gas. It can be shown, however, that the speed of sound for a given gas is relatively independent of ambient pressure, since $\gamma$ is, to a first approximation, independent of pressure and temperature over the range of values survivable in the diving environment. Furthermore,

$$\rho = \rho_0 P/P_0 \qquad (5.2a)$$

or $$\rho/\rho_0 = P/P_0 \qquad (5.2b)$$

where $\rho_0$ is the density measured at some pressure $P_0$.

Substitution of equations 5.2a,b into equation 5.1 illustrates that the speed of sound is independent of pressure for a given gas. Thus, it might be expected that the mechanism and perception of human speech is relatively unaffected at high ambient air pressure: experiments have shown, however, that there is a degradation of the intelligibility of speech uttered in such an environment (Fant and Sonesson 1964; Fant and Sonesson 1967; Fant *et al.* 1971).

It has been demonstrated that unvoiced consonant sounds suffer a relative attenuation of the order of 10dB with respect to voiced vowel sounds at high pressure (Fant and Sonesson 1964); therefore, in consideration of the above discussion relating to the effect of the vowel-consonant intensity ratio, an increase in this ratio will inherently disfavour intelligibility. However, this drop in intelligibility has been attributed in the main to (a) the increased nasality of high-pressure air speech and (b) to a nonlinear shift of the lower formant frequencies (Hollien, Thompson and Cannon 1973). These effects have been related theoretically to the acoustic impedance properties of the respiratory gas and the tissue of the face and vocal tract (Berg 1955). Specifically, at normal temperatures and pressures the vocal tract walls can be essentially regarded as perfect reflectors of acoustic energy due to the large mismatch of impedance between the transmission medium and the skin tissue.

However, the characteristic acoustic impedance, $Z_0$, of each air-filled vocal tract cavity increases with density, $\rho$, i.e. with ambient pressure. At high pressures, therefore, the vocal tract walls may in fact absorb acoustic energy as the impedance mismatch is reduced, thereby producing resonatory vibrations of the wall tissue which will, in turn, affect the pressure wave produced by the vocal mechanism. This effect has been quantised by a theoretical consideration of the total volume of air enclosed by the vocal tract and acoustic properties of the vocal tract walls (Fant and Sonesson 1964). The result is an equation relating the resonance frequency of the closed vocal tract, $F_w$, to the ambient pressure $P$:

$$F_w = F_{w0}\sqrt{(P/P_0)} \qquad (5.3a)$$

and substituting equation 5.2b gives:

$$F_w = F_{w0}\sqrt{(\rho/\rho_0)} \qquad (5.3b)$$

where $F_{w0}$ is the closed tract resonance frequency at pressure $P_0$. At normal atmospheric pressure $F_{w0}$ is considered to be of the order of 150–200 Hz (Fant 1972).

The resonating vocal tract walls, under conditions of high ambient pressure, can be considered to act as a side chamber shunting the main vocal tract, and would therefore be expected to add a low-frequency pole and zero to the vocal tract transfer function in a similar manner to the effect of nasal cavity resonance (Fujimura 1962).

The resulting deterministic relationship between formant

centre frequency in high-pressure air, $F_{pn}$, and original formant frequency, $F_n$, at some pressure $P_0$ is given by:

$$F_{pn} = \sqrt{(F_n^2 + F_w^2)} \tag{5.4}$$

with the proviso that equation 5.4 is valid only on an average basis (Fant and Lindqvist 1968), since the shunting effect of the vocal tract walls is not likely to be uniformly distributed. Published results (Suzuki, Ooyama and Kido 1974) have been roughly in agreement with equation 5.4, and an illustrative example is shown in figure 5.5a. Notice that the nonlinearity of the formant frequency translation characteristic is most pronounced in the low frequency region of the speech spectrum. In addition, it has been argued that the bandwidths of the vocal tract resonances are not radically changed with increasing pressure. Coupling of a shunt element across the vocal tract would, of course, be expected to produce an increased damping and hence increased bandwidth of those vocal tract poles in the vicinity of the shunt element poles and zeros. Here, however, the effects of the high-pressure atmosphere are such that the low-frequency region most affected in this respect is translated upwards in frequency with the lowest frequencies shifted by the greatest amount, the effect of which is to nullify any low-frequency formant damping.

A consequence of this trend is that certain voiced sounds having an otherwise very low first formant will increase in intensity at high pressure because the low first formant $Q$-factor is effectively increased by the upwards frequency shift. The combined effect of this, in addition to acoustic radiation idiosyncrasies related to the mechanism of production of voiced and voiceless sounds, is to promote an increase in the sound pressure level of voiced to voiceless sounds by a factor roughly proportional to $\sqrt{P}$ (Fant and Sonesson 1964, Fant and Lindqvist 1968).

As regards the increased nasality of high-pressure air speech, this has been directly attributed to the increase in frequency of the low-frequency formants (Fant and Sonesson 1967). The degradation of the intelligibility of speech at high ambient pressure has been correlated to this increase in nasality, and also due to the rise in the first formant (F1) frequency towards that of the second formant (F2), in that auditive distinction between F1 and F2 is impaired, thereby affecting the perceived phoneme. The increase in the vowel-consonant intensity ratio also deteriorates the intelligibility of high-pressure speech, and plosive and fricative consonants are found to be worst affected in this respect (Sergeant 1967).

In a similar manner to speech in high-pressure air, attempts

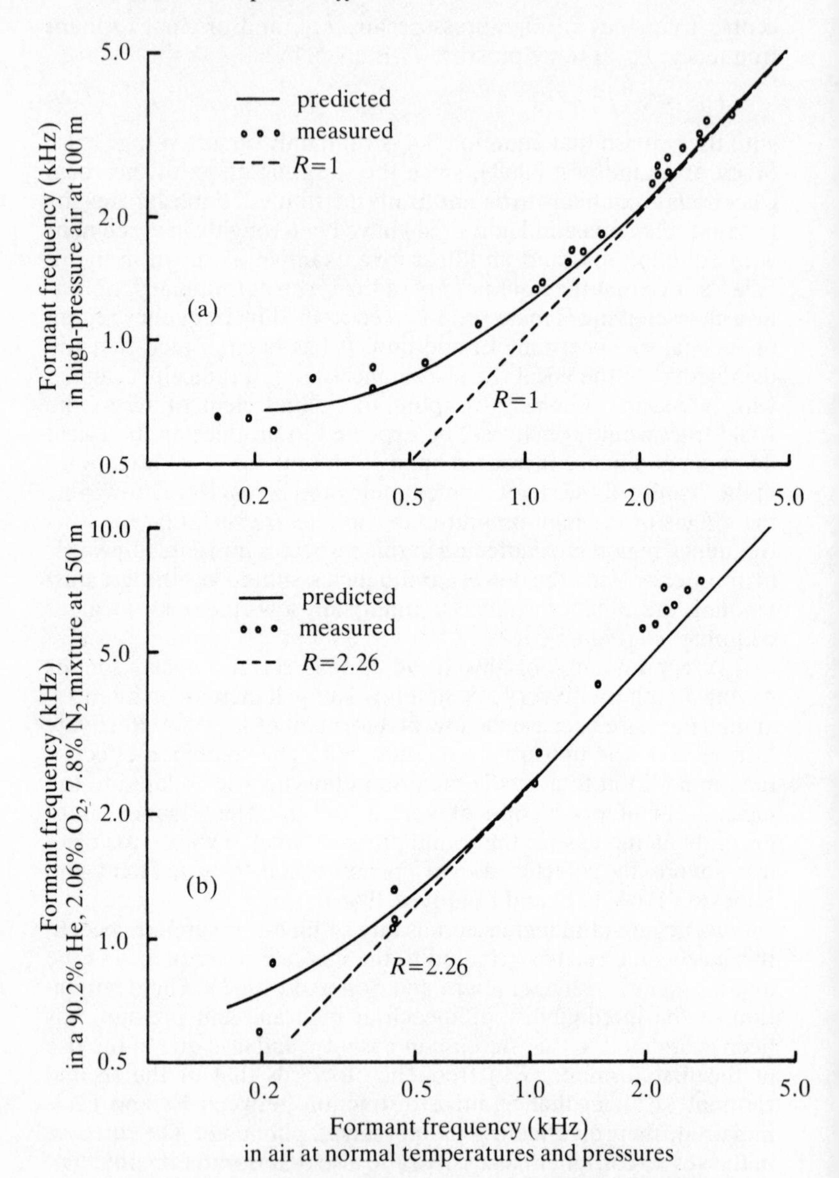


*Figure* 5.5. Formant transposition for one subject (a) in high-pressure air, and (b) in a high-pressure helium-oxygen-nitrogen mixture (after Fant 1968).

have been made to quantify the effect of a pressured atmosphere containing a high percentage volume of helium on the speech spectrum, with particular emphasis on the characteristics as applied to voiced speech (Fant and Lindqvist 1968, Morrow 1971). Recalling equation 5.1 relating the velocity of sound, $c$, of a gas to its pressure $P$, density $\rho$ and adiabatic constant $\gamma$, then the ratio $R$ of the velocity of sound of a respiratory mixture containing helium gas, $c_h$, to the velocity of sound in air at the same pressure, $c_a$, is given by:

$$R = c_h / c_a = \surd(\gamma_h \rho_a / \gamma_a \rho_h) \qquad (5.5)$$

where $\gamma_h$ is the adiabatic constant for the mixture with helium, given by:

$$\gamma_h = \sum_i Q_i \gamma_i \qquad (5.6)$$

where $Q_i$ and $\gamma_i$ are the percentage by volume and corresponding adiabatic constant for the $i$th gas in the mixture, and similarly:

$$\rho_h = \sum_i Q_i \rho_i \qquad (5.7)$$

Since the frequency of resonance of any rigid resonator is proportional to the velocity of sound, then the closed vocal tract resonance $F_{wh}$ for the mixture with helium at pressure $P$ can be found by combining equation 5.3 and equation 5.5, that is:

$$F_{wh} = F_{wa} R \surd(\rho_h P_h / \rho_a P_a) \qquad (5.8)$$

where $F_{wa}$ is the closed tract resonance in air at normal pressures. Furthermore, substituting equation 5.2b and equation 5.5 into equation 5.8 gives:

$$F_{wh} = F_{wa} \surd(\gamma_h P_h / \gamma_a P_a) \qquad (5.9)$$

Thus, assuming $F_{wa}$ is measured in air at one atmosphere pressure, then for a given gas mixture $F_{wh} \propto \surd P$, and the general relationship relating formant frequency in a high-pressure helium mixture, $F_{hn}$, to original formant frequency in air, $F_n$, is therefore given (Fant and Lindqvist 1968) by:

$$F_{hn} = \surd(R^2 F_n^2 + F_{wh}^2) \qquad (5.10)$$

Notice that since $\gamma$ varies only slightly with gas composition, then $F_{wh}$ is approximately independent of gas mixture and varies as a function of pressure only. Basic assumptions relating to equation 5.10 are (a) that the formant frequency in air $F_n$ is that as observed by a technique such as spectrographic analysis, and (b) that the vocal tract walls act as perfect acoustic reflectors. However, since the closed tract resonance is measurable in air at normal pressures, then there is no reason to suggest that equation 5.4 should not apply equally to normal speech in air (Morrow 1971), that is:

$$F_n^2 = F_{rn}^2 + F_{wa}^2 \qquad (5.11a)$$

or $$F_{rn}^2 = F_n^2 + F_{wa}^2 \qquad (5.11b)$$

where $F_{rn}$ is the resonance (not directly observable) assuming the vocal tract to be a perfect rigid resonator, and $F_{wa}$ is the closed tract resonance frequency as measured, the implication being that the term $F_n$ in equation 5.10 should actually be replaced by $F_{rn}$ as defined by equation 5.11b. Thus, replacing the variable $F_n$ in equation 5.10 by $F_{rn}$ as defined in equation 5.11b and replacing explicitly the term $F_{wh}$ by equation 5.9 gives:

$$F_{hn} = R\sqrt{(F_n^2 + (c_a^2 v_h P_h / c_h^2 v_a P_a - 1) F_{wa}^2)} \qquad (5.12)$$

Figure 5.5b demonstrates the application of equation 5.12 to formant data, measured for one subject breathing a pressured helium-oxygen-nitrogen mixture at depth. The value for $F_{wa}$ is 180 Hz in this case.

While certain published results regarding formant frequency analysis in helium (Nakatsui and Suzuki 1971; Nakatsui *et al.* 1973; Tanaka, Nakatsui and Suzuki 1974) are claimed to approximate closely to the theoretical relationship of equation 5.12, there is no general accord as regards the relative effect this has on speech perception. Those whose analytical results approximate to equation 5.12 have postulated that the nonlinear formant shift characteristic, which is most pronounced below approximately 700 Hz in a pressured helium environment, has a large role to play in the decrease of intelligibility of voiced helium speech (Gerstman, Gamertsfelder and Goldberger 1966). However, it has also been argued that it is the large shift of formant frequencies upwards in the spectrum which most affects intelligibility, since the formant shift characteristic is grossly linear (Sergeant 1963, Holywell and Harvey 1964, Nixon and Sommer 1968). There has also been some criticism that estimates of the closed tract resonance in air of between 150 and 200 Hz are too high, and that it is only at pressured helium to normal air velocity ratios of $R$ greater than 2.0 that intelligibility is impaired (Morrow 1971).

There is also disagreement as to the effect of the spectral shift characteristic in respect of formant bandwidths. Some results (Gerstman, Gamertsfelder and Goldberger 1966; Morrow 1971; Giordano, Rothman and Hollien 1973) have found that formant bandwidth is unaffected by the formant frequency shift characteristic, whereas more recent research (Belcher and Hatlestad 1982,

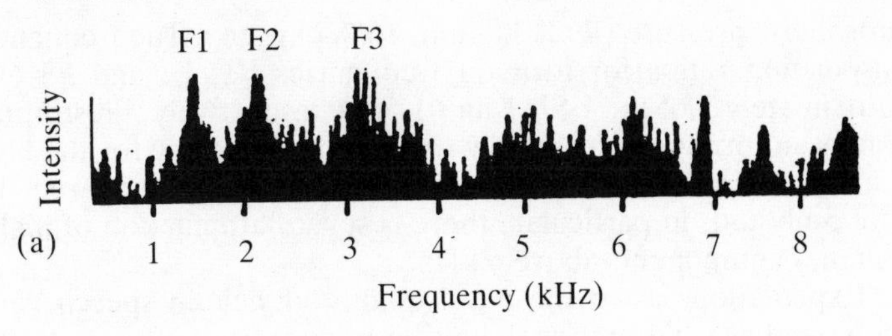


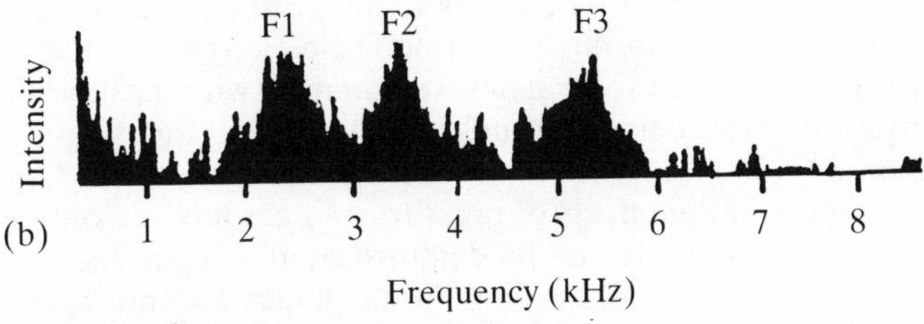


*Figure* 5.6. Vowel spectra for one subject (a) in air, and (b) in a helium-oxygen mixture at sea level (after Sergeant 1963).

Richards 1982) has found that the lower formant bandwidths increase by a ratio of the order of $R^2$ with the bandwidths of higher formants increasing by the ratio $R$ or even less.

There is, however, overall agreement as regards the impairment of intelligibility due to high-frequency attenuation in the voiced spectrum (Sergeant 1963, Flower and Gerstman 1971) and the nasal quality of helium speech (Sergeant 1963, Maclean 1966, Fant and Lindqvist 1968). High-frequency attenuation, effective beyond approximately 5–7 kHz, has been attributed in the main to the fact that, whereas the vocal tract formant frequencies are shifted upwards in frequency, the vocal tract excitation source spectral characteristic remains unaltered (Takasugi, Nakatsui and Suzuki 1971; Tanaka *et al.* 1974). Therefore, since this characteristic exhibits a power spectrum roll-off of −12 dB/octave (Miller 1959, Fant 1979), then high-frequency spectral events in the voiced speech waveform are relatively more attenuated than in a normal air environment. It has also been suggested that acoustic loading of the vocal cords themselves may indeed affect the high-frequency region of the vocal cord source characteristic (Morrow 1971). Figure 5.6 shows two speech spectra of the same vowel uttered (a) by a speaker in air and (b) in a helium environment at normal atmospheric pressure (79% helium, 16% oxygen). The frequency transposition ratios for formant frequencies F1, F2 and F3 are approximately 1.65:1, 1.57:1 and 1.56:1 respectively, illustrating the non-uniformity of frequency translation especially for the first formant. An increase of formant bandwidth is also apparent in figure 5.6b and, in particular, there is severe attenuation of high-frequency components above 6 kHz.

Explanations as to the nasal quality of helium speech vary widely, and as yet a precise definition has not been offered. It has been tentatively related to (a) the increase in overall formant frequency, and in particular nonlinear shifts of low-frequency formants (Nakatsui *et al.* 1973; Tanaka, Nakatsui and Suzuki 1974); and (b) increased transmission of acoustic energy into the nasal cavity itself, resulting in actual nasal resonance, with the most likely transmission path being through the tissue of the soft palate (Morrow 1971).

In respect of the effects of pressure and gas mixture on fundamental frequency, early results demonstrated a slight decrease in pitch period when breathing helium gas under laboratory conditions at normal ambient temperatures and pressures (Beil 1962). However, this effect has been attributed to a physical contraction of the larynx muscles, since the helium gas was colder than room temperature.

Acoustic theory would predict no increase in pitch period as a result of increased helium concentration and little change, if any, due to an increase in ambient pressure, as indeed might be expected from what is in the main a muscularly controlled event. Published data (Copel 1966) relating the pitch period for the same utterances made in air and in a pressured helium-oxygen (heliox) environment found that there was a close correlation between the pitch distribution in both atmospheric air and a heliox atmosphere, simulated at pressure in a compression chamber, on the same day. In contrast, the pitch distribution measured some three months later for the same utterance in air at sea level showed a noticeable change. Conversely, other results demonstrated that pitch period in a pressured heliox atmosphere was reduced in comparison to the pitch period in air (Copel 1966). However, the reduction in pitch period was not found to vary directly with depth since the reduction was greater for a depth of 70 ft than for 200 ft, and in any case the observed changes fell within the expected range of pitch variations for normal speech in air. Several possible causes for these observed changes in pitch period at depth have been suggested. First, at depth and especially in a diving habitat, it has been observed that

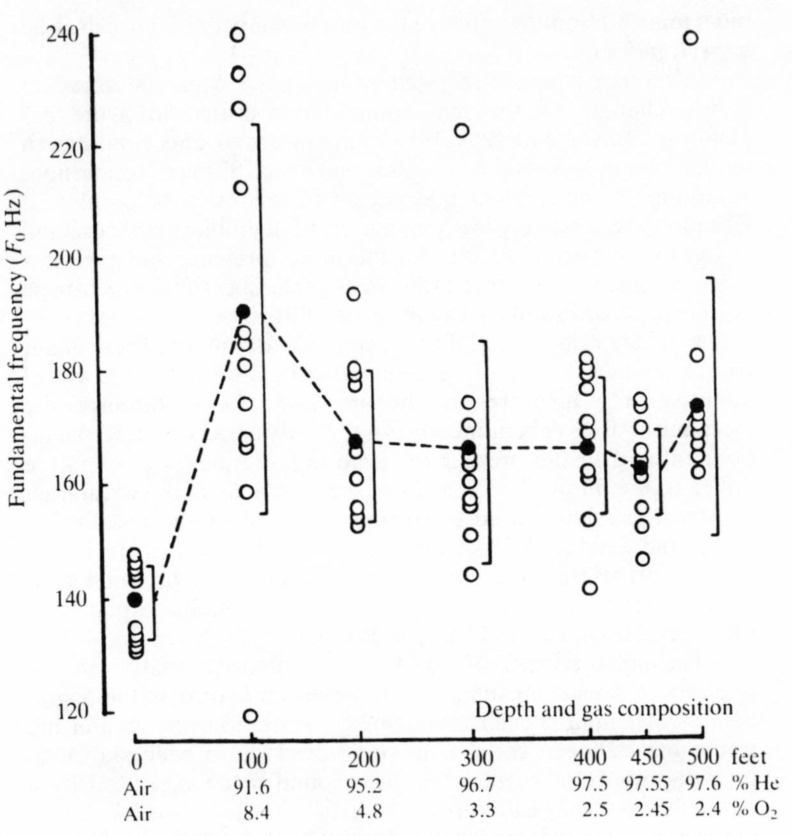


*Figure* 5.7. Fundamental frequency versus depth for one subject in heliox, analysed over 11 different vowels. • = mean $F_0$ over vowel space ; ---- = trace of mean $F_0$ from depth to depth ; ] = standard deviation of $F_0$ about the mean at each depth (after Duncan 1983b).

divers tend to speak with increased vocal intensity in an attempt to overcome background noise levels experienced in the diving chamber environment (Hollien, Shearer and Hicks 1977). Such increases in vocal intensity are normally accompanied by an increase in fundamental frequency. Secondly, environmental effects on the diver's speech mechanism, such as changes in the acoustic loading of the vocal cords due to increased gas density, is expected to produce some slight change in pitch period (Morrow 1971). Finally, the diver may invoke modifications to his speech to alter his pitch in an attempt to enhance the intelligibility of his speech as he himself judges.

Although changes in pitch period have been measured at depth, changes of the magnitude demonstrated in figure 5.7 (Duncan, Laver and Jack 1983) have minimal effect on speech intelligibility (Nakatsui *et al.* 1973, Nakatsui 1974). In conclusion, variations of the fundamental period of repetition of vocal cord vibration are relatively independent of both ambient pressure and respiratory gas composition. Furthermore, pressure and gas composition appear to produce little effect on the spectral characteristic of the vocal cord source (Tanaka *et al.* 1974).

It is generally agreed that the intensity of unvoiced consonant speech sounds is severely attenuated compared to voiced vowel intensity in a high-pressure helium atmosphere, although the mechanism of this phenomenon is not clearly understood. Explanations offered to the present relate to high-frequency attenuation linked to the physical properties of the air stream in the vocal tract constriction as compared to voicing produced by the vocal cords (Fant and Lindqvist 1968, Speakman 1968). An increase in the vocal effort of the diver, which would alter the vowel-consonant intensity ratio, has also been proposed as a possible explanation (Hollien, Thompson and Cannon 1973).

The enhancement of vowel sound intensity relative to the intensity of consonant sounds is an important feature in the degradation of helium speech intelligibility, since consonants and the transitions between consonants and vowels have been shown to provide important cues to the next sound to be produced by a particular speaker (Liberman *et al.* 1954). Hence, if such cues are missing or degraded, the listener's perceptual system may be caught unawares and will effectively suffer a time delay in attempting to assimilate information from the helium speech waveform, and may misinterpret and confuse vowel formant transitions, thereby reducing intelligibility (Sergeant 1967, Suzuki and Nakatsui 1971).

An example of consonant attenuation in helium speech (Brubaker and Wurst 1968) is demonstrated in figure 5.8, which compares (a) vocal intensity as a function of time for the word *fish* spoken at the surface in air and then (b) at 300 ft in a pressured helium atmosphere.

The quantification of the effects of a high-pressure helium atmosphere on the self-intelligibility of the diver is a somewhat complex problem. Whilst it has been estimated that overall spectral sensitivity in the aural canal is attenuated (Fluur and Adolfson 1966), it has also been shown that the diver's auditory characteristic

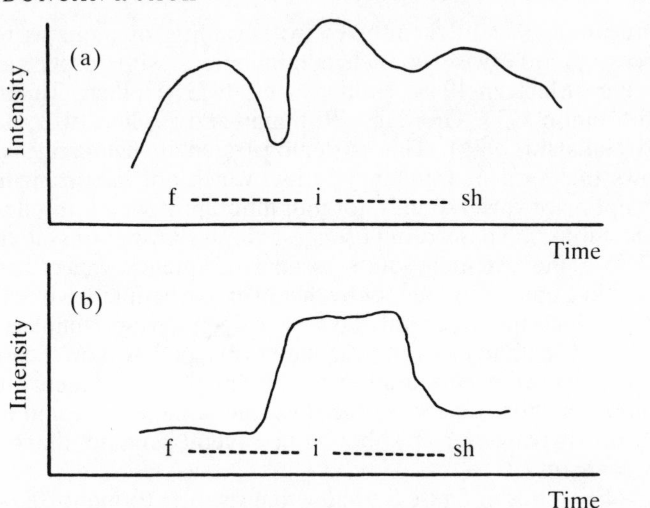


*Figure* 5.8. The word 'fish' spoken by one subject (a) in normal air, and (b) in heliox at a depth of 300 ft (after Brubaker 1968).

has a tendency to emphasise by +10 dB/octave those frequencies above 5 kHz, and attenuate those below 5 kHz (Morrow 1971), which in some respects compensates for the attenuation of high-frequency voiced formants due to the source spectral fall-off characteristic and may also help boost unvoiced consonant intensity. Auditory feedback, however, does not depend solely on the speech signal path through the surrounding respiratory medium, but is a rather convoluted process involving acoustic pathways through the facial tissue and bone, and the processing characteristics of the human brain (Fairbanks 1954). Under normal circumstances in speech in air, the interaction of the signals transmitted through each of these several acoustic pathways produces recognisable signal patterns which indicate to the talker that the sound he has uttered was produced in the manner he desired (Verzeano 1950). Experimentation (Lee 1950) has shown that disruption of the auditory feedback mechanism causes the talker to change his voice output in a manner which, presumably, renders the speech signal more intelligible to himself as he judges, but can be detrimental in terms of the effect upon intelligibility to the listener. In addition, different subjects alter their speech in an individual manner to the same change in the auditory feedback system (Fairbanks 1955).

It has been reported that some divers appear to voluntarily adapt their voices (usually over a matter of days) in order to sound more intelligible to themselves, with varying opinions as to the success or otherwise of such manipulations with respect to the listener (Maclean 1966; Hollien *et al.* 1973; Hollien, Thompson and Cannon 1973; Giordano, Rothman and Hollien 1973; Suzuki and Nakatsui 1974). This is demonstrated in figure 5.9, which shows the formant frequency ratios, which are important in the perception of vowel sounds, for four different vowels uttered by the same subject. These results suggest that relative formant ratios, and to some extent absolute formant frequency values, can be varied in a conscious manner by the diver in a helium environment. The subject spent several days in a deep diving chamber in a pressured, helium environment, and his respective vowel formant frequencies were measured in air before the experiment, in the chamber a short time after the dive had commenced, and finally prior to leaving the chamber after several days in the helium environment.

The trends of figure 5.9 show that the first formant frequency F1 has ultimately tended towards its value in air, and that the relative formant frequency ratios (F1:F2, F1:F3) have also demonstrated a consistent trend towards their values in air.

### *Signal-Processing Techniques for Improving the Intelligibility of Deep Submergent Speech*

The earliest attempts to unscramble helium speech involved the use of tape recorders. The helium speech was recorded at a fast speed and was subsequently played-back at a lower speed, at which the resulting speech was more intelligible (Holywell and Harvey 1964, Geil 1974). A modified version of this technique which enabled real-time processing employed a continuous tape loop on which the helium speech signal was recorded, and was then read-off by a rotating capstan which contained 4 pick-up heads. The speed of rotation was such that the relative velocity of any one head was one-half that of the tape, and so the played-back speech was time-expanded by a factor of two, so rendering it more intelligible. However, these tape methods inherently involved bulky, moving mechanisms and, more importantly, were limited in their ultimate fidelity since, in addition to shifting the formant frequencies of the speech signal, the fundamental frequency was also shifted.

The search to produce unscrambling methods which would enable correction of only the vocal tract frequency response and leave the pitch information intact has led to the development of helium speech unscramblers based on a variety of processing strategies. These approaches can be divided into analogue pro-

| | Vowel 1 | | | Vowel 2 | | | Vowel 3 | | | Vowel 4 | | |
|---|---|---|---|---|---|---|---|---|---|---|---|---|
| | F1 | F2:F1 | F3:F1 | F1 | F2:F1 | F3:F1 | F1 | F2:F1 | F3:F1 | F1 | F2:F1 | F3:F1 |
| Early in experiment | 1000 | 1.9 | 3.25 | 1100 | 1.91 | 3.27 | 950 | 2.3 | 3.3 | 1450 | 1.93 | 2.97 |
| | ↓ | ↓ | ↓ | ↓ | ↓ | ↓ | ↓ | ↓ | ↓ | ↓ | ↓ | ↓ |
| Late in experiment | 900 | 2.0 | 3.28 | 950 | 2.0 | 3.47 | 800 | 2.6 | 3.56 | 1100 | 2.09 | 3.36 |
| In air at sea level | 550 | 3.09 | 4.45 | 600 | 3.0 | 4.17 | 400 | 3.0 | 5.75 | 700 | 2.0 | 3.36 |

*Figure* 5.9. Self-adaptation of formant frequency ratios in a high-pressure heliox environment (after Maclean 1966). F1 in Hz.

cessing techniques, which include unscrambling by time-domain methods; modulation/bandpass filtering and vocoder techniques; and waveform coding techniques based on digital signal processing (DSP). Advances in real-time array processing systems have permitted the realisation of an advanced unscrambling technique within the latter category, but most, although proven in off-line simulation, remain uncommitted to real-time architectures. However, the principal attraction of such techniques relates to their potential to perform nonlinear spectral relocation of the speech signal. Of paramount importance too is their ease of adaptation to adjustments in the algorithms which provide the unscrambling mechanism, both as the knowledge of the nonlinearities affecting the helium speech signal is refined, and indeed as new respiratory mixtures are introduced to allow deeper penetrations of saturation diving.

The system architecture of the most prevalent unscrambler system currently in use, based on time-domain processing (Hicks, Braida and Durlach 1981) is outlined below, and functional singularities which are likely to affect the intelligibility of the unscrambled speech are reviewed. The time-domain technique, due to its convenient realisation as a real-time processing system, can be considered to furnish a reference basis for a comparison of performance of more complex unscrambler systems.

*Helium Speech Unscrambling by Direct Processing of the Time-Domain Waveform*

Helium speech unscramblers in the category of time-domain processing (Stover 1967, Gill 1970, Flower and Gerstman 1971, Dildy 1976, Suzuki *et al.* 1977, Jack *et al.* 1981) achieve improvements in speech intelligibility by direct time-expansion of the helium speech waveform, in synchronism with each pitch period of voiced speech. This strategy maintains voiced fundamental frequency, but each pitch waveform undergoes an overall linear bandwidth compression by the ratio $R$, corresponding exactly to the amount of applied time-base expansion, $R$ (see equation 5.5).

The helium speech unscrambler system based on time expansion is arranged such that the signal is sampled at some rate $f_s$. A means to detect the point on the waveform from which the subsequent waveform segment is to be expanded, which will normally be from the start of the voiced pitch period, is required, and the signal for expansion is stored first in a storage channel and then subsequently read out at a slower rate to achieve effective time expansion of the waveform. This ostensibly implies that in addition to providing a pitch detector (Rabiner *et al.* 1976; McGonegal, Rabiner and Rosenberg 1977) as a means of achieving pitch-synchronous waveform expansion, there may also be a requirement to provide a very large number, $M$, of storage channels since, if voiced fundamental frequency were considered capable of varying instantaneously over a large frequency range, then a number of subsequent pitch periods may occur during the time to expand any single pitch period. From the same consideration, if all of the pitch waveform were to be expanded in time, then each of the $M$ channels would require a large amount $N$ of variable storage space. However, assuming that the pitch period is approximately constant over long periods of time, then the number of storage channels required is fixed at $M = R_{max} + 1$, where $R_{max}$ is the maximum possible *integer* time expansion ratio to be encountered. This is because for a constant pitch period, there may be $R_{max}$ possible pitch periods detected in the time required to expand any existing stored waveform. In the case of helium speech, assuming a worst-case frequency expansion of $R_{max} = 3$ corresponding to a 100% helium atmosphere, then the number of channels required, $M = 4$. However, an infinite amount of variable channel storage space $N$ would cause the resulting time-expanded waveform to be discontinuous in places. Thus, requirements for continuity of the speech signal demand a regular output of time-expanded speech from the unscrambler device, which therefore limits the amount of storage space $N$ in each channel store to vary up to some maximum length. Thus, for long-duration inter-pitch waveforms, only a certain maximum portion of the signal can be stored; that is, the trailing portion of the inter-pitch waveform must be discarded. The requirements for a real-time cost-effective unscrambler system likewise demand a fixed amount of channel storage space $N$, which in turn limits the minimum acceptable pitch period to a value dependent upon both $N$ and the signal sample rate, $f_s$.

The block structure of a miniature helium speech unscrambler system architecture for diver-borne use (Jack, Milne and Virr 1979; Jack *et al.* 1981) is shown in figure 5.10. Note that most systems apply a preamplification filter with invariant characteristics to the input helium speech signal both to improve (reduce) dynamic range and to provide some correction for the attenuation of the high-frequency spectral components of helium speech. In the miniature system, which employs analogue charge transfer devices for waveform storage and CMOS digital circuitry for control logic functions, the channel length is $N = 256$ points with a sample frequency of $f_s = 80$ kHz, which gives a waveform capture time of

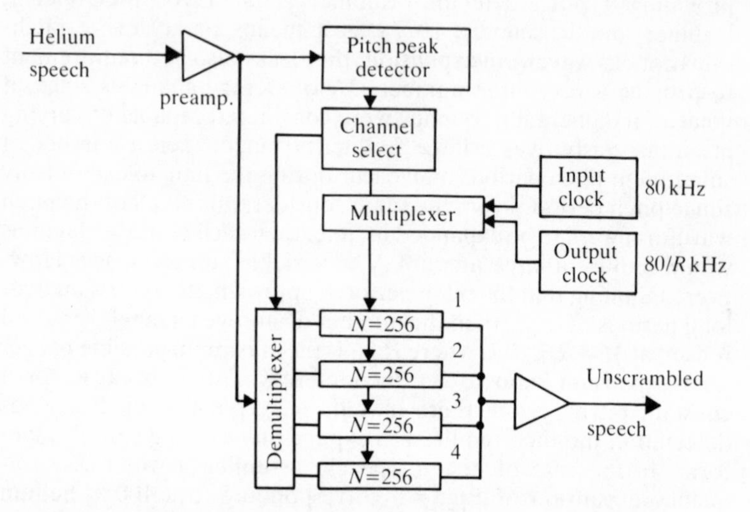


*Figure* 5.10. Block diagram of the time-domain-based pitch-synchronous real-time helium speech unscrambler system.

3.2 ms, and since no other pitch periods may trigger the device during this time, then the unscrambler is capable of operating with diver fundamental frequencies up to 310 Hz. However, it is the design of the pitch detector circuitry which is the most crucial aspect in the operation of this system, since it not only detects the start of a pitch period, but controls the multiplexed input and output clocks to individual storage channels.

The pitch detector in this device is based on waveform amplitude peak detection with hysteresis (Jack, Milne and Donaldson 1979). The upward shift in spectral components, due to the increased speed of sound in the heliox mixture in comparison to air, produces a faster decay time constant in the helium speech pitch waveform envelope, and therefore pitch peaks become more pronounced than in air, so that a peak detector can be successfully employed. When no voiced speech is detected, the waveform segmentation device relies on random excursions of the waveform to provide triggering. An alternative unscrambling strategy involves retriggering at regular intervals, in this case according to default timing intervals supplied by an internal clock.

In respect of factors affecting the performance of the time-domain unscrambling technique, the most important criterion relates to the operation of the pitch detection device. Considering the peak-detection method with hysteresis, the crucial parameters have been found to be the choice of peak trigger voltage $V_g$ and hysteresis decay time constant $\tau_d$. A trigger voltage $V_g$ that is set too close to the quiescent (no input) level of the system may cause repeated triggering of the peak detector within any single pitch period, as indeed may a hysteresis decay constant $\tau_d$ that is short compared to the expected time constant $\tau_e$ of the inter-pitch waveform envelope decay. Conversely, the peak detector may fail to identify the start of a pitch period if $V_g$ is set too high such that the full-wave rectified input speech signal never crosses the required trigger threshold, and/or if the decay constant $\tau_d$ is too long compared to $\tau_e$.

Investigation of these systems invariably shows that the resultant unscrambled speech exhibits audible discontinuities, to the detriment of intelligibility. This effect cannot directly be related to the acoustic properties of helium speech, but is attributable to the processing mechanics of the unscrambler system. There are several possible causes for this fragmented speech output.

Assuming meantime that all voiced pitch intervals are correctly detected, then the fragmentory nature of the unscrambled speech may be due to the time expansion of only a finite portion of the pitch waveform, in that there may be significant perceptual information residing in the discarded portion of the pitch waveform. However, experimental results (Duncan 1983) show that the signal energy of the pitch waveform of air speech falls off sharply beyond some 3–4 ms after the start of each pitch period. Given that each channel store in the device retains the first 3.2 ms of the pitch waveform of helium speech, and since the inter-pitch helium speech waveform decay is much faster than in air, then it is therefore not expected that significant perceptual information is being rejected.

Another possible reason for the fragmentation of the speech may be due to the choice of pitch detector trigger threshold, in that not all pitch periods are detected. Investigation has found that for strongly-voiced speech sounds the triggering of the detector corresponds well to the pitch intervals detectable by phonetic experts, whereas pitch in weakly-voiced segments is less reliably detected, thereby accounting for part of the broken effect of the unscrambled speech. The same problem, however, occurs with unvoiced speech, for which there is no defined pitch interval. The operation of the device under these conditions depends on random excursions of the speech waveform beyond the peak detector trigger threshold, and therefore also depends on the dynamic range of the helium speech. Subjective evaluation of the unscrambler system shows that the

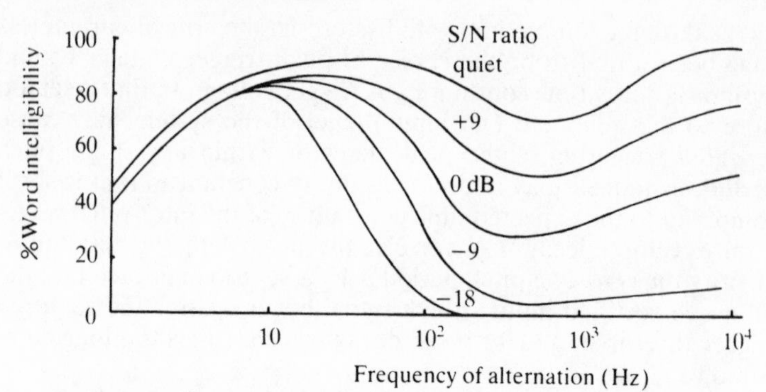


*Figure* 5.11. Word intelligibility as a function of the frequency of alternation between speech and noise, with signal-to-noise ratio as the parameter (after Miller 1950).

worst breaks in the unscrambler output, in fact, occur during periods of unvoiced speech. It has been shown that the disruption of continuity of speech in this manner degrades intelligibility. Increasing the frequency of disruption beyond 10 Hz produces a progressive impairment of intelligibility by effectively distracting the human perceptual mechanism, thereby masking the speech with intermittent noise. Figure 5.11 demonstrates the effect on intelligibility of alternating intervals of equal length of speech and silence or white noise (Miller and Licklider 1950). The intelligibility is measured in terms of the percentage of words heard correctly, and is plotted against frequency of interruption, which ranged from 0.1 Hz to 10 kHz, for signal-to-noise power ratio (SNR) which varied from +9 dB to −18 dB during the interruptions. Shown also is the response, marked 'quiet', where the intervals between speech were silent, with no added noise.

It can be seen from figure 5.11 that for an interruption rate in the range 10–560 Hz, and for each SNR curve, intelligibility is increasingly impaired with interruption rate. The pitch frequencies of most male and female talkers fall within this range (Laver and Trudgill 1979); hence if a succession of unvoiced pitch decisions are made in error of actual voiced speech sounds, then the voiced speech will be effectively masked in noise as the pitch detector malfunction causes either no output to be produced or produces output at default intervals, dependent on the unscrambler system under consideration. Thus, signal continuity is an important aspect of intelligibility.

However, notwithstanding the above-mentioned theoretical

limitations on the performance of time-domain-based unscrambling methods using waveform segmentation and expansion, systems based on this technique remain the most prevalent real-time helium speech unscrambler systems in day-to-day heliox diving operations. The nature of the signal processing involved in time domain methods lends itself well to the production of inexpensive electronic unscrambler systems, together with permitting low power consumption and miniaturisation leading to the inception of diver-borne unscrambler systems (Jack *et al.* 1981). Such diver-borne miniature systems engender an economisation of speech transmission bandwidth to the surface, hence reducing the effects of additive noise on the communications channel – a significant advantage in consideration of the problems of crossfeed from high-frequency, high-current switched signals to control heating/lighting, which share the same umbilical cable as the millivolt signals of transmitted speech.

### *Helium Speech Unscrambling using Modulation and Vocoder Techniques*

The first technique to appear in this category employs spectral relocation by applying amplitude modulation, where the input baseband helium speech is first converted upwards in frequency by a balanced modulator. This is followed by a down-conversion stage consisting of a balanced demodulator of slightly higher but variable frequency. Use of such a variable-frequency mixing oscillator allows some form of fine tuning of spectral relocation and hence caters for variations in helium mixture.

A variant of this technique allows for separate relocation characteristics to be applied to individual segments of the helium speech baseband through the inclusion of a dual-band heterodyning arrangement (Copel 1966). The spectral effects of this technique are illustrated in figure 5.12. Note that in either technique, the helium speech effect is corrected by arithmetic subtraction of a frequency relocation constant. That is, each component of the baseband helium speech is shifted wholesale down in frequency by $\delta f = F_{mo} - F_{dmo}$, where $F_{mo}$ is the frequency of the modulation oscillator and $F_{dmo}$ that of the demodulation oscillator.

A second approach to unscrambling using modulation techniques employs analysis-synthesis (vocoder) schemes (Holmes 1972). This method offers unscrambling by scaling the frequency spectrum of the helium speech by ratio multiplication as opposed to the modulation techniques above which effectively employ arithmetic subtraction.

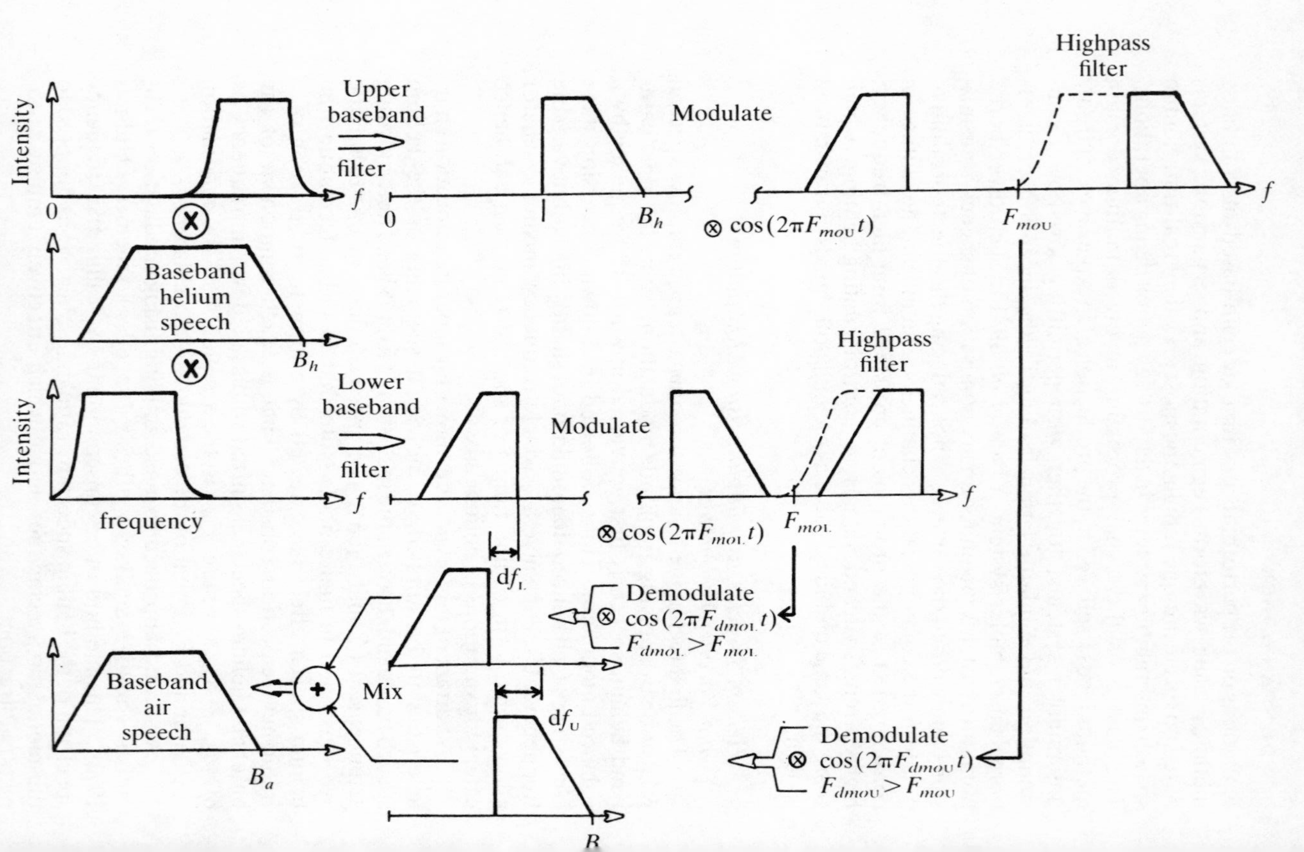


*Figure* 5.12. Sequence of modulation-demodulation spectral manipulations in unscrambling helium speech by frequency subtraction (after Copel 1966).

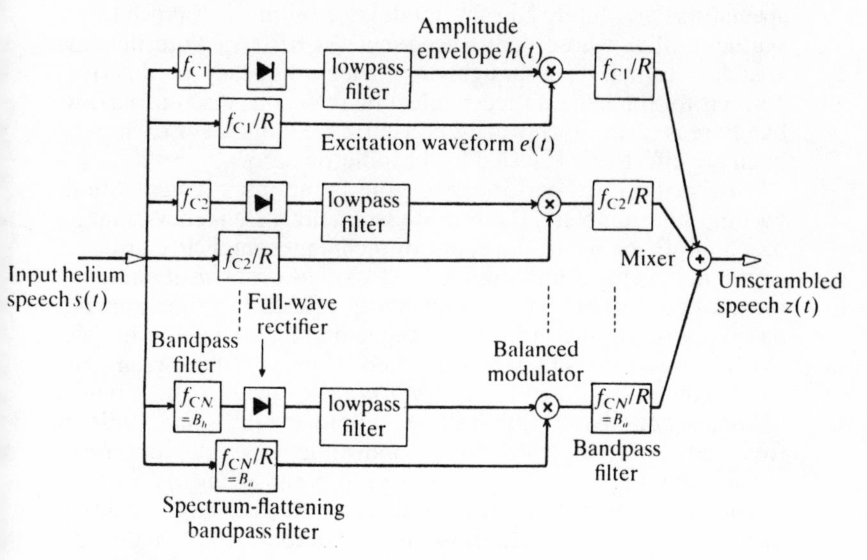


*Figure* 5.13. Block diagram of the self-excited vocoder system proposed for use as a helium speech unscrambler system (after Golden 1966).

In general, the input helium speech with bandwidth $B_h$, say, is decomposed in the analysis operation into two sets of components (Golden 1966, Zurcher 1980). Let there be $N$ contiguous bandpass filters in any given analysis/synthesis section of the device, and let $R$ be the required gross spectral compression ratio required to restore intelligibility. Initially, the full helium speech bandwidth $B_h$ is resolved into $N$ components. Each analysis filter in the filter bank is followed by a full-wave rectifier and smoothing circuitry to provide a slowly-varying time-amplitude (or spectral energy) envelope of the helium speech characteristic in the given passband (see figure 5.13). At the same time, a second set of $N$ analysis filters, which are constrained to cover the range $B_h/R$, provide estimates of the glottal waveform over a set of frequency bands which cover the required bandwidth of air speech, $B_a$, say. This operation is valid since the glottal excitation is considered to be invariant from air to heliox. In order to relocate the spectral energy envelope of any $n$th component of helium speech in the passband centered on $nB_h/N$ to its air-equivalent spectral location at $nB_h/NR$ $(=nB_a/N)$, then the output of the $n$th glottal waveform analysis filter (centered on $nB_a/N$) is modulated by the time envelope from the $n$th helium speech analysis filter, and the modulator output is followed by a synthesis filter whose centre frequency is $nB_a/N$. Note that the vocoder technique, although employing frequency modulation, corrects for the helium speech effect by relocating discrete narrowband areas of the spectrum. The centre frequencies of each passband are effectively rescaled by a geometric ratio.

In respect of performance considerations of modulation/vocoder unscramblers, the use of essentially a frequency-domain-based approach in this category of techniques implicitly provides continuity of output unscrambled speech. The problem of randomly interrupted output, which particularly affects the time-domain-based approach due to its dependency on pitch detection for successful operation, is therefore avoided. However, in unscrambling by frequency subtraction, every frequency within the original helium speech bandwidth will be shifted by the same constant frequency, thus, original formant bandwidths are totally conserved. Since helium speech is by nature very nasal in quality with an attendant increase in formant bandwidth, then the nasal quality is likely to be conserved in the unscrambled output. Additionally, this technique does not attempt to deconvolve glottal source characteristics from the composite helium speech waveform prior to application of any spectral correction. Therefore both pitch frequency and the overall spectral characteristic of the glottal characteristic are adversely affected.

The vocoder-based techniques explicitly seek to conserve the glottal excitation characteristic during translation from heliox to air speech. Additionally, spectral correction by geometric ratio compression of narrowband segments can be expected to relieve some of the nasal quality of helium speech. However, this is a result of applying synthesis filters, with bandwidth $B_a/N$, to each relocated passband, rather than any explicit compression of formant bandwidth. Since each synthesis (air) passband relies on modulation by the energy envelope of its corresponding heliox passband, then formant bandwidths and amplitudes will be totally conserved through the use of the same modulating signal applied to both heliox and air carrier waveforms (narrowband glottal excitation). Note that the geometric scaling properties of this unscrambling method relate uniquely to the relationship between corresponding analysis and synthesis filter passband centre frequencies. Any apparent formant compression is a fortuitous consequence of applying a narrower spectral window, of bandwidth $B_h/NR$, to each modulated synthesis passband as opposed to the original analysis bandwidth of $B_h/N$. The vocoder-based techniques theoretically have the ability to cater for nonlinear frequency translation by varying the geometrical relationships between corresponding analysis and synthesis filter centre frequencies, but analogue implementations of these devices have not included this feature. Additionally, correction for high frequency formant attenuation, although possible by manipulation of gain within selected passbands, has not been realised in practical implementations. Finally, note that all of the modulation/vocoder techniques employ a piecewise decomposition of the speech spectrum through their dependence on bandpass filtering. That is, individual frequency components, given any required arbitrary spectral resolution $\delta f < B_h/N$, cannot be appropriately corrected for the helium speech effect.

It is interesting to note that whilst several patents relating to unscrambler devices based on modulation/vocoder techniques have been lodged, there are no known helium speech unscrambler systems of this type currently in service. However, use of computer-controlled filter centre frequency adjustments and digital filtering techniques now make systems of this type attractive to produce as real-time, robust unscrambler systems.

*Unscrambling Techniques based on Waveform Coding*

For the general class of linear systems, then by the superposition principle (Kuo 1966), if $[e_1(t), s_1(t)]$ and $[e_2(t), s_2(t)]$ are excitation-response pairs, then if the excitation were $e(t) = e_1(t) + e_2(t)$, the response $s(t) = s_1(t) + s_2(t)$. Also, if the system impulse response is described by $h(t)$, then $s(t)$ is the convolution (D'Azzo 1966) of $h(t)$ and $e(t)$:

$$s(t) = h(t) * e(t) \tag{5.13}$$

In the case of helium speech correction, one or all of the system waveforms $h(t)$ (vocal tract impulse response), $e(t)$ (glottal excitation waveform) or $s(t)$ (helium speech) must be manipulated according to some predetermined correction algorithm. One method to achieve the required correction might be to solve for the required parameter by expressing equation 5.13 as a series of linear differential equations. However, since the problem here relates to manipulation of the speech spectrum, which is a function of frequency, then it is natural to encode the time-varying system waveforms of equation 5.13 into an equivalent frequency-domain representation. By the convolution theorem (Jenkins and Watts 1968):

$$s(t) = h(t) * e(t) \rightarrow S(f) = H(f)E(f) \tag{5.14}$$

where $V(f)$ is some frequency transform of the respective time-varying signal $v(t)$, and $t$ and $f$ represent continuously variable time and frequency respectively.

Waveform encoding to a frequency-domain representation is particularly useful since the convoluted relationships of time-varying quantities transform into algebraic relationships in the frequency domain and *vice versa.* This simplifies considerably the mathematical manipulation of the various component signals. Notice at this point that whereas the helium speech correction algorithms may be nonlinear as a function of frequency, the application of frequency transformations generally demands that the system under consideration is linear and conforms to the principle of superposition. This is exemplified for the speech signal by a consideration of the model of speech production presented earlier and in figure 5.2. A further general assumption relating to the use of waveform coding is that the speech system is time-invariant (Oppenheim, Willsky and Young 1983), that is:

$$s(t-\tau) = h(t) * \delta(t-\tau) \qquad (5.15)$$

where $\delta(t-\tau)$ is a Dirac impulse occuring at some random time $t=\tau$. This general condition is required since the theoretical definition of virtually all frequency transformations involves signal integration over infinite time. From the discussion of the speech mechanism presented earlier, it was seen that the configuration of the vocal tract, corresponding to $h(t)$ in equation 5.15, could only be considered time-invariant over short periods of time of up to 30 ms or so, and therefore special precautions are required to ensure that any waveform-coding-based correction algorithm will process only those portions of the speech signal which can be considered as time-invariant in the short term. This involves application of time windows (Harris 1978) to the speech waveform, prior to processing, which effectively force the signal to exist only for finite time but with minimal detrimental effect on the processing mechanics of the appropriate signal transform. The complexity of these signal processing techniques also demands their implementation in digital computer architectures.

The advantages of the digital waveform coding approach to unscrambling relate principally to (a) improved spectral resolution allowing correction of within-formant frequency components; (b) ease of adaptation to nonlinear spectral correction characteristics; and (c) improved deconvolution of the spectral and temporal characteristics of the glottal source waveform and vocal tract impulse response.

Advances in array processing technology have allowed the implementation in real time of a highly complex helium speech unscrambler based on the waveform coding technique of the short-time Fourier transform (STFT) (Allen 1977, Portnoff 1981, Richards 1982, Belcher and Andersen 1983).

It is a well-established result that the Fourier spectrum $E(f)$ of a periodic impulse train is itself periodic in frequency. If this impulse train excites a linear time-invariant system whose transfer function is $h(t)$ and whose spectral response is given by $H(f)$, then the resulting time signal $s(t)$ is given by the convolution integral (Healy 1969)

$$s(t) = \int \delta(\tau - n\tau) h(t-\tau) d\tau \qquad (5.16)$$

$$= h(t - n\tau) \qquad (5.17)$$

That is, $s(t)$ is a periodic function in time with the impulse response $h(t)$ repeated every $\tau$ seconds. Furthermore, from the convolution theorem expressed in equation 5.14, then the two spectra $H(f)$ and $E(f)$ are simply multiplied in the frequency domain, therefore the signal spectrum $S(f)$ is similar to the spectrum of $E(f)$ in that $S(f)$ is also periodic in frequency with the same spectral line spacing, $F_0 = 1/\tau$. If the signal $s(t)$ is assumed to be the result of the periodic impulse excitation of a linear time-invariant system, then if $H(f)$ is known or can be estimated, then $E(f)$ can be recovered (Richards 1982, Belcher 1982) from equation 5.14:

$$E(f) = S(f)/H(f) \qquad (5.18)$$

This is an important feature in the use of the short-time Fourier transform for helium speech unscrambling. As has been discussed above the fundamental frequency of voiced speech is unaltered by a pressured heliox atmosphere, and therefore the periodicity of normal and helium speech voiced waveforms is identical. If $S(f)$ in equation 5.18 is the spectrum of the periodic helium speech voiced waveform, which is directly observable and calculable, and if the helium vocal tract frequency response $H_h(f)$ can be obtained, which is unobservable and therefore must be estimated, then the excitation spectrum $E(f)$ can be derived, thereby obviating the need to maintain pitch information by pitch extraction in the time domain (Harris and Weiss 1963). Correction for the helium speech distortion is then applied directly to $H_h(f)$ to produce a new estimate of the vocal tract frequency response $H_a(f)$, which is then remultiplied together with the excitation spectrum $E(f)$ to produce a new spectrum $Z(f)$ corresponding to the unscrambled speech signal $z(t)$:

$$Z(f) = H_a(f)E(f) \tag{5.19}$$

and $z(t)$ is obtained from the inverse Fourier transform of $Z(f)$:

$$z(t) = \int Z(f)e^{j2\pi ft}df \tag{5.20}$$

Note, however, that special precautions are necessary to ensure that the spectral periodicity (line spacing $F_0$) remains intact on manipulation of $H_h(f)$. In the case of speech, the waveform can at best be considered quasiperiodic in the short term (Rabiner and Schafer 1975). Additionally, conditioning for application of the FFT demands a truncated time series. The combined effect of these two conditions is to introduce non-zero values of sidelobe amplitude between the (otherwise perfectly periodic) line spectrum peaks located at $f=n/\tau$, where $n$ is any integer value, and $\tau$ is the mean pitch period of voiced speech. Thus, not only does the frequency response of the vocal tract filter multiply the principal peaks of the line spectrum, but the inter-line sidelobes too are affected (Stewart 1960). Therefore the combined effects of signal truncation and quasiperiodicity render direct calculation of the helium vocal tract frequency response $H_h(f)$, even given information regarding the statistical characteristics of the waveform jitter, very difficult and tedious, and such a solution is not propitious for real-time implementation. Spectral envelope estimation algorithms (Makhoul 1976) are therefore necessary, and the STFT unscrambler method employs a piecewise-linear curve-fitting technique (Richards 1982) in order to separate helium vocal tract frequency response from the glottal excitation spectrum.

In the real-time STFT unscrambler architecture, the helium speech signal is initially band-limited to 7.4 kHz and sampled at 14.8 kHz. The samples are windowed by a symmetrical triangular (Bartlett) window and transformed with an analysis frame length $N=512$ points, therefore the spectral resolution is 29 Hz. The fast Fourier transform (FFT) frequency transformation (Brigham and Morrow 1967, Bergland 1979, Brigham 1974, Arambepola 1980) is then applied to aid with real-time operation. Contiguous analysis frames overlap by 50% such that the data in the last half of any analysis frame $A_n$ also constitutes the data in the first half of the subsequent analysis frame $A_{n+1}$. Notice that the window is structured such that addition of the frames given by $\Sigma A_n$ would restore the original signal $s(t)$. This overlapping is required for two reasons. First, the loss of information due to the deliberate signal attenuation towards each edge of the window in frame $A_n$ is restored by the emphasis at the window apex in frames $A_{n-1}$ and $A_{n+1}$. Secondly, whilst the window length is 34.6 ms, new updates in the signal variations can be processed every 17.3 ms, thereby improving the trade-off between spectral resolution and length of time window (Crochiere 1980).

Once each frame is transformed into the frequency domain, magnitude and phase spectra are obtained from the complex FFT frequency spectrum. In all of the STFT unscrambler system architectures reported to date, correction for the helium speech distortion is confined to the magnitude spectrum alone, with the phase spectrum being left unchanged. After magnitude spectrum correction for the helium speech effect, the new warped magnitude is recombined with the retained original helium speech phase spectrum, and the new air-equivalent resulting complex spectrum is then inverse-transformed to produce a signal representing the unscrambled intelligible speech (see figure 5.14).

The spectral envelope is therefore derived solely from the magnitude spectrum, and the envelope estimation algorithm is based here on a piecewise-linear curve-fitting method (Richards 1982). The basic principle is very similar to that of peak detection with hysteresis in the time domain, as outlined earlier except that the method used here employs a linear threshold decay characteristic in the frequency spectrum whose slope is dependent on the amplitude of the last spectral pitch line detected. That is, the normalised slope connecting some candidate spectral line peak to the previously detected spectral line peak (assuming a peak search from 0 Hz increasing in frequency up to half the sample frequency, $f_s/2$) cannot be less than some threshold given by a minimum negative number $\epsilon$. This constraint is intended to restrain the spectral envelope from descending too rapidly to include sidelobe peaks due to truncation and pitch jitter as discussed above. There is no limit on the maximum rising slope of the envelope in order to allow a fast recovery should a sidelobe or noise peak be detected.

Notice that if the same sampling frequency $f_s$ is conserved at the system output, then overall spectral compression by a ratio $R$ results in an undefined region of the spectrum between $f_s/2R \rightarrow f_s/2$ and its image from $f_s/2 \rightarrow (1-1/2R)f_s$. Simply setting this spectral region to zero amplitude is reported to produce spurious artefacts in the resulting waveform (Richards 1982), and to avoid this effect, the spectrum is tapered linearly to zero across this region from the last known amplitude value at $H(f_s/2R)$ to $H(f_s/2)$.

The resulting unscrambled speech frame is not in a suitable state for immediate output, however. In consideration of the overlapped-frame method of analysis, then only the first half of the

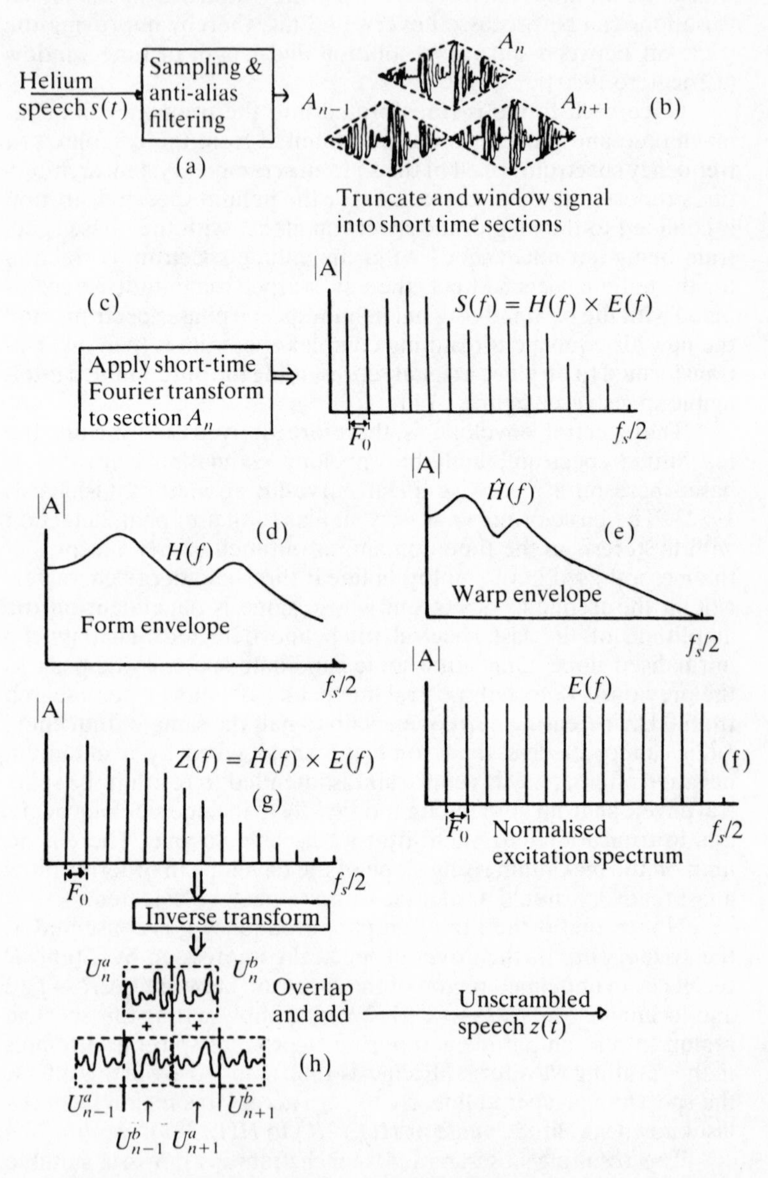


*Figure* 5.14. Sequence of operations in helium speech unscrambling using the short-time Fourier transform.

current frame of unscrambled speech, $U_n$, is used and added to the (stored) last half of the previous frame, $U_{n-1}$, with $U_n$ being stored to await the arrival of frame $U_{n+1}$, and so the resulting unscrambled speech output is produced by a continuous overlap-and-add process (Crochiere 1980).

Signal continuity and homogeneity of processing are a feature of the short-time Fourier transform unscrambler system (Richards 1982, Belcher and Andersen 1983) and, additionally, signal integrity is conserved with no requirement to discard portions of the input speech signal. Although the real-time implementation of this complex system excludes presently a diver-borne unscrambler system and so necessitates conservation of helium speech bandwidths for transmission, the frequency-domain-based rationale of this technique permits ease of nonlinear spectral correction of both frequency and amplitude for the helium speech distortion. The voiced/unvoiced nature of the speech is intrinsically conserved with no explicit pitch detection required whatsoever, so preserving signal continuity.

The principal enigma of this system, however, relates to the retention of the original phase spectrum and its subsequent recombination with the (co-phase) magnitude spectrum prior to execution of the inverse Fourier transform to produce the unscrambled speech. The reason offered for retention of the original phase spectrum relates to the apparent insensitivity of the ear to spectral phase in that human speech perception is considered to be based mainly on magnitude spectrum information (Fant 1960), and therefore as far as the phase information is concerned, '. . . since the ear is known to be relatively insensitive to moderate phase distortion, the simple approach of doing nothing has been taken' (Richards 1982).

There is an important dichotomy of principles here, however. Whilst the speech perceptual system within the human brain appears indeed to depend mainly on spectral magnitude (Fant 1960), transduction of the acoustic waveform at the ear (Flanagan 1965), does however, depend on the signal phase characteristic (Mathes and Miller 1947). The maxim of phase insensitivity is an oft-misquoted version of de Boer's rule, (de Boer 1961) which asserts that '. . . the timbre (or apparent frequency content) of a sound does not change when the phases of the components are shifted by a constant amount and/or by amounts that are linearly dependent on frequency.'

In respect of the STFT unscrambler architecture, the processing mechanics can equivalently be viewed as producing a signal with the correct air-speech phase spectrum as required, but which is then passed through a phase-shifting network, prior to output, whose phase characteristic is non-uniform with frequency such that the original helium speech phase characteristic is restored. The following extract summarises the importance of relative phase on the perceived timbre of the acoustic signal (de Boer 1961):

> . . . Thus when the phase $\phi(f)$ of the component with frequency $f$ is changed by an amount:
>
> $$\phi(f) = \alpha + \beta f \tag{5.21}$$
>
> where $\alpha$ and $\beta$ are constants, no change of the timbre will occur. The evidence for this rule comes from experiments from signals of which the components have equal spacings. Such signals, be they harmonic or not, have a steady sound quality.
>
> The term $\beta f$ in the formula is easily understood. In fact it represents a simple shift in time equal for all components. For a steady sound this is, of course, not noticeable. The important term is the constant $\alpha$, expressing a constant phase shift of all components. The phases of the components can all be changed by the same amount without bringing about a change in timbre. . . . If such a common phase shift produced an audible change, we would hear beats (in this case slow variations of timbre). Nothing of the kind occurs, however, and we must conclude that such a uniform phase shift is not detectable. We have studied this phenomenon for many frequency ranges, and the rule seems to hold always. The exceptions could all be traced down to artefacts giving rise to non-uniform phase shifts.

Complementary research has supported this phenomenon (Schouten, Ritsma and Cardozo 1962; Ritsma and Engel 1964; Buunen *et al.* 1974), and the consensus of opinion is that changes of phase in a signal which are nonlinear with frequency alter the envelope of the time waveform, which creates extraneous frequencies when transduced by the ear, thereby generating an internal magnitude spectrum (Goldstein 1967) in the human perceptual system which may be very different to that obtained by a straightforward machine-generated Fourier analysis of the signal.

Although approached from a different point of view, dependency of the time waveform on spectral phase is reinforced by experiments involving signal reconstruction from either phase or magnitude spectra alone (Hayes, Lim and Oppenheim 1980; Nawab, Quatieri and Lim 1983; Yegnanarayana and Dhayalan 1983), which implies that the phase spectrum contains significant information relating to the frequency content of the signal

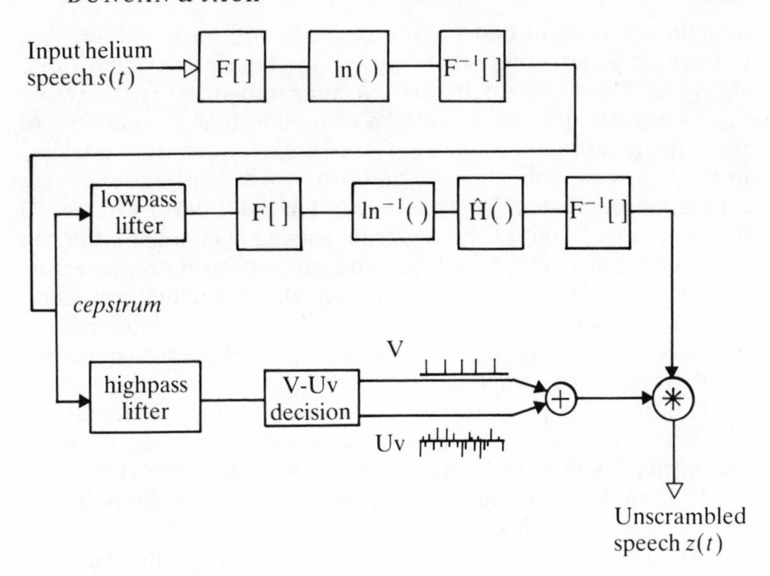


*Figure* 5.15. Helium speech unscrambler employing cepstrum-based digital signal processing.

(Oppenheim and Lim 1981), and therefore that magnitude and phase spectra of linear time-invariant systems are in essence unique to each other (Cook and Bernfeld 1967).

In cepstrum-based helium speech unscrambling, the principal objective remains that of deconvolving the vocal tract frequency response from the glottal excitation waveform. As outlined previously, the short-term view of the speech signal as originating from a linear time-invariant filter system leads to multiplicative relationships for $H_h(f)$ and $E(f)$ in the frequency domain. The cepstrum technique initially applies a complex Fourier transform to short-time windowed segments of the speech waveform. The complex logarithm of this Fourier spectrum is then obtained and the resulting sequence is inverse Fourier-transformed to give a pseudo-time series known as the 'cepstrum' (Childers, Skinner and Kemerait 1977; see figure 5.15):

$$\mathrm{F}^{-1}[\ln(H_h(f)E(f))] = \mathrm{F}^{-1}[\ln(H_h(f)) + \ln(E(f))] \quad (5.22)$$

$$\rightarrow h_h(t') + e(t') \quad (5.23)$$

where $\mathrm{F}^{-1}$[ ] is the inverse Fourier transform operator, ln( ) is the logarithmic operator, and $t'$ is a pseudo-time variable known as 'quefrency'. That is, convolution relationships in the original time domain are transformed to additive relationships in the cepstral domain. In general, note that $H_h(f)$ in the Fourier spectrum varies slowly as a function of frequency compared to $E(f)$, therefore $h_h(t')$ is a low-quefrency function of pseudo-time $t'$ compared to the cepstral representation of $e(t')$. Thus, low-quefrency 'liftering' in the cepstrum will separate the vocal tract contribution to the cepstrum from that of the glottal excitation (Noll 1967). Once $e(t')$ has been liftered out of the cepstrum leaving $h_h(t')$, the latter can be inverse-transformed back to produce a spectrum representing an estimate of $H_h(f)$ alone, to which any desired helium correction characteristic can be applied (Quick 1970). The corrected spectrum then becomes the frequency response $H_a(f)$ of the air-equivalent vocal tract which, when inverse-transformed back to the time-domain, is the impulse response of the vocal tract. Treating this time waveform as a finite impulse response (FIR) filter, it can be reconvolved with an estimate of the glottal excitation waveform to produce an air-translated version of the original helium speech waveform over the short-time analysis interval.

Similarly to the STFT unscrambler system (Richards 1982), the foremost criterion affecting system performance of the cepstrum-based approach relates to considerations of signal phase. In this case, the problem is two-fold and encompasses both spectral and cepstral phase characteristics. The theoretical definition of the cepstrum presumes a complex logarithm operator to be applied to the spectrum. In practice, however, it is difficult to generate a complex logarithm as a direct result of using a sampled signal because of the uncertainty of a multiple of $\theta = 2\pi j$ in the discrete Fourier spectrum phase. Additionally, cepstral phase is sensitive to time-shifting of the waveform segment under analysis, and indeed can be dominated by this factor. Although there are procedures for direct generation of the cepstrum without recourse to logarithmic operators, ceptrum analysis techniques generally reject spectral phase and apply the logarithmic operator solely to the magnitude spectrum (Oppenheim and Schafer 1968). In purely waveform analysis scenarios, this does not represent much of an inconvenience. However, where the cepstrum is to provide a basis for correcting a time waveform, the lack of accurate cepstral phase information can cause distortions in the resulting spectrum and hence in the time waveform. The problem is compounded by the consideration of retained spectral phase when correction is applied only to the magnitude spectrum, as discussed earlier in de Boer's rule. Additionally, the problem of undefined areas of the corrected spectrum still exists, and the resulting time waveform is sensitive to the assumptions and values ascribed to these undefined spectral regions.

The resulting corrected time waveform resulting from cepstrum processing is an FIR representation of the vocal tract impulse response in air, and hence improves on the STFT method in that, assuming that the glottal excitation waveform has been correctly detected, any signal distortion is confined to the inter-pitch waveform. The STFT method, on the other hand, may potentially distort the entire composite speech waveform since it inverse Fourier-transforms both pitch and vocal tract response into the time domain using a nonlinear phase characteristic. Also, although the original helium speech segment is time-windowed in the cepstrum approach to helium speech unscrambling, the added complexity and indeterminate effects of an overlap-and-add approach (Crochiere 1980) are avoided by treating the corrected spectrum as a filter frequency response. The effects of windowing can then be considered to merely improve the estimate of this filter response by reducing spectral leakage, as opposed to affecting the time envelope of the entire composite speech waveform. However, accurate estimation of the glottal waveform is a nontrivial processing task, and no automatic means of cepstrum pitch determination as applied to helium speech has been reported, although pitch estimation algorithms do exist and are applied regularly to problems of vocal-fold pathology (Laver, Hiller and Mackenzie 1984) and to speech recognition (Tucker and Bates 1978).

Helium speech unscrambling by autoregressive signal processing generally employs linear prediction coding (LPC) analysis (Makhoul 1975, Atal and Hanauer 1971) as the main signal-processing vehicle. This technique explicitly views the speech signal as being the result of a filter (the vocal tract), characterised by an assumed all-pole transfer function, which has been excited by a train of unit Dirac impulses or by white noise (i.e. the excitation is assumed to be spectrally white).

In the LPC approach, the present signal value is assumed to be calculable from a weighted linear sum of several past values of the signal waveform, i.e.

$$s(n) = a_1 s(n-1) + a_2 s(n-1) + \ldots + a_p s(n-p) \qquad (5.24)$$

although the problem is more often written as:

$$\sum_{k=0}^{p} a_k s(n-k) = \epsilon \qquad (5.25)$$

where the coefficient $a_0 = 1$, and the least mean squares minimisation conditions are applied to each value of $a_k$ to force the prediction error, or residual, $\epsilon$, to some arbitrarily small value.

From a filtering standpoint, the coefficient sequence

$$a_0 + a_1 z^{-1} + a_2 z^{-2} + \ldots + a_p z^{-p} \quad (5.26)$$

can be viewed as the tap weights of a $p+1$ length FIR filter, $i(t)$, which is applied to the speech signal, $s(t)$. Specifically, $i(t)$ is viewed as an inverse or prediction error filter of order $p$ in that, if the output residual waveform $\epsilon(t)$ is spectrally white, then the frequency response of the inverse filter, $I(f)$, must be the exact inverse of that of the vocal tract, and the residual signal itself then represents the glottal excitation waveform. Applying this technique to the helium speech waveform, then $I(f) = 1/H_h(f)$, and hence simply inverting each component of the inverse filter frequency response will yield an estimate of $H_h(f)$, which is now deconvolved from the pitch characteristic.

Similarly to the cepstrum unscrambling technique (Quick 1970), any desired nonlinear correction characteristic can now be applied to produce an air-equivalent vocal tract response, $H_a(f)$, which is then inverse-transformed to produce a new FIR filter sequence:

$$b_0 + b_1 z^{-1} + b_2 z^{-2} + \ldots + b_p z^{-p} \quad (5.27)$$

where $b_0 = 1$. This filter represents the inverse of the required air vocal tract, and therefore must be inverted to produce an infinite impulse response (IIR) filter. The inversion of the filter is structurally simple to achieve, and merely requires a single interconnection change on one tap weight. This synthesis filter can then be excited with the conserved glottal excitation (residual) signal to produce air-equivalent speech.

Several architectures have been proposed which employ LPC processing. The earliest implementation (Suzuki, Ooyama and Kido 1974) merely time-expanded the original FIR sequence in equation 5.26 to produce the new filter sequence of equation 5.27, whose tap weights $b_k$ were linearly interpolated from the original $a_k$ sequence, dependent on the value, $R$, of the overall spectral compression required (see figure 5.16).

A later technique (Beet and Goodyear 1983a,b) has attempted explicitly to solve for the $z$-plane transfer function of the inverse filter by using an approximate root estimating transform implemented as a digital adaptive filter consisting of a cascade of second-order sections (see figure 5.17). Each section of the adaptive filter effectively solves for the positions of one pole-pair in the vocal tract transfer function, and produces a $z$-plane pole position as defined

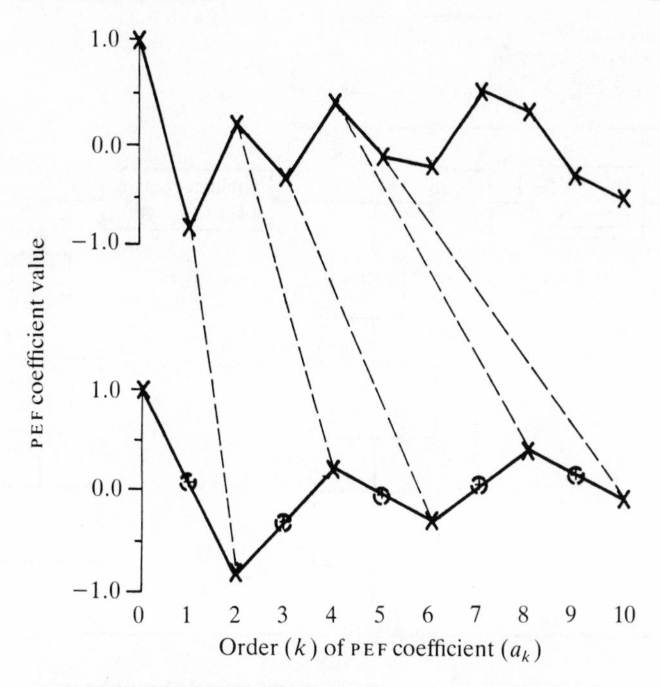


*Figure* 5.16. Correction of a 10th-order prediction error filter (PEF) for the helium speech effect by linear interpolation. $R$=2. × = original PEF coefficients, ○ = interpolated values (after Suzuki, Ooyama and Kido 1974).

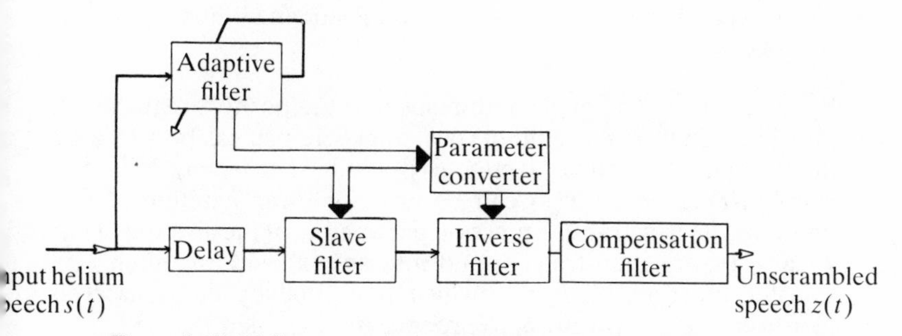


*Figure* 5.17. Helium speech unscrambling using an adaptive filtering approach (after Beet 1983a).

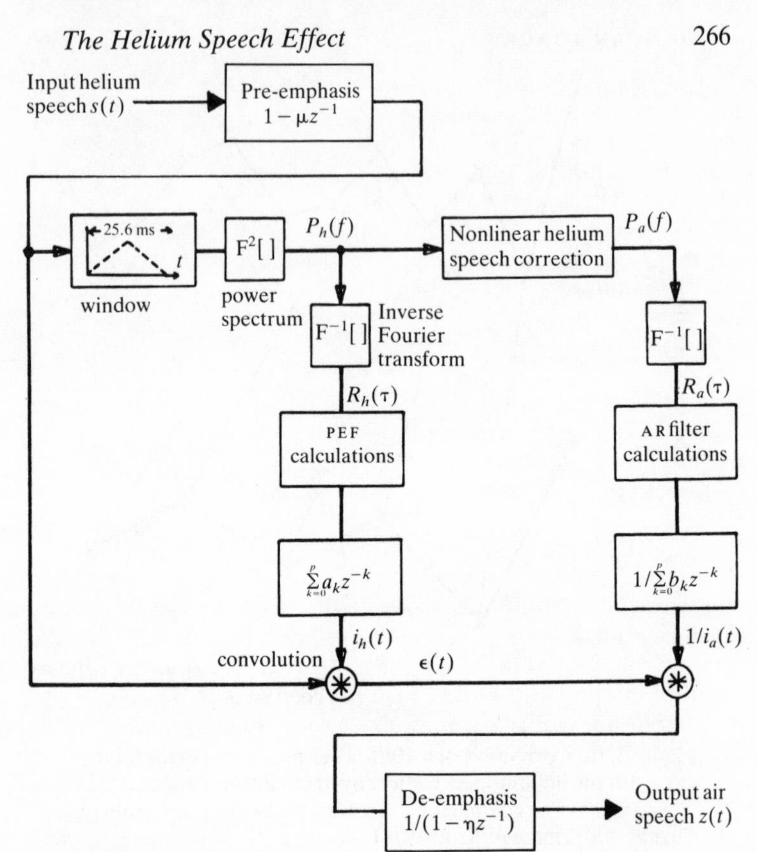


*Figure* 5.18. Block diagram of the spectrum-based LPC helium speech unscrambler system. F[ ] is the Fourier transform operator.

by its radius, $r$, and angle with respect to the positive real axis, $\theta$. Any individual $z$-plane helium vocal tract pole-pair can be relocated in frequency by some factor $R(f)$ such that $r_a = r_h^{1/R(f)}$, and $\theta_a = \theta_h / R(f)$, where $R(f)$ can be any nonlinear function of frequency. The new transfer function parameters of the air-equivalent vocal tract are then transformed into an IIR synthesis filter with transfer function $1/\Sigma b_k z^{-k}$, which is excited by the conserved glottal excitation waveform to produce unscrambled speech.

Explicit spectral-domain helium unscrambling permitting nonlinear correction is afforded by a third LPC-based unscrambler system (Duncan 1983, Duncan and Jack 1983; see figure 5.18). Similarly to the above techniques, the glottal waveform is estimated by calculating short-time inverse filter parameters and applying each filter to its corresponding short-time segment of helium speech. However, this system differs in its approach by using the Weiner-Kinchine theorem (Weiner 1930) in estimating both heliox and air-equivalent filter structures.

The solution for the coefficient set $a_k$ of equation 5.25 using least mean squares techniques necessitates the calculation of the autocorrelation sequence of the signal. From the Weiner-Kinchine theorem:

$$P(f) = \int A(\tau) e^{-j2\pi f\tau} d\tau \qquad (5.28)$$

where $P(f)$ is the power spectrum of the signal and $A(\tau)$ is the autocorrelation sequence. For a real-valued signal such as speech, then the power spectrum can simply be calculated by applying a short-time Fourier transform to the signal and squaring and adding real and imaginary frequency components at each discrete frequency. Calculation of the helium speech autocorrelation function $A_h(\tau)$ is then possible by applying the inverse transform to that in equation 5.28 above. Note however that special precautions are necessary in digital signal processing to force a linear autocorrelation from the cyclic autocorrelation afforded by the discrete Fourier transform (Cooley, Lewis and Welch 1967; Rader 1970).

The strategy of using the power spectrum as a route to the autocorrelation function is particularly convenient since the power spectrum implicitly contains spectral information relating to vocal tract formants and can therefore itself be directly corrected for the helium speech effect.

The resulting power spectrum $P_a(f)$ corresponding to normal air speech can be used to form the autocorrelation function $A_a(\tau)$ of the spectrally corrected signal, permitting the calculation of the new FIR filter sequence of equation 5.27, from which the synthesis filter is constructed and re-excited by the conserved residual excitation. Note that in this method, the power spectrum, prior to spectral correction, still contains information regarding the glottal excitation, since no attempt is made to deconvolve this when applying the Fourier transform to the short-time segment. Therefore, applying spectral compression to the power spectrum of voiced speech necessarily entails an effective increase in the fundamental frequency of the resulting speech. However, this apparent imperfection is redressed here by a consideration of an implicit but rarely explicitly expressed property of autoregressive signal processing. Namely, although any single prediction error filter is constructed from a signal having, say, a well-defined fundamental frequency, the resulting inverse filter will produce a spectrally-white residual when applied to any similar signal originating from the same, fixed linear time-invariant system, but which possesses a different fundamental frequency.

As with all waveform coding techniques, LPC-based helium speech unscramblers offer continuity of the unscrambled speech waveform. In linear prediction modelling in general, the optimum fit to the vocal tract transfer function assumes an all-pole model. Therefore, those features of the signal spectrum which are directly due to the influence of transfer function poles, that is, the areas of resonance (formants) in the spectrum, will be modelled best, whereas features due to the influence of transfer function zeros, i.e. spectral antiresonances, will not be faithfully represented. This is in contrast to STFT and cepstrum techniques, which make no assumptions other than that the system under analysis is linear time-invariant in the short term, and which therefore treat all components (poles and zeros) of the vocal tract transfer function. However, although the characteristic transfer function of certain speech sounds, such as nasals, do indeed contain zeros of transmission, the important aspects of speech perception still relate to formant information, that is, the relative locations, amplitudes and bandwidths of spectral peaks, so vindicating the use of AR modelling techniques. However, whereas this is acceptable in an analysis-only scenario, the explicit application of the prediction error filter to the helium speech is unlikely to result in a residual which is spectrally white, since it will contain those elements in the short-time spectrum which were not well matched by the all-pole model, specifically, areas of antiresonance.

LPC-based unscramblers employing simple time expansion of the prediction error filter coefficients (Suzuki, Ooyama and Kido 1974) afford only linear spectral compression with no ability to correct individually either formant bandwidth or amplitude. It is also likely that the corrected vocal tract frequency response is sensitive to the interpolation algorithm employed to obtain the new coefficient set of equation 5.27. This is because small changes in prediction error coefficients are known to produce greater changes in the spectrum when compared to the same percentage changes in, say, cepstral coefficients representing the same vocal tract frequency response. Indeed, for this reason, the latter are preferred to LPC coefficients as a data reduction mechanism in many speech recognition systems based on template matching (Davis and Mermelstein 1980).

The technique based on the matched-$z$ transform (Beet and

Goodyear 1983a,b) offers overall nonlinear correction of the helium vocal tract transfer function. However, formant bandwidth and amplitude correction is difficult to control, since although individual pole-pairs can be manipulated, the gross spectral characteristic of any formant depends on all components of the system transfer function, except in the unlikely case where a formant has a sufficiently high $Q$-factor to let it be considered decoupled from the remaining system poles. The helium speech effect is best understood in terms of its spectral and temporal characteristics (Bond and Myatt 1969, Maitland and Thomas 1974), and the adoption of a transfer function approach represents an abstract view of the problem from which the effects of transfer function manipulation are currently difficult to quantify. The use of the matched-$z$ transform in the formulation of the corrected transfer function is also reported to adversely affect intelligibility by imparting a low-frequency boost to the unscrambled speech, although this can be attenuated by applying a post-processing bass-cut filter.

The unscrambling technique based on frequency-domain correction (Duncan and Jack 1983, Duncan 1987) also offers nonlinear correction of all aspects of the short-time vocal tract frequency response. Although the time waveform requires windowing to avoid spectral leakage, no overlap-and-add (Crochiere 1980) approach is required since, similarly to the cepstrum approach, the spectrum is used to estimate a filter impulse repsonse which operates on an unwindowed version of the waveform. Additionally, preemphasis, which imparts a +6 dB/octave emphasis to the speech waveform, is a well-established requirement to increase the accuracy of the modelling process (Markel and Gray 1967), and helps implicitly to correct for the high-frequency attenuation of helium speech. A matching de-emphasis filter cascaded to the synthesis filter restores the usual air speech formant characteristic. Precautions necessary to ensure linear autocorrelation from the inverse power spectrum (Rader 1970) effectively double the spectral resolution and hence improve the accuracy of spectral correction using index remapping and interpolation. Similarly to the STFT method, areas of the spectrum are undefined after spectral compression. In contrast to the STFT method, however, spectral tapering has an adverse effect on the synthesis filter. Indeed, here, the solution is much simpler and consists of simply repeating the last known frequency value (which must be non-zero) in the power spectrum over the undefined region. This satisfies conditions of maximum entropy which the technique must satisfy. Most importantly, the use of the co-phase power spectrum and the Wiener-Kinchine relationship avoids any explicit consideration of spectral phase whatsoever. Figure 5.19 illustrates the spectrum-based LPC unscrambler applied to the correction of helium speech spoken at a depth of 250 ft (pressure = 8.5 bar) in an atmosphere consisting of 96 % helium and 4 % oxygen. Shown inset is the Hamming-windowed waveform segment for analysis in each case. Note the fundamental excitation periodicity of approximately 6 ms, which is also apparent in each power spectrum due to the 167 Hz spacing of spectral lines. The vocal tract frequency response is estimated by the dotted-line envelope which has been overlaid on each spectrum, and formants F1, F2 and F3 have been identified.

### *Summary, Conclusions and Future Developments*

Correction of the helium speech effect by signal processing demands an *a priori* knowledge of the desirable characteristics of the end-product waveform which will result from the application of processing to the input signal. Here, the resulting acoustic waveform must conform to the needs of the human speech perception mechanism such that the information which the talker wishes to convey is in a form which is intelligible to the listener. In normal speech in air, the human interpretation of the speech waveform depends upon a complex interplay of temporal and acoustic events, the most salient of which have been detailed here. The known effects of a pressured helium-oxygen atmosphere on the acoustic waveform have also been outlined and related to those perceptual factors which are considered to directly affect the intelligibility of helium speech.

Notwithstanding the technological complexity of systems available for helium speech unscrambling, which embrace a variety of signal processing strategies as presented above, there has still been cause to vilify the quality of the helium speech they produce (Hollien and Rothman 1974, Brown and Feinstein 1976). There are two categories into which the root causes for the poor performance of these systems may be divided. Firstly, there may be assumptions pertaining to the acoustic events in helium speech which these systems make but which are false, or indeed the systems may be deficient in their provisions for processing certain acoustic attributes of the signal which have an important bearing on intelligibility. This may certainly be true in that, whilst there is general agreement that the glottal excitation characteristic is invariant and that the vocal tract frequency response is nonlinearly affected, most published research differs in the precise details regarding the nonlinear characteristics. Secondly, the processing algorithms which

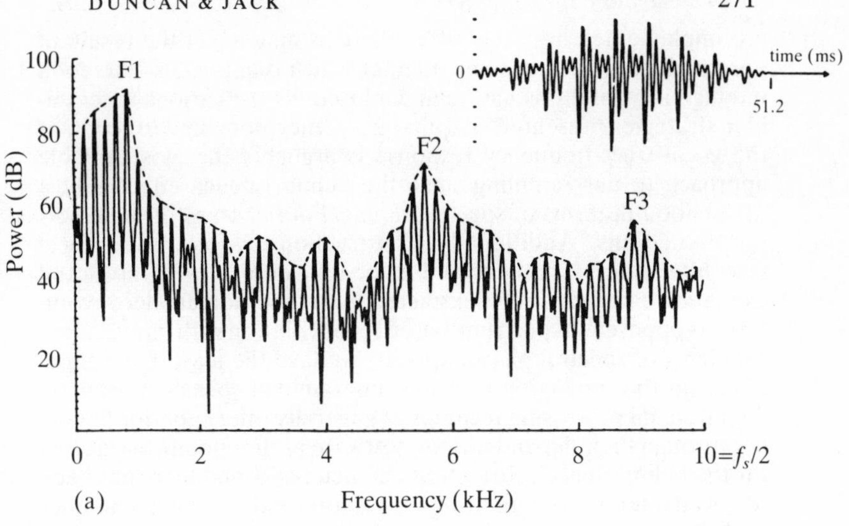


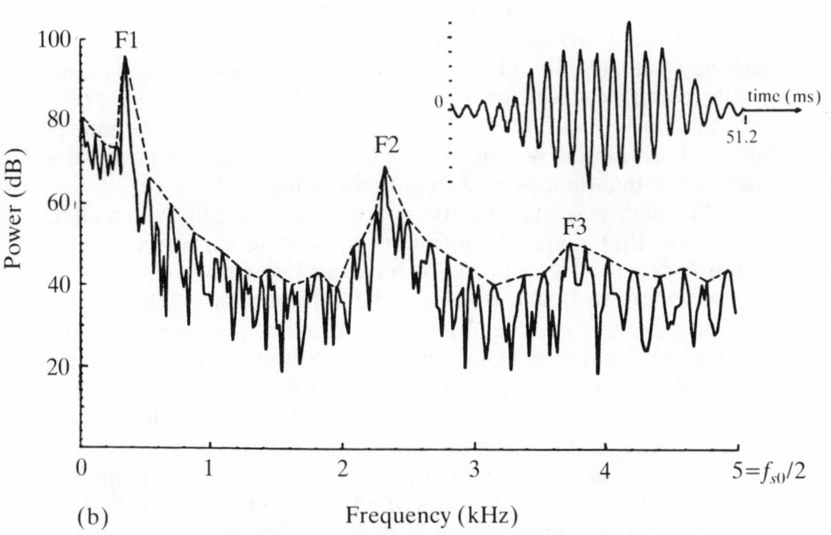


*Figure* 5.19. (a) Helium speech power spectrum for the vowel 'ee' (51.2 ms) before processing by the spectrum-based LPC unscrambler system. (b) Power spectrum of corrected speech segment using a linear compression of $R$=2.4. (Note: the power spectrum of (b) is obtained from a post-processing analysis of the output unscrambled speech waveform.)

are implemented in unscrambler systems may affect the resultant unscrambled speech in some manner which is antagonistic to good intelligibility, and this has been explored for the various unscrambler strategies presented. Explicit frequency-domain correction of the vocal tract frequency response is arguably the most suitable approach to unscrambling since the helium speech effect is best understood in terms of spectrographic/Fourier transform analysis representations. Additionally, construction of a corrected vocal tract filter which is reconvolved in the time domain with the glottal excitation estimate also as extracted from time-domain deconvolution, as opposed to spectrum-based reconvolution with its attendant problems of spectral phase, appears to have the least detrimental effect on the intelligiblity of the unscrambled speech. Certainly, digital signal processing techniques generally offer superior flexibility through their dependence on software programming in catering for the helium speech correction characteristic, and there has been renewed interest in committing modulation and vocoder techniques to high-speed digital architectures (Jelinek 1986).

A problem which afflicts all unscrambler systems is the inadequacy with which the glottal excitation is estimated, since although assumed to be a Dirac impulse train or white noise for most processing strategies, the glottal waveform is in reality a highly composite complex waveform (Laver 1980). However, to the extent that the vocal tract filter function is unobservable, then estimation of the true glottal waveform is likely to remain an intractable problem in one-dimensional signal processing.

Ancillary research into the nature of the helium speech effect has found that there are formants present in certain sounds in helium speech which appear to have no known correlates in air speech (Duncan 1983). It has also been found that certain sounds are shifted uniquely in helium speech, exhibiting phoneme-specific formant shift profiles (Duncan 1983), certain of which are very different in form compared to a deterministic formant shift characteristic. Such findings support the theory that the helium speech effect depends not only on the properties of gas mixture and pressure on the vocal tract, but also on the talker's perception of his own speech, in that he may be affecting his articulation such that his speech is more intelligible as he perceives it. The reinterpreted data from earlier research (Maclean 1966) supports this view. Indeed, future developments in unscrambler technology may include a limited application for speech recognition techniques, and this method is already under trial (Trehern and Jack 1986).

Several important factors relating to the diving environment also indirectly affect the performance of helium speech unscramblers, although these factors remain relatively unquantified. The acoustic properties of the cavities in the mask worn by a diver can worsen the speech distortion by selectively amplifying areas of the speech spectrum (Morrow and Brouns 1971a,b). Microphone type and placement with respect to the lips and nose can also play an important role, particularly in attenuating noise from the surrounding environment (Murry and Sergeant 1971, Trukken and Pelton 1984). Bubble noise released from the gas demand valve and broadband noise from carbon dioxide extraction equipment is particularly troublesome in this respect (Duncan 1983).

Man's excursion into the deep-sea environment necessitates the use of extraordinary respiratory mixtures in order to sustain life. Exploitation of the ocean floor to extract organic fuels and minerals is at present confined to working depths of some 600 ft or so, and therefore helium-oxygen mixtures are physiologically suited to the well-being of man at these depths. It has already been proposed, however, that current oil resources in the known sites will be exhausted by the mid-twenty-first century, and therefore exploration will become necessary at depths far in excess of those commonly encountered at present. Hydrogen-oxygen mixtures are in limited use in order to allow deeper penetration into the ocean environment (Sergeant 1972; Macedo, Strohl and Brecher 1985). It has also been shown (Bennett *et al.* 1974) that the nitrogen which is narcotic at present working depths is in fact physiologically necessary for excursions beyond 800 ft or so, which changes the physical properties of the respiratory gas, and therefore the associated speech distortions.

It is indeed a paradox to reflect that man may extend his hand to other worlds whilst still enjoying instantaneous and effective interlocution with his fellow man, and yet may scarcely explore 200 ft inside his own world without encountering severe difficulties in communication.

## REFERENCES

Allen, J. B. (1977) Short term spectral analysis, synthesis and modification by discrete Fourier transform. *IEEE Trans. ASSP-25*, 235-8.

Arambepola, B. (1980) General discrete Fourier transform and fast Fourier transform algorithm, in *Signal Processing: Theories and Applications* (eds M. F. Kunt & F. de Coulon). North Holland Publishing Company.

Atal, B. S. & S. L. Hanauer (1971) Speech analysis and synthesis by linear prediction of the speech wave. *J. Acoustical Society of America 50*, 637-55.

Bacon, F. (Viscount St. Albans) (1620) lib.II, cap.L, *Novum Organum, sive Indicia Vera de Interpretatione Naturae,* 339-40. London.

Baume, A. D., D. R. Godden & J. R. Hipwell (1982) Procedures and language for underwater communication. *UEG Technical Note 26,* Underwater Engineering Group, London.

Beet, S. & C. C. Goodyear (1983a) Making helium speech intelligible. *IEE Colloquium on Digital Processing of Speech,* 1983/31, 11/1-5.

—— (1983b) Helium speech processor using linear prediction. *Electronics Letters 19,* no.11, 408-10.

Behnke, A. R., R. M. Thompson & E. P. Motley (1935) The psychologic effects from breathing air at 4 atmospheres. *American J. Physiology 112,* 554-8.

Beil, R. G. (1962) Frequency analysis of vowels produced in a helium-rich atmosphere. *J. Acoustical Society of America 34,* no.3, 347-9.

Belcher, E. O. (1982) The cause and enhancement of helium speech. *Publication of Norwegian Underwater Technology Centre,* Bergen, 1982.

Belcher, E. O. & K. Andersen (1983) Helium speech enhancement by frequency-domain processing. *Proc. 1983 IEEE Conference on Acoustics, Speech and Signal Processing (ICASSP-83) 3,* 1160-63.

Belcher, E. O. & S. Hatlestad (1982) Analysis of isolated vowels in helium speech. *Report 26-82, Norsk Undervannsteknologisk Senter,* March 1982.

Bennett, P. B., G. D. Blenkarn, J. Roby & D. Youngblood (1974) Suppression of the high pressure nervous syndrome in human deep dives by He-$N_2$-$O_2$. *Undersea Biomedical Research 1,* 221-37.

Berg, J. (1955) Transmission of the vocal cavities. *J. Acoustical Society of America 27,* 161-68.

—— (1958) Myoelastic/aerodynamic theory of voice production. *J. Speech and Hearing Research 1,* 227-44.

Bergland, G. D. (1969) A guided tour of the Fast Fourier Transform. *IEEE Spectrum,* 41-52.

Bert, P. (1876) La pression de l'air et les êtres vivants. *La Revue Scientifique de La France et de l'Etranger 3,* 49-55.

Bjork, L. (1961) Velopharyngeal function in connected speech. Studies using tomography and cineradiography synchronized with speech spectrography (book review). *Acta Radiographica 56,* 399.

de Boer, E. (1961) A note on phase distortion in hearing. *Acustica 11,* 182.

Bond, W. H. & J. M. Myatt (1969) Investigation of distortion of diver's speech using power spectral estimates based on the fast Fourier transform. Master's thesis, US Naval Postgraduate School, Monterey, June 1969.

Borelli, G. A. (1685) Pt.I, propos.CCXXII. Machinae constructio, qua homines, demersi intra aquam, possent per plures horas respirare et vivere, in *De Motu Animalium,* 270-2 (+tab.XIV). Lugduni in Batavis.

Borovickova, B., V. Malac & S. Pauli (1970) F2-transitions in the perception of Czech stops. *Quarterly Progress and Status Report, Speech Transmission Laboratory, STL-QPSR-1,* Royal Institute of Technology, Stockholm, 14-15.

Brigham, E. O. (1974) *The Fast Fourier Transform.* Prentice-Hall, Eaglewood Cliffs, NJ.

Brigham, E. O. & R. E. Morrow (1967) The fast Fourier transform. *IEEE Spectrum,* 63-70.

Brown, D. D. & S. H. Feinstein (1976) An evaluation of three helium speech unscramblers to a depth of 1000 feet. *J. Sound Vibration 48,* no.1, 123-35.

Brubaker, R. S. & J. W. Wurst (1968) Spectrographic analysis of divers' speech during decompression. *J. Acoustical Society of America 43,* no.4, 798-802.

Buunen, T. J. F., J. M. Festen, F. A. Bilsen & G. van den Brink (1974) Phase effects in a three-component signal. *J. Acoustical Society of America 55,* no.2, 297-303.

Carlson, R. & B. Granstrom (1977) Perception and synthesis of speech. *Quarterly Progress and Status Report, Speech Transmission Laboratory, STL-QPSR-1,* Royal Institute of Technology, Stockholm, 1-16.

Carlson, R., B. Granstrom & S. Pauli (1972) Perceptive evaluation of segmental cues. *Quarterly Progress and Status Report, Speech Transmission Laboratory, STL-QPSR-1,* Royal Institute of Technology, Stockholm, 18-24.

Case, E. M. & J. B. S. Haldane (1941) Human physiology under high pressure: I. Effects of nitrogen, carbon dioxide and cold. *J. Hygiene 41,* 225-49.

Childers, D. G., D. P. Skinner & R. C. Kemerait (1977) The cepstrum: A guide to processing. *Proc. IEEE 65,* no.10, 1428-43.

Clark, J. M. (1982) Oxygen toxicity, in *The Physiology and Medicine of Diving* (3rd edition, ed. D. H. Elliott). Baillere & Tyndall, London.

Cook, G. E. & M. Bernfeld (1967) *Radar Signals: an introduction to theory and application.* Academic Press, New York.

Cooley, J. W., P. A. Lewis & P. D. Welch (1967) Application of the fast Fourier transform to computation of Fourier integrals, Fourier series, and convolution integrals. *IEEE Trans. AU-15,* no.2, 79-84.

Copel, M. (1966) Helium voice unscrambling. *IEEE Trans. AU-14,* no.3, 122-6.

Crochiere, R. E. (1980) A weighted overlap-add method of short-time Fourier analysis/synthesis. *IEEE Trans. ASSP-28,* 99-102.

D'Azzo, J. J. & C. H. Houpis (1966) *Feedback Control Systems Analysis and Synthesis* (2nd ed.). McGraw-Hill.

Davis, S. B. & P. Mermelstein (1980) Comparison of parametric representations for monosyllabic word recognition in continuously spoken sentences. *IEEE Trans. ASSP-28,* 357-66.

Delattre, P. C., A. M. Liberman & F. S. Cooper (1955) Acoustic loci and transitional cues for consonants. *J. Acoustical Society of America 27,* no.4, 769-73.

Denes, P. B. & E. N. Pinson (1963) The acoustic characteristics of speech, in *The Speech Chain.* Bell Telephone Laboratories.

Diehl, C. F. & E. T. McDonald (1956) Effect of voice quality on communication. *J. Speech and Hearing Disorder 21,*233-7.

Dildy, C. A., Jr (1976) Helium speech unscrambler with pitch synchronisation. *US Patent Specification 3 950 617,* 13th April 1976.

Duncan, G. (1983) Analysis and correction of the helium speech effect by autoregressive signal processing. PhD thesis, University of Edinburgh, 1983.

—— (1987) Correction of the helium speech effect by short-time autoregressive signal processing. *Proc. 5th Int. Conf. Electronics for Ocean Technology,* Institution of Electronic and Radio Engineers Publication, 125-30.

Duncan, G. & M. A. Jack (1983) Residually excited LPC processor for enhancing helium speech intelligibility. *Electronics Letters 19,* no.18, 710-11.

Duncan, G., J. Laver & M. A. Jack (1983) A psycho-acoustic analysis of divers' fundamental frequency variations in a high-pressure helium-oxygen atmosphere. *Work in Progress, Dept of Linguistics, University of Edinburgh, 16,* 9-16.

Fairbanks, G. (1954) A theory of the speech mechanism as a servosystem. *J. Speech and Hearing Disorders 19,* 133-9.

—— (1955) Selective vocal effects of delayed auditory feedback. *J. Speech and Hearing Disorders 20,* 333-46.

Fairbanks, G. & P. Grubb (1961) A psychophysical investigation of vowel formants. *J. Speech and Hearing Research 4,* 203-19.

Fairbanks, G. & A. S. House (1953) The influence of consonant environment upon the secondary acoustical characteristics of vowels. *J. Acoustical Society of America 25,* 105-13.

Fairbanks, G. & M. S. Miron (1957) Effects of vocal effort upon the consonant-vowel ratio within the syllable. *J. Acoustical Society of America 29,* 621-6.

Fairbanks, G., A. S. House & K. N. Stevens (1950) An experimental study of vowel intensities. *J. Acoustical Society of America 22,* 457-9.

Fant, G. (1959) The acoustics of speech. *Proc. 3rd Int. Cong. Acoustics,* 188-201.

—— (1960) *Acoustic Theory of Speech Production.* Mouton, The Hague (Netherlands).

—— (1967) Sound, features and perception. *Quarterly Progress and Status Report, Speech Transmission Laboratory, STL-QPSR-2,* Royal Institute of Technology, Stockholm, 1-14.

—— (1972) Vocal tract wall effects, losses and resonance bandwidths. *Quarterly Progress and Status Report, Speech Transmission Laboratory, STL-QPSR-2-3,* Royal Institute of Technology, Stockholm, 28-52.

—— (1979) Glottal source excitation analysis. *Quarterly Progress and Status Report, Speech Transmission Laboratory, STL-QPSR-1,* Royal Institute of Technology, Stockholm, 85-107.

Fant, G. & J. Lindqvist (1968) Pressure and gas mixture effects on diver's speech. *Quarterly Progress and Status Report, Speech Transmission Laboratory, STL-QPSR-1,* Royal Institute of Technology, Stockholm, 1-17.

Fant, G. & B. Sonesson (1964) Speech at high ambient air pressure. *Quarterly Progress and Status Report, Speech Transmission Laboratory, STL-QPSR-1,* Royal Institute of Technology, Stockholm, 9-21.

—— (1967) Diver's speech in compressed air atmosphere. *Military Medicine 132,* 434-6.

Fant, G., J. Lindqvist, B. Sonesson & H. Hollien (1971) Speech distortion at high pressure, in *Underwater Physiology: Proc. 4th Int. Cong. Diver Physiology,* 293-99.

Flanagan, J. L. (1955) Difference limen for vowel formant frequency. *J. Acoustical Society of America 27,* 613-7.

—— (1965) *Speech Analysis, Synthesis and Perception.* Academic Press, New York.

Flower, R. A. & L. J. Gerstman (1971) Correction of helium speech distortions by real-time electronic processing. *IEEE Trans. COM-19,* no.3, 362-4.

Fluur, E. F. & J. Adolfson (1966) Hearing in hyperbaric air. *Aerospace Medicine 37,* 783-5.

Fujimura, O. (1962) Analysis of nasal consonants. *J. Acoustical Society of America 34,* 1865-75.

Fujisaki, H. & T. Kawashima (1968) The roles of pitch and higher formants in the perception of vowels. *IEEE Trans. AU-16,* 73-7.

Geil, F. G. (1974) Instant replay helium speech unscrambler using slowed tape for correction. *US Patent Specification 3 813 687,* 28th May 1974.

Gerstman, L. J., G. R. Gamertsfelder & A. Goldberger (1966) Breathing mixture and depth as separate effects on helium speech. *J. Acoustical Society of America 40,* no.5, 1283 (abstract).

Gill, J. S. (1970) British patent 1321313, June 1970.

Giordano, T. A., H. B. Rothman & H. Hollien (1973) Helium speech unscramblers – a critical review of the state of the art. *IEEE Trans. AU-21,* no.5, 436-44.

Golden, R. M. (1966) Improving naturalness and intelligibility of helium speech, using vocoder techniques. *J. Acoustical Society of America 40,* no.3, 621-4.

Goldstein, J. L. (1967) Auditory nonlinearity. *J. Acoustical Society of America 41,* no.3, 676-89.

Halley, E. (1716) The art of living underwater: or, a discourse concerning the means of furnishing air at the bottom of the sea, in any ordinary depths. *Phil. Trans. Royal Soc. London 29,* no.349, 492-9.

Harris, C. M. & M. R. Weiss (1963) Pitch extraction by computer processing of high-resolution Fourier analysis data. *J. Acoustical Society of America 35,* 339.

Harris, F. J. (1978) On the use of windows for harmonic analysis with the discrete Fourier transform. *Proc. IEEE 66-1,* 51-83.

Harris, K. S., H. Hoffman, A. M. Liberman, P. C. Delattre & F. S. Cooper (1958) Effect of third-formant transitions on the perception of the voiced stop consonants. *J. Acoustical Society of America 30,* 122-6.

Hayes, M. H., J. S. Lim & A. V. Oppenheim (1980) Signal reconstruction from phase or magnitude. *IEEE Trans. ASSP-28,* no.6.

Healy, T. J. (1969) Convolution revisited. *IEEE Spectrum 6,* no.4, 87-93.

Heinz, J. H. & K. N. Stevens (1961) On the properties of voiceless fricative consonants. *J. Acoustical Society of America 33,* 589-96.

Hicks, B. L., L. D. Braida & N. I. Durlach (1981) Pitch invariant frequency lowering with nonuniform spectral compression. *Proc. IEEE Conf. Acoustics, Speech and Signal Processing ICASSP-81,* 121-4.

Holbrook, A. & G. Fairbanks (1962) Diphthong formants and their movements. *J. Speech and Hearing Research 5,* 38-58.

Hollien, H. & H. B. Rothman (1974) Evaluation of helium speech unscramblers under controlled conditions. *MTS Journal 8,* no.9, 35-44.

Hollien, H., W. Shearer & J. W. Hicks, Jr (1977) Voice fundamental frequency levels of divers in helium-oxygen speaking environments. *Undersea Biomedical Research 4,* no.2, 199-207.

Hollien, H., C. L. Thompson & B. Cannon (1973) Speech intelligibility as a function of ambient pressure in a helium-oxygen atmosphere. *Aerospace Medicine 44,* 249-53.

Hollien, H., H. B. Rothman, S. H. Feinstein & P. Hollien (1973) The speech characteristics of divers in He-$O_2$ breathing mixtures at high pressures. *Florida University Gainesville Communication Sciences Lab., CSL/ONR-TR-49.*

Holmes, J. N. (1972) *Speech Synthesis.* Mills & Boon Ltd, London.

Holywell, K. & G. Harvey (1964) Helium speech. *J. Acoustical Society of America 36,* 210-11.

House, A. S. (1957) Analog studies of nasal consonants. *J. Speech and Hearing Disorders 22,* no.2, 190-204.

House, A. S. & K. N. Stevens (1956) Analog studies of the nasalization of vowels. *J. Speech and Hearing Disorders 21,* no.2, 218-31.

Hunter, E. K. (1968) Problems of diver communication. *IEEE Trans. AU-16,* no.1, 118-120.

Jack, M. A., A. D. Milne & W. Donaldson (1979) Final report and operating manual for compact helium speech unscrambler equipment. *Wolfson Microelectronics Institute,* University of Edinburgh.

Jack, M. A., A. D. Milne & L. E. Virr (1979) Compact helium speech unscrambler using charge transfer devices. *Electronics Letters 15,* no.18, 548-50.

Jack, M. A., A. D. Milne, L. E. Virr & R. Hicks (1981) Miniature helium speech unscrambler for diver-borne use. *Conference on Electronics for Ocean Technology, IERE Conference Proceedings 51,* 13-18.

Jaminet, A. (1871) *Physical effects of compressed air and of the causes of pathological symptoms produced on man by atmospheric pressure employed for the sinking of piers in the construction of the Illinois-St Louis bridge over the Mississippi River at St Louis, Missouri.* R. & T. A. Enious, St Louis.

Jelinek, H. J. (1986) Automated recognition of helium speech: Phase I. Investigation of microprocessor based analysis/synthesis system. *Electronic Design Associates (Final Report)* Costa Mesa, USA.

Jenkins, G. M. & D. G. Watts (1968) *Spectral Analysis and its Applications.* Holden-Day, San Francisco.

Kuo, F. F. (1966) *Network Analysis and Synthesis* (2nd ed.). J. Wiley & Sons Inc., New York.

Laver, J. (1980) Phonatory settings, in *The Phonetic Description of Voice Quality.* Cambridge University Press.

Laver, J., S. Hiller & J. Mackenzie (1984) Acoustic analysis of vocal fold pathology. *Proc. Institute of Acoustics 6,* no.4, 425-30.

Laver, J. & P. Trudgill (1979) Phonetic and linguistic markers in speech, in *Social Markers in Speech,* (eds K. Scherer & H. Giles). Cambridge University Press.

Lee, B. S. (1950) Effects of delayed speech feedback. *J. Acoustical Society of America 22,* 824-6.

Liberman, A. M., P. C. Delattre, F. S. Cooper & L. J. Gerstman (1954) The role of consonant-vowel transitions in the perception of the stop and nasal consonants. *Psychological Monographs 68,* no.8, 1-13.

Licklider, J. C. R. & I. Pollack (1948) Effects of differentiation, integration, and infinite peak clipping upon the intelligibility of speech. *J. Acoustical Society of America 20,* no.1, 42-51.

Lindblom, B. & A. Floren (1965) Estimating short-term context-dependance of formant pattern perception: stimulus specification. *Quarterly Progress and Status Report, Speech Transmission Laboratory, STL-QPSR-2,* Royal Institute of Technology, Stockholm, 24-6.

Lindqvist, J. & J. Lubker (1970) Mechanisms of stop consonant production. *Quarterly Progress and Status Report, Speech Transmission Laboratory, STL-QPSR-1,* Royal Institute of Technology, Stockholm, 1-2.

Lintz, L. B. & D. Sherman (1961) Phonetic elements and perception of nasality. *J. Speech and Hearing Research 4,* 381-96.

Lisker, L., J. Martony, B. Lindblom & S. Ohman (1960) F-pattern approximations of voiced stops and fricatives. *Quarterly Progress and Status Report, Speech Transmission Laboratory, STL-QPSR-1,* Royal Institute of Technology, Stockholm, 20-2.

McGonegal, C. A., L. R. Rabiner & A. E. Rosenberg (1977) A subjective evaluation of pitch detection methods using LPC synthesised speech. *IEEE Trans. ASSP-25,* 221-9.

Maclean, D. J. (1966) Analysis of speech in a helium-oxygen mixture under pressure. *J. Acoustical Society of America 40,* no.3,625-7.

Macedo, J. F., P. Strohl & M. Y. Brechet (1985) *Etude technico-économique de la plongée à l'hydrogène.* B+, Marseille, France (internal report).

Maitland, G. & J. R. Thomas (1974) Frequency spectrum analysis of speech changes during open sea habitat and hyperbaric chamber dives. *Aerospace Medicine 45,* no.8, 855-9.

Makhoul, J. (1975) Spectral linear prediction: properties and applications. *IEEE Trans. ASSP-23,* no.3, 283-96.

—— (1976) Methods for nonlinear spectral distortion of speech signals. *Proc. IEEE Int. Conf. Acoustics, Speech and Signal Processing, ICASSP-76,* 87-90.

Markel, J. D. & A. H. Gray, Jr (1967) *Linear Prediction of Speech.* Springer-Verlag, New York.

Martony, J. (1962) On the synthesis and perception of voiceless fricatives. *Quarterly Progress and Status Report, Speech Transmission Laboratory, STL-QPSR-1,* Royal Institute of Technology, Stockholm, 17-22.

—— (1964a) On the vowel source spectrum. *Quarterly Progress and Status Report, Speech Transmission Laboratory, STL-QPSR-1,* Royal Institute of Technology, Stockholm, 3-4.

—— (1964b) The role of formant amplitudes in synthesis of nasal consonants. *Quarterly Progress and Status Report, Speech Transmission Laboratory, STL-QPSR-3,* Royal Institute of Technology, Stockholm, 28-31.

Mathes, R. C. & R. L. Miller (1947) Phase effects in monaural perception. *J. Acoustical Society of America 19,* 780.

Mermelstein, P. (1978) Difference limens for formant frequencies for steady-state and consonant-bound vowels. *J. Acoustical Society of America 63,* no.2, 572-80.

Miller, G. A. & J. C. R. Licklider (1950) The intelligibility of interrupted speech. *J. Acoustical Society of America 22,* no.2, 167-73.

Miller, R. L. (1959) Nature of the vocal cord wave. *J. Acoustical Society of America 31,* no.6, 667-77.

Morrow, C. T. (1971) Speech in deep submergence atmospheres. *J. Acoustical Society of America 50,* no.3(1), 715-28.

Morrow, C. T. & A. J. Brouns (1971a) Speech communications in diving masks: I. Acoustics of microphones and mask cavities. *J. Acoustical Society of America 50,* no.1(1), 1-9.

—— (1971b) Speech communications in diving masks: II. Communication in mask cavities. *J. Acoustical Society of America 50,* no.1(1), 10-22.

Morse, P. M. (1948) *Vibration and Sound* (2nd ed.). McGraw-Hill, New York.

Moser, H. M., J. J. Dreher & S. Adler (1955) Comparison of hyponasality, hypernasality and abnormal voice quality on the intelligibility of two-digit numbers. *J. Acoustical Society of America 27,* 872-4.

Murry, T. & R. L. Sergeant (1971) Response variations of three types of microphones pressurized to 19 ATA. *Report No. NSMRL-690,* Naval Submarine Medical Research Lab., Groton, Connecticut, USA.

Nakatsui, M. (1974) Comment on helium speech – insight into speech event needed. *IEEE Trans. ASSP-22,* 472-3.

Nakatsui, M. & J. Suzuki (1971) Observations of speech parameters and their daily variations in a helium-nitrogen-oxygen mixture at a depth of 30 m. *J. Radio Research Laboratories of Japan 18,* no.97, 221-5.

Nakatsui, M., J. Suzuki, T. Tagasugi & R. Tanaka (1973) Nature of helium speech and its unscrambling. *Record of IEEE Conference on Engineering in the Ocean Environment,* 137-40.

Nawab, S. H., T. F. Quatieri & J. S. Lim (1983) Algorithms for signal reconstruction from short-time Fourier transform magnitude. *Proc. IEEE Conf. Acoustics, Speech and Signal Processing, ICASSP-83,* 800-3.

Nixon, C. W. & H. C. Sommer (1968) Subjective analysis of speech in a helium environment. *Aerospace Medicine 39,* 139-44.

Noll, A. M. (1967) Cepstrum pitch determination. *J. Acoustical Society of America 41,* no.2, 293-309.

Öhman, S. E. G. (1966) Perception of segments of VCCV utterances. *J. Acoustical Society of America 40,* 979-88.

Oppenheim, A. V. & J. S. Lim (1981) The importance of phase in signals. *Proc. IEEE 69,* no.5, 529-41.

Oppenheim, A. V. & R. W. Schafer (1968) Homomorphic analysis of speech. *IEEE Trans. AU-16,* no.2, 221-6.

Oppenheim, A. V., A. S. Willsky & I. T. Young (1983) *Signals and Systems.* Prentice-Hall, New Jersey.

Peterson, G. E. (1952) The information-bearing elements of speech. *J. Acoustical Society of America 24,* no.6, 629-37.

Peterson, G. E. & H. L. Barney (1952) Control methods used in the study of vowels. *J. Acoustical Society of America 24,* 175-84.

Pinson, E. N. (1963) Pitch-synchronous time-domain estimation of formant frequencies and bandwidths. *J. Acoustical Society of America 35,* 1264-73.

Portnoff, M. R. (1981) Short-time Fourier analysis of sampled speech. *IEEE Trans. ASSP-29,* no.3.

Quick, R. F. (1970) Helium speech translation using homomorphic techniques. *J. Acoustical Society of America 48,* 130(a).

Rabiner, L. R. & R. W. Schafer (1975) *Digital Processing of Speech Signals.* Prentice-Hall.

Rabiner, L. R., M. J. Cheng, A. E. Rosenberg & C. A. McGonegal (1976) A comparative performance study of several pitch detection algorithms. *IEEE Trans. ASSP-24,* no.5, 399-417.

Rader, C. M. (1970) An improved algorithm for high speed autocorrelation with applications to spectral estimation. *IEEE Trans. AU-18,* no.4, 439-41.

Richards, M. A. (1982) Helium speech enhancement using the short-time Fourier transform. *IEEE Trans. ASSP-30,* no.6, 841-53.

Ritsma, R. J. & F. L. Engel (1964) Pitch of frequency modulated signals. *J. Acoustical Society of America 36,* no.9, 1637-44.

Rosenberg, A. E. (1971) The effect of glottal pulse shape on the quality of natural vowels. *J. Acoustical Society of America 49,* no.2(2), 583-90.

Scherer, K. R. (1981) Vocal indicators of stress, in *The Evaluation of Speech in Psychiatry* (ed. J. Darby). Grune & Stratton, New York.

Schott, G. (1664) Lib.VI. Mirabilia Mechanica : cap.IX. Cacabus aquaticus et aquatica Lorica, qua quis tectus sub aquis ambulat. *Technica Curiosa, sive Mirabilia Artis,* 393-6.

Schouten, J. F., R. J. Ritsma & B. L. Cardozo (1962) Pitch of the residue. *J. Acoustical Society of America 34,* no.8, 1418-24.

Scott, B. J. (1976) Temporal factors in vowel perception. *J. Acoustical Society of America 60,* no.6, 1354-65.

Sergeant, R. L. (1963) Speech during respiration of a mixture of helium and oxygen. *Aerospace Medicine 34,* 826-8.

—— (1967) Phonemic analysis of consonants in helium speech. *J. Acoustical Society of America 41,* no.1, 66-9.

—— (1972) The intelligibility of hydrogen-speech at 200 feet of seawater equivalent. *Report No. 701,* Naval Submarine Medical Center, Groton, Connecticut, USA.

Speakman, J. D. (1968) Physical analysis of speech in helium environments. *Aerospace Medicine 39,* 48-53.

Stevens, K. N. & A. S. House (1956) Studies of formant transitions using a vocal tract analog. *J. Acoustical Society of America 28,* no.4, 578-85.

Stevens, K. N. & D. H. Klatt (1974) Role of formant transitions in the voiced-voiceless distinction for stops. *J. Acoustical Society of America 55,* 653-9.

Stewart, J. L. (1960) *Fundamentals of Signal Theory.* McGraw-Hill.

Stover, W. R. (1967) Technique for correcting helium speech distortion. *J. Acoustical Society of America 41,* 70-4.

Suzuki, J. & M. Nakatsui (1971) Articulation of mono-syllables uttered in a helium-nitrogen-oxygen mixture at a depth of 30 m. *J. Radio Research Laboratories of Japan 18,* no.97, 233-7.

—— (1974) Perception of helium speech uttered under high ambient pressures. *Publication SCS-74, Speech Communication Seminar,* Stockholm.

Suzuki, J., G. Ooyama & K. Kido (1974) Analysis-conversion-synthesis system for improving naturalness and intelligibility of speech at high pressure in a helium gas mixture. *Publication SCS-74, Speech Communication Seminar,* Stockholm, 97-105.

Suzuki, J., M. Nakatsui, T. Takasugi & R. Tanaka (1977) Translation of helium speech by the method of segmentation, partial-rejection and expansion. *J. Radio Research Laboratories of Japan 24,* no.113, 1-16.

Takasugi, T., M. Nakatsui & J. Suzuki (1971) Long-term speech spectrum in a helium-nitrogen-oxygen mixture at a depth of 30 m. *J. Radio Research Laboratories of Japan 18,* no.97, 227-31.

Tanaka, R., M. Nakatsui & J. Suzuki (1974) Formant frequency shifts under high ambient pressures. *J. Radio Research Laboratories of Japan 21,* no.105, 261-7.

Tanaka, R., M. Nakatsui, T. Takasugi & J. Suzuki (1974) Source characteristics of speech produced under high ambient pressure. *J. Radio Research Laboratories of Japan 21,* 269-73.

Trehern, J. & M. A. Jack (1986) Helium speech recognition system. *Internal Report,* Centre for Speech Technology Research, University of Edinburgh.

Trukken, S. J. & J. D. Pelton (1984) Evaluation of commercially available underwater communicators for use with the Mk.I MOD 0 and Mk.12 SSDS (Surface Supplied Diving System) in the air mode. *Report No.NEDU-9-84,* US Navy Experimental Diving Unit, Panama City.

Tucker, W. H. & R. H. T. Bates (1978) A pitch estimation algorithm for speech and music. *IEEE Trans. ASSP-26,* no.6, 597-604.

Verzeano, M. (1950) Time-patterns of speech in normal subjects. *J. Speech and Hearing Disorders 15,* 197-201.

Webster, D. H. F. (1978) in *The Danger Game.* Hale Publications, London.

Weiner, N. (1930) Generalized harmonic analysis. *Acta Mathematica 55,* 117-258.

Williams, C. E. & K. N. Stevens (1972) Emotions and speech: some acoustical correlates. *J. Acoustical Society of America 52,* 1238-50.

Wood, S. (1977) A radiographic analysis of constriction locations for vowels. *Working papers 15,* Phonetics Laboratory, Lund University, 103-31.

Yegnanarayana, B. & A. Dhayalan (1983) Noniterative techniques for minimum phase signal reconstruction from phase or magnitude. *Proc. IEEE Conf. Acoustics, Speech and Signal Processing, ICASSP-83,* 639-42.

Zurcher, J. F. (1980) Voice transcoder. *UK Patent specification 1561918,* 5th March 1980.

# INDEX